Lecture Notes in Computer Sc

T0238087

Commenced Publication in 1973
Founding and Former Series Editors:
Gerhard Goos, Juris Hartmanis, and Jan van Leeuwen

Violet R. Syrotiuk   Edgar Chávez (Eds.)

Ad-Hoc, Mobile, and Wireless Networks

4th International Conference, ADHOC-NOW 2005
Cancun, Mexico, October 6-8, 2005
Proceedings

 Springer

Volume Editors

Violet R. Syrotiuk
Arizona State University
Computer Science and Engineering
P.O. Box 878809, Tempe, AZ 85287-8809, USA
E-mail: syrotiuk@asu.edu

Edgar Chávez
Universidad Michoacana de San Nicolás de Hidalgo
Facultad de Ciencias Físico-Matemáticas
México
E-mail: elchavez@fismat.umich.mx

Library of Congress Control Number: 2005932993

CR Subject Classification (1998): C.2, D.2, H.4, H.3, I.2.11, K.4.4, K.6.5

ISSN   0302-9743
ISBN-10   3-540-29132-6 Springer Berlin Heidelberg New York
ISBN-13   978-3-540-29132-9 Springer Berlin Heidelberg New York

Springer is a part of Springer Science+Business Media

springeronline.com

© Springer-Verlag Berlin Heidelberg 2005
Printed in Germany

Typesetting: Camera-ready by author, data conversion by Scientific Publishing Services, Chennai, India
Printed on acid-free paper   SPIN: 11561354   06/3142   5 4 3 2 1 0

Preface

The 4th International Conference on Ad-Hoc Networks and Wireless (ADHOC-NOW 2005) was held October 6–8, 2005 in Cancun, Mexico. Adhoc Now started as a workshop in 2002 and was held at the Fields Institute in Toronto. In 2003, it was held in Montreal, and in 2004 it was held in Vancouver. 2005 was the first year for the conference to move outside of Canada. The purpose of the conference is to create a collaborative forum between mathematicians, computer scientists, and engineers for research in the field of mobile ad hoc and sensor networks.

In 2005, we received over 100 submissions from 22 different countries: Australia, Canada, China, France, Germany, Greece, India, Ireland, Italy, Japan, Korea, Malaysia, Mexico, Nepal, Nigeria, Pakistan, Poland, Spain, Sweden, Tunisia, the UK, and the USA — a true international conference. Of the papers submitted, we selected 27 for presentation at the conference and publication in the proceedings.

We are grateful to our Technical Program, Organizing, and Steering Committees; without their help, expertise, and experience we could not have selected such a fine program. We thank Jorge Urrutia of the Instituto de Matemáticas, Universidad Nacional Autónoma de México, and J.J. Garcia-Luna-Aceves of the Computer Engineering Department, University of California, Santa Cruz for accepting our invitation to speak at the conference. Special thanks are due to the crew from the Facultad de Ciencias Físico-Matemáticas Universidad Michoacana for handling the local arrangements, and to the Mobile Adhoc Research Lab at Arizona State University for handling odd jobs on a moment's notice.

August 2005

Violet R. Syrotiuk
Edgar Chávez
Program Co-chairs
ADHOC-NOW 2005

Organization

Organizing Committees

Steering Committee　　　　Michel Barbeau
　　　　　　　　　　　　　　　　Evangelos Kranakis
　　　　　　　　　　　　　　　　Ioanis Nikolaidis
　　　　　　　　　　　　　　　　S.S. Ravi
　　　　　　　　　　　　　　　　Violet R. Syrotiuk
Publicity Chair　　　　　　Pedro M. Ruiz   .
Panel Chair　　　　　　　　S.S. Ravi
Demo Chair　　　　　　　　Jose A. García Macias

Technical Program Committee

E. Altman, INRIA

M. Barbeau, Carleton Univ.

P. Bose, Carleton Univ.

R. Bazzi, Arizona State Univ.

T. Camp, Colorado School of Mines

E. Chávez, Univ. Michoacana

J. Cobb, Univ. Texas at Dallas

M. Conti, CNR-IIT

S. Dobrev, Univ. Ottawa

A. Faragó, Univ. Texas at Dallas

L. Feeney, SICS

A. Garcia-Macias, CICESE

S. Giordano, SUPSI

A. Gomez-Skarmeta, Univ. of Murcia

A. Jukan, UIUC

G. Konjevod, Arizona State Univ.

E. Kranakis, Carleton Univ.

D. Krizanc, Wesleyan Univ.

S. Krumke, Univ. Kaiserslautern

T. Kunz, Carleton Univ.

J. Misic, Univ. Manitoba

A. Mielke, LANL

P. Morin, Carleton Univ.

L. Narayanan, Concordia Univ.

I. Nikolaidis, Univ. Alberta

J. Opatrny, Concordia Univ.

R. Prakash, Univ. Texas at Dallas

S. Rajsbaum, UNAM

S.S. Ravi, SUNY Albany

M. Reisslein, Arizona State Univ.

F. Rousseau, LSR-IMAG

P.M. Ruiz, Univ. of Murcia

L. Stacho, Simon Fraser Univ.

M. Steenstrup, Stow Research LLC

V.R. Syrotiuk, Arizona State Univ.

G. Toussaint, McGill Univ.

D. Turgut, Univ. of Florida

P. Widmayer, ETH Zürich

G. Záruba, Univ. Texas at Arlington

J. Zhao, ICSI-Berkeley

R. Zheng, Univ. Houston

Adhoc Reviewers

A. Agah

K. Anna

M. Biagi

J. Boleng

L. Boloni

W.S. Chan

Y. Chu

M. Colagrosso

F. Galera Cuesta

M. Cui

V. Govindaswamy

R. Fonseca

S. Gandham

H. Hirst

C. Hu

R. Huang

E. Krohne

S. Kuppa

S. Kurkowski

D.W. McClary

M.P. McGarry

M. Moshin

R. Musunuri

Y. Moon

L. Ritchie

C.M. Yago Sánchez

P. Seeling

C. Sengul

A. Sobeih

M. Thoppian

S. Tixeuil

B. Turgut

K.K. Vadde

G. van der Auwera

H.-S. Yang

H. Zhang

Table of Contents

Another Look at Dynamic Ad-Hoc Wireless Networks[*]

J.J. Garcia-Luna-Aceves[1,2]

[1] Computer Engineering, University of California, Santa Cruz,
1156 high street, Santa Cruz, CA 95064, U.S.A.
Phone:1-831-4595436, Fax: 1-831-4594829
jj@soe.ucsc.edu
[2] Palo Alto Research Center (PARC), 3333 Coyote Hill Road,
Palo Alto, CA 94304, U.S.A.

Abstract. The price, performance, and form factors of sensors, processors, storage elements, and radios today are enabling the development of network-supported applications for a wide range of environments, including the monitoring of disruptive phenomena, object tracking, establishment of on-demand network infrastructure for disaster relief or military purposes, and peer-to-peer vehicular or interpersonal networks. However, while in theory ad hoc networks are the ideal vehicle for such applications, the practice today is far from this theory. In this talk, I argue that many of the limitations of ad hoc networks today stem from the fact that the architectures and protocols used for them are in many ways a derivative of the Internet architecture, and describe a research agenda that considers developing ad hoc networks without having to adhere to many of the design choices that, until now, have proven so successful for internetworking of wired networks.

Biography

J.J. Garcia-Luna-Aceves received the B.S. degree in electrical engineering from the Universidad Iberoamericana in Mexico City, Mexico in 1977, and the M.S. and Ph.D. degrees in electrical engineering from the University of Hawaii, Honolulu, HI, in 1980 and 1983, respectively. He holds the Jack Baskin Chair of Computer Engineering at the University of California, Santa Cruz (UCSC). He is also a Principal Scientist at the Palo Alto Research Center (PARC). Prior to joining UCSC in 1993, he was a Center Director at SRI International (SRI) in Menlo Park, California. He has been a Visiting Professor at Sun Laboratories and a Principal of Protocol Design at Nokia.

Dr. Garcia-Luna-Aceves has published a book, more than 290 papers, and seven U.S. patents. He has directed 21 Ph.D. theses and 19 M.S. theses at UCSC over the past 11 years. He is the General Chair for the IEEE SECON 2005 Conference. He has also been Program Co-Chair of ACM MobiHoc 2002 and ACM

---

[*] This work was supported in part by the Palo Alto Research Center and by the Baskin Chair of Computer Engineering at University of California, Santa Cruz.

V.R. Sirotiuk and E. Chávez (Eds.): ADHOC-NOW 2005, LNCS 3738, pp. 1–2, 2005.

Mobicom 2000; Chair of the ACM SIG Multimedia; General Chair of ACM Multimedia '93 and ACM SIGCOMM '88; and Program Chair of IEEE MULTIMEDIA '92, ACM SIGCOMM '87, and ACM SIGCOMM '86. He has served in the IEEE Internet Technology Award Committee, the IEEE Richard W. Hamming Medal Committee, and the National Research Council Panel on Digitization and Communications Science of the Army Research Laboratory Technical Assessment Board. He has been on the editorial boards of the IEEE/ACM Transactions on Networking, the Multimedia Systems Journal, and the Journal of High Speed Networks. He received the SRI International Exceptional-Achievement Award in 1985 and 1989, and is a senior member of the IEEE.

Routing in Wireless Networks and Local Solutions for Global Problems

Jorge Urrutia

Instituto de Matematicas,
Universidad Nac. Aut. de Mexico,
Area de la Inv. Cientifica,
Circuito Exterior,
Ciudad Universitaria,
Mexico D.F. C.P. 04510

Abstract. Let P_n be a set of points. The unit distance graph of P_n is the graph with vertex set P_n, in which two points are connected if their distance is at most one. Unit distance graphs of point sets can be used to model wireless networks in which the elements of P_n represent the location the broadcast stations of our wireless networks. The stations are assumed to broadcast with the same power.

In recent years, it has been proved that many global problems for this type of networks can be solved by means of local algorithms, that is algorithms in which a node needs to communicate only with its neighbours. The first example of this, was the extraction of a planar connected subgraph of a unit distance wireless network, which was then used for a local type routing algorithm. In this talk we will survey several results in this area of research, and present recent results related to approximations of minimum weight spanning trees, snapshots of networks, etc.

Biography

Jorge Urrutia obtained his B.Math. in UNAM, Mexico in 1975, and Ph.D. in Waterloo 1980. His main area of research is Discrete and Computational Geometry. Founder and editor-in-Chief of Computational Geometry, Theory and Applications 1990-2000. In 1998 he joined the Instituto de Matematicas, Universidad Nacional Autonoma de Mexico. Previously he was at the Department of Computer Science at the University of Ottawa.

He has written many papers in Discrete and Computational Geometry and in several areas of Combinatorics. He has delivered numerous plenary talks in conferences in Europe, Asia and the Americas and organized and participated in the organization committees of many conferences. He is well respected in the Mexican Research Community, and is member of the Mexican Sistema Nacional de Investigadores level III.

V.R. Sirotiuk and E. Chávez (Eds.): ADHOC-NOW 2005, LNCS 3738, p. 3, 2005.

Equilibria for Broadcast Range Assignment Games in Ad-Hoc Networks

Pilu Crescenzi[1], Miriam Di Ianni[2], Alessandro Lazzoni[1], Paolo Penna[3,*],
Gianluca Rossi[2,*], and Paola Vocca[4,**]

[1] Dipartimento di Sistemi e Informatica, Università di Firenze, Florence, Italy
piluc@dsi.unifi.it, alex@email.it
[2] Dipartimento di Matematica, Università degli Studi di Roma "Tor Vergata",
Rome, Italy
{diianni, rossig}@mat.uniroma2.it
[3] Dipartimento di Informatica ed Applicazioni "R.M. Capocelli",
Università di Salerno, Salerno, Italy
penna@dia.unisa.it
[4] Dipartimento di Matematica, Università di Lecce, Lecce, Italy
paola.vocca@unile.it

Abstract. Ad-hoc networks are an emerging networking technology, in which the nodes form a network with no fixed infrastructure: each node forwards messages to the others by using the wireless links induced by their power levels. Generally, energy-efficient protocols heavily rely on cooperation. In this paper, we analyze from a game-theoretic point of view the problem of performing a broadcast operation from a given station s. We show both theoretical and experimental results on how the existence of (good) Nash equilibria is determined by factors such as the transmission power of the stations or the payment policy that stations can use to enforce their reciprocal cooperation.

1 Introduction

Ad-hoc networks do not need any fixed infrastructure for communication: nodes consist of radio stations that are able to communicate by sending messages with a certain power. This feature is particularly attractive for users since they do not have to rely on a service provider for building/using the network.

Tipically, stations are located in a two-dimensional Euclidean space and are connected by *wireless links* that are induced by their power levels. Each station v is equipped with an *omnidirectional antenna* and, depending on the environmental conditions, a signal transmitted with power P_v can be received by every other station t such that

$$d(v,t)^\alpha \le \frac{P_v}{\gamma}, \tag{1}$$

---

* Supported by the European Union under the Project IST-2001-33135 "Critical Resource Sharing for Cooperation in Complex Systems" (CRESCCO).
** Partially supported by the Italian Research Project PRIN 2003 "Optimization, simulation and complexity of the design and management of communication networks"

where $d(v,t)$ is the Euclidean distance between v and t, $\alpha \geq 1$ is the distance-power gradient, and $\gamma \geq 1$ is the transmission quality parameter. In an ideal environment (i.e., in empty space) it holds that $\alpha = 2$, but it may vary from 1 to more than 6 depending on the environment conditions at the location of the network (see [16]). According to the previous equation, when a station v transmits with power P_v, it covers an area consisting of all points at distance at most $r_v \geq (P_v/\gamma)^{1/\alpha}$ from v. The value r_v is the *transmission range* of v, i.e., the maximum distance at which station v can transmit in one hop with power P_v. Hence, assigning transmission ranges to the stations is equivalent to decide their transmission powers. In the remaining of this work, we assume $\gamma = 1$, although all of our results easily apply to any constant γ.

The set of all transmission ranges yields a *range assignment* that is a function $r : S \to \mathbb{R}^+$, where S denotes the set of stations and $r(v) = r_v$. We consider *broadcast range assignments*, that is, range assignments which, given a source station $s \in S$, allow this station to transmit to all other stations (via a multi-hop communication). Formally, consider a *transmission graph* $G_r = (S, E_r)$, such that $(v, t) \in E_r$ if and only if $d(v, t) \leq r(v)$. Then r is a broadcast range assignment if G_r contains a directed spanning tree rooted at s.

The *social cost* (or, simply, the cost) of a (broadcast) range assignment is measured as the overall energy that all stations in the network spend to implement these ranges, that is,

$$\mathsf{cost}(r) = \sum_{v \in S} r(v)^\alpha.$$

If ranges are assigned to stations by a central authority, then it is possible to get broadcast range assignments whose cost do not differ to much from the optimum cost (see Subsection 1.2). Implicit in this approach is the assumption that each station will actually transmit with the range specified by the authority. This assumption cannot be take for granted in a (more realistic) scenario in which stations are managed by different (potentially selfish) users. This is indeed the case of ad-hoc networks for which it is fundamental to develop mechanisms that enforce stations cooperation.

In this work, we consider a game-theoretic setting in which each station corresponds to a different player (or agent) of a game named *broadcast range assignment game*. The strategy of each player v is to decide its transmission range $r(v)$ and/or to provide some payment to some other players in order to convince them to transmit with a given range.

The range assignment r derived by the strategies of all players can induce a *benefit* $b_v(r)$ to every station v. The benefit can represent, for example, the interest of station v in guaranteeing the given connectivity, the sum of the payments received/provided from/to the other stations, or a combination of these two things. Since implementing the range $r(v)$ induces a cost of $r(v)^\alpha$, we can define a *utility function*

$$\mu_v(r) = b_v(r) - r(v)^\alpha \tag{2}$$

that station v aims at maximizing. Observe that $\mu_v(r)$ depends on the strategies of *all* the stations. In particular, if station v' changes its transmission range from $r(v')$ to $r'(v')$, then we obtain a range assignment r' and the utility of all stations v will change to $\mu_v(r')$.

We are interested in sets of "stable" strategies for which no player has an incentive in unilaterally switching to a different strategy. These configurations are known as Nash equilibria [15]. A range assignment r is a *Nash equilibrium* if $\mu_v(r) \geq \mu_v(r')$, for every station v and for every r' obtained from r by changing $r(v)$ into $r'(v)$. We are interested in *good* Nash equilibria, that is, strategies that minimize the overall power consumption (the social cost). Sometimes, it can be convenient to consider ϵ-*approximate* Nash equilibria, that is, range assignments r that guarantee $\epsilon \cdot \mu_v(r) \geq \mu_v(r')$, for all v and r'.

As already noticed, some station v may be interested in guaranteeing the given connectivity requirement (i.e. broadcast) and may be willing to pay some other station in order to maintaining the needed connectivity property. We model the connectivity requirement of a station v by saying that v is *penalized* if its connectivity requirement is not satisfied by the range assignment r. In this case, we define $b_v(r) = -\infty$ and thus $u_v(r) = -\infty$ as well (see Eq. 2). Otherwise, we define $b_v(r)$ as the "balance" derived from all money exchanged with the other stations that is,

$$b_v(r) = \sum_{u \in S} \left(p_u^v(r) - p_v^u(r) \right),$$

where $p_u^v(r)$ is the payment from station u to station v when range assignment r is implemented.

The simplest games we consider are the *Payments-free* games. Here, no payments are allowed (as in [9]). Clearly, a broadcast range assignment will be a Nash equilibrium if at least one station is penalized.

We next consider payment games in which a pricing policy is defined that will depend on which transmission ranges are "used" by a station v, given a range assignment r. In particular, payments provided by a station v are used to allow this station to increase the transmission range of other stations in order to create a path from s to v. In this case, the strategy of node v consists in specifying a path from s to v. Our payment policies differ with respect to which stations in the chosen path will receive a payment from v. For every station u, we define $\text{Used}_r(u)$ as the set of stations having to pay u.

- *Edge-payment.* Here we consider a simple local policy in which payments associated to the range assignment r are only provided to neighbor stations. In particular, assume u is the last station in the path from s to v. Then, station u is the only station that receives a payment from v. For every such pair of stations u and v, $v \in \text{Used}_r(u)$.
- *Path-payment.* In this case, station v can be required to pay for (some part of) all the ranges $r(s), r(u_1), \ldots r(u_k)$, where $\langle s = u_0, u_1, \ldots, u_k, v \rangle$ is the strategy of v. In this case, $v \in \text{Used}_r(u_i)$ for all $i = 0, \ldots, k$.

We then consider two possible payment policies:

- *No-profit.* The cost $r(u)^\alpha$ is divided among all stations $v \in \text{Used}_r(u)$ (similarly to [2,3]).
- *Profit.* Every station v using station u pays exactly $p_v^u(r) = r(u)^\alpha$. Clearly, station u may have a profit if $|\text{Used}_r(u)| > 1$.

The connectivity requirements we consider are of two types: (i) *Reachability*, that is, station v has $b_v(r) = -\infty$ whenever r does not allow s to transmit to v, and (ii) *B-Broadcasting*, that is, for all $v \in B \subseteq S$, $b_v(r) = -\infty$ whenever r is not a broadcast range assignment. In particular, we consider s-Broadcasting and S-Broadcasting, i.e., only the source s or all stations in S are interested in the information dissemination, respectively.

1.1   Paper Contribution

We investigate the existence of Nash equilibria, the computational complexity of finding (a good) one, and convergence properties to Nash equilibria under the natural *best response*[1] assumption. For the latter, we take into account both the convergence time and the quality of the final Nash equilibrium. A significant measure of the quality of a Nash equilibrium is the *price of stability*, that is, the ratio between the cost of the equilibrium and the cost of the optimal solution (that, in general, is not a Nash equilibrium).

We also consider a weaker notion of ϵ-approximate Nash equilibria since we observe that, for some of the games we consider, ϵ-approximate Nash equilibria are difficult to obtain, even when agents changing their strategies can only attain a very small gain. We thus introduce the concept of *Payment ϵ-approximate Nash equilibrium*, which takes into account these aspects.

Regarding the Payment-free games, we prove that the s-Broadcasting, the s-Broadcasting with Reachability and the S-Broadcasting have Nash equilibria. In particular, the *unique* equilibrium for the s-Broadcasting and for the s-Broadcasting with Reachability can be arbitrary more expensive than the optimal solution, whereas *all* the Nash equilibria for the S-Broadcasting game are optimal range assignments (NP-hard to be computed).

Table 1 summarizes our main results for the Reachability problem in the profit models for the Edge- and Path-payment policies.

Finally, we experimentally evaluate the behavior of an algorithm that looks for a Nash equilibrium for the Reachability game in the No-profit model for both the payment policies we have introduced (Edge- and Path-payment). We test this algorithm on thousands of random instances and instances derived by the mobility model described in [11]. We obtain the following results: (i) the algorithm converges to a Nash equilibrium for all the generated instances; (ii) the convergence of the algorithm is guaranteed in a bounded number of steps that weakly depends on the size of the instances; (iii) the Nash equilibrium

---

[1] Best response strategies assume each player to select the strategy that currently maximize its utility [15].

Table 1. Results for the Reachability problem in the profit models for the Edge- and Path-payments policies

Payment	Profit	No-Profit
Edge-	A polynomial time computable Nash equilibrium that is a 6 approximation of the optimal solution	Experimental evaluation + A polynomial time computable Payments 6-approximated Nash equilibrium that is a 6 approximation of the optimal solution
Path-	A polynomial time computable Payments ϵ-approximated Nash equilibrium that is a $6(1 + \frac{2}{1-\epsilon})$ approximation of the optimal solution	Experimental evaluation

created by the algorithm is a constant approximation of the optimal solution and (iv) the intermediate configurations are feasible solutions whose cost is an approximation of the optimal cost.

1.2   Related Works

(Broadcast) Range Assignment. The broadcast range assignment problem has been deeply investigated from the point of view of centralized/distributed algorithms. In both cases, the underlying assumption is that stations will always implement the solution computed by such algorithms, even if this solution will not be advantageous for themselves. Several heuristics for this problem version have been proposed [10,12,17]. Among those, the MST-based[2] algorithm has been proved to achieve, for $\alpha \geq 2$, a *constant* approximation ratio [5,7]. A tight bound of 6 has been achieved in [1] (the lower bound of 6 is due to [5]). Interestingly, the (analysis of the) MST-based algorithm turns out to be useful for studying other heuristical approaches: indeed, several of these can be proved to produce a cost which is bounded from above by the cost of MST-based solutions [5], or to be only a constant factor away from the latter [17]. No approximation algorithm is known for $1 < \alpha < 2$. The problem is known to be NP-hard for all $\alpha > 1$ [6], while the case $\alpha = 1$ is trivially in P.

Nash Equilibria and Network Design Games. In [2] the authors introduce *network design games*: each agent offers to pay for an *arbitrary* fraction of the cost of building/maintaining a link of a network, and the corresponding link "exists" if and only if enough money is collected from all agents. Also the agents have a *connectivity requirement* and Nash equilibria correspond to those strategies for which no agent can reduce its payments still having its connectivity fulfilled. In this game, (pure) Nash equilibria may not exist for point-to-point connectivity requirements [2]. (Notice that mixed – i.e., randomized – strategies are meaningless for these games due to the fact that $u_v = -\infty$ if the graph does not support v's connectivity requirement.)

---

[2] This algorithm is denoted as BLiMST algorithm in [10].

Fixing a "fair" pricing policy in which the cost of an edge is evenly divided among all agents using it ensures the existence of pure Nash equilibria [3]. The result is an application of *potential functions* [13], which the authors use to bound the *price of stability* – i.e., the loss of performance due to this "strict" pricing policy ensuring Nash equilibria. Indeed, given k agents, the *best* Nash equilibria attains a cost of at most $O(\log k)$ the optimum [3]. For directed graphs, this bound is tight [3].

Network design games in ad-hoc wireless networks have been first considered in [9] for point-to-point and strong connectivity requirements. In this game every station has to choose its own transmission range. For point-to-point connectivity, the problem admits pure Nash equilibria and there exists an algorithm to find one of them of cost at most twice the optimum [9]. Conversely, strong connectivity games do not always have Nash equilibria, and not even ϵ-approximate Nash equilibria, for any $\epsilon > 1$. In [4], the authors deal with the multicast games in general ad-hoc networks introducing a pricing policy similar to the one introduced in [3] and they prove that the games induced by these payments have a Nash equilibrium but, finding such equilibrium is NP-hard.

2   Analytic Results

Due to the lack of space, the proofs of the results in this section will be given in the full version of the paper (see [8] for a preliminary draft).

2.1   Payments-Free Games

In the payments-free games messages are forwarded for free. This means that only stations that are penalized when broadcast cannot occur have positive ranges. Hence, broadcast is not supported in the model in which only the stations that do not receive the broadcasted message are penalized.

Proposition 1. *For the s-Broadcasting and the s-Broadcasting with Reachability games (i.e. s and the non-receiving stations are penalized) the only Nash equilibrium is the range assignment in which* $r(s) = \max\{d(s,v) : v \in S - \{s\}\}$ *and* $r(v) = 0$ *for any* $v \in S - \{s\}$.

This result implies that the cost-quality ratio is unbounded.

Proposition 2. *For the S-Broadcasting game the only Nash equilibria are the minimum cost broadcast range assignments.*

Proposition 3. *Consider an s-Broadcasting game in which s could pay other stations v an amount* $p_s^v = r(v)^\alpha$ *for implementing a certain range* $r(v)$. *Then, only minimum cost broadcast range assignments are Nash equilibria for this game.*

By comparing Proposition 1 and 3 we observe that the introduction of payments may reduce the cost-quality ratio while, on the other hand, it makes the computation of Nash equilibria to become NP-hard [7].

2.2   Payments-Games for Reachability Games

Profit Models: Both Edge- and Path-payment models do admit Nash equilibria which can be found in polynomial time.

Proposition 4. *There exists an Edge-payment policy based on the profit models such that any range assignment yielded by a minimum cost spanning tree of the complete Euclidean graph G derived from the instance is a Nash equilibrium.*

From [1] we can obtain the following result:

Theorem 1. *The cost-quality ratio of the Reachability games under the Edge-payment policy is 6.*

Similarly to Proposition 4 we can prove the following:

Proposition 5. *There exists a Path-payment policy based on the profit models such that any range assignment yielded by a shortest path tree rooted at s of the complete Euclidean graph G derived from the instance is a Nash equilibrium.*

Unfortunately, the shortest path tree does not guarantee any approximation of the optimal solution. Moreover, even ϵ-approximate Nash equilibria are difficult to obtain since in the utility function

$$\mu_v(r) = \sum_{u \in S} \left(p_u^v(r) - p_v^u(r) \right) - r(v)^\alpha$$

station v can only affect the payments it provides to the others, while the money received depends only on the other stations' strategies. This means that a considerable change in the station strategy may result in a negligible change in the station utility. However, if we limit our requirements to some weaker notion of approximate equilibria, then, for the Path-payment games the social optimum can be approximated by such equilibria. In the following we define the notion of *payments ϵ-approximate Nash equilibrium*.

Definition 1 (Payments ϵ-approximate Nash equilibria). *A range assignment r is a* Payments ϵ-approximate Nash equilibrium *if, for any station v, and any range assignment r' derived from r by changing only v's strategy, it holds that $\sum_{u \in S} p_v^u(r) \leq \epsilon \sum_{u \in S} p_v^u(r')$.*

Remark 1. Let r be a Payments ϵ-approximate Nash equilibrium. Then, r is an ϵ-approximate Nash equilibrium for the game in which (i) station v cannot refuse to implement a transmission range $r(v)$ if receiving an amount of money not smaller than $r(v)^\alpha$, (ii) a station strategy is to choose a path (thus providing the corresponding money) for being reached, and (iii) the utility of station v is the inverse of the sum of all payments provided to the other agents (or $-\infty$ if not reached).

Theorem 2. *For the Reachability Path-payment game it is possible to compute in polynomial time a Payments ϵ-approximate Nash equilibrium r such that* $\mathsf{cost}(r) \leq 6\,(1 + 2/(1 - \epsilon))) \cdot OPT$, *where OPT is the optimum social cost, for any* $\epsilon > 1$.

As a final remark, notice that the profit models introduced in this section require some form of encryption. Actually, a station transmitting with some range r reaches *all* stations at distance r while only those stations having paid their fee must be reached.

No-Profit Models. We consider both the Edge-payment model and the Path-payment model.

We now define our specific Edge-payment policy. Suppose stations $u_1, \ldots, u_k$ receive from station v and suppose that $d(v, u_1) \leq \ldots \leq d(v, u_k)$, that is, station v transmits at range $d(v, u_k)$. Let $r_1 < \ldots < r_h$ be the set of distinct distances between v and any station $u_1, \ldots, u_k$ ($h \leq k$). Let $N_v(r_j)$ be the set of stations in $\{u_1, \ldots, u_k\}$ at distance exactly r_j from v.

The payments in the Edge-payment model are defined as follows:

$$p_u^v(r) = \sum_{i=1}^{\substack{j:r_j=d(v,u)}} \frac{r_i^\alpha - r_{i-1}^\alpha}{|N_v(r_i)|}$$

Intuitively speaking, each increment $(r_i^\alpha - r_{i-1}^\alpha)$ in the transmission power of v is equally shared among all the stations using the new range (r_i). Hence, the total amounts of payments received by v equals the energy spent by v for implementing the range $r(v)$.

The Edge-payment model guarantees the existence of an easy to compute Payment ϵ-approximate Nash equilibrium, as stated in the next theorem.

Theorem 3. *Let T be a minimum spanning tree of the complete geometric graph induced by S and let Δ be the maximum out-degree of T. Then it is possible to compute in polynomial time a Payments $(\Delta + 1)$-approximate Nash equilibrium in the Edge-payment model game.*

Since every geometric spanning tree T can be transformed in polynomial time into a spanning tree T' such that T' has the same cost of T and every node in T' has at most 5 neighbors [14], we can conclude with the following result.

Corollary 1. *For the Edge-payment model game it is possible to compute in polynomial time a Payments 6-approximate Nash equilibrium.*

We now define our specific Path-payment model. Suppose stations $u_1, \ldots, u_k$ receive from v and suppose that $d(v, u_1) \leq \ldots \leq d(v, u_k)$, that is, station v transmits at range $d(v, u_k)$. Let $r_1 < \ldots < r_h$ be the set of distances between v and any station $u_1, \ldots, u_k$ ($h \leq k$). Let T be the directed tree rooted at s induced by the range assignment r and T_v be the subtree of T rooted at v.

Define $T_v(r_j)$ as the tree obtained by T_v by removing all the subtrees T_{u_i} such that $d(v, u_i) \neq r_j$.

Let $P_v = \{v_0 \equiv s, v_1, \dots, v_\ell \equiv u\}$ be the path in T from s to u, then for $i = 1, \dots, \ell$

$$p_u^{v_i}(r) = \sum_{h=1}^{j:r_j=d(v_h,v_{h-1})} \frac{r_h^\alpha - r_{h-1}^\alpha}{|T_{v_i}(r_h) - \{v_i\}|}$$

Intuitively speaking, each increment $(r_h^\alpha - r_{h-1}^\alpha)$ in the transmission power of v_i is equally shared among all the stations using the new range (r_j) in their paths. Hence, the total amounts of payments received by v_i equals the energy spent by v_i for implementing the range $r(v_i)$.

These payments are introduced in [4]. The authors of this paper prove that this kind of payments always induce a Nash equilibrium, however computing such equilibrium is NP-hard. Moreover, from the analysis in [3], it is possible to derive a upper bound of the cost-quality ratio that is logarithmic on the number of stations. This upper bound is not necessarily tight.

In the next section we experimentally test an algorithm that provides empirical evidence on the existence of a an algorithm that converges in polynomial time to a Nash equilibrium with constant cost-quality ratio.

3   Experimental Evaluation of the No-Profit Models

We conjecture that the Reachability game under the No-profit model (both Edge- and Path- payment policies) admits a Nash equilibrium and that there exists an equilibrium having a social cost within a constant factor of the cost of an optimal solution.

Some evidence in favor of the conjecture has been obtained by our experimental evaluations. We have tested the behavior of the algorithm described in Fig. 1. Procedure findNE takes as inputs the set of stations S and the broadcast source $s \in S$. As a first step, it computes a directed minimum spanning tree of S rooted at s and having arcs oriented towards the leaves (the MST-based algorithm mentioned in the introduction). Then, every station, in turn, tries to decrease the amount of its payments. This last step continues till a Nash equilibrium is found. Notice that, the findNE algorithm can be seen as a "simulation" of a distributed protocol for the construction of a broadcast range assignment by selfish stations that adjust a solution computed by the well known MST-based algorithm.

We have applied the algorithm findNE to two different kinds of instances: random instances and mobility instances. For the first ones, experiments have been carried out for several sizes n of the instances (between 10 and 2,000) and for each n, one thousand instances have been randomly generated according to the uniform distribution. For the second ones, the instances have been generated by using a recently proposed mobility model whose main objective is to take into account the existence of obstacles and of pathways [11]. This mobility model tries to simulate this behavior as follows. Given a set of polygonal obstacles, it first computes the Voronoi diagram determined by the vertices of the

procedure findNE(S, s)
   $T_0 \leftarrow$ mst(S);
   compute T by rooting T_0 at s and by orienting all its edges towards the leaves;
   for $v \in S - \{s\}$ **do**
        $p_T(v) \leftarrow$ the sum of all payments due by v according to T
            and to the payment model;
     while T does not represent a Nash equilibrium **do**
        choose $v \in S - \{s\}$; $m \leftarrow p_T(v)$; $T_2 \leftarrow T$;
        for $x \in S - \{s\}$ **and** x is not in the subtree of T rooted at v
           let u be the father of v in T;
           $T_1 \leftarrow E(T) - \{(u, v)\} \cup \{(x, v)\}$
           if $p_{T_1}(v) < m$ **then** $m \leftarrow p_{T_1}(v)$; $T_2 \leftarrow T_1$;
        if $p_T(v) < m$ **then** $T \leftarrow T_2$;
   return T;

Fig. 1. The findNE algorithm

polygons: the edges of the diagram are the pathways that a mobile user has to follow. Subsequently, for each user, the source node and the destination node are randomly chosen among all the vertices of the Voronoi diagram. Finally, the user is moved along the minimal path (with respect to the diagram) between the source and the destination node with a randomly chosen speed. Once the user arrives at destination, a waiting time is randomly chosen: after this time, the movement process is repeated. By using this mobility model, experiments have been carried out for two obstacle scenarios and for different numbers n of users (between 10 and 2,000): for each n, 100 instances have been generated according to the obstacle mobility model.

Remarkably, in all the experiments the algorithm findNE has been able to end in a Nash equilibrium in a very small number of rounds (a round is an iteration of the **while** loop).

Let $\langle S, s \rangle$ be an input of algorithm findNE. In what follows, the algorithm performance will be discussed by using the following parameters: the cost $SC(S)$ of the kick-off configuration (that is, the cost of the minimum spanning tree); the maximum cost $WC(S)$ between all configurations reached by the algorithm; the cost $FC(S)$ of the final configuration (that is the cost of the solution representing the Nash equilibrium); the executed number of rounds $rnds(S)$ before reaching the final state ($rnds(S) = 1$ if the kick-off configuration is a Nash equilibrium).

Convergence Speed. The necessary number of rounds for random instances is summarized in Table 2. This table shows that for the majority of the instances the convergence is within 6 rounds (1 round means that the starting solution is a Nash equilibrium). Moreover, only for a negligible number of instances the required rounds are in the interval $7 - 12$. No instance require more than 13 rounds.

The necessary number of rounds for mobility instances is summarized in Table 3. This table shows that for the majority of the instances the convergence is within 5 rounds. Notice that there are some differences between the two considered scenarios.

Table 2. Percentage of instances that converge to the Nash equilibrium in a given step. The rows indicate the cardinality of the instances and the columns the number of steps. Each column is divided in two sub-columns: the ones labelled with e refer to the Edge-payment model and the ones labelled with p refer to the Path-payment model.

n	1		2		3		4		5		6		7		...
	e	p	e	p	e	p	e	p	e	p	e	p	e	p	...
10	40.9	12.0	50.9	69.5	7.5	16.8	0.6	1.5	0.0	0.1	0	0	0	0	...
100	0	0	46.4	5.2	48.9	65.9	4.6	25.4	0.1	3.3	0	0.2	0	0	...
200	0	0	24.1	0.1	67.9	50.5	7.8	40.8	0.2	7.2	0	1.3	0	0.1	...
300	0	0	10	0	77.2	33.9	12.3	54	0.4	9.6	0.1	1.7	0	0.4	...
400	0	0	4.4	0	79.6	23.8	15.5	55.4	0.5	16.5	0	3.7	0	0.5	...
500	0	0	3.1	0	76.9	15.5	19.1	61.6	0.9	17.8	0	3.5	0	1.3	...
1000	0	0	0.1	0	62.4	2.6	34.7	58.1	2.7	30.3	0.1	6.9	0	1.7	...
1500	0	0	0	0	50.9	1.3	46.3	41.8	2.7	45.3	0.1	10.4	0	0.9	...
2000	0	0	0	0	41.3	0.2	54	33.4	4.3	45.7	0.4	14.9	0	3.4	...

Table 3. Number of mobility instances that converge to the Nash equilibrium in a given round. The rows indicate the cardinality of the instances for the two scenarios and the columns the number of steps. Each column is divided in two sub-columns: the ones labelled with e refer to the Edge-payment model and the ones labelled with p refer to the Path-payment model. The rows are divided in two subrows, the first one refers to the first scenario, the second one to the second scenario.

n		1		2		3		4		5		6		7		8	
		e	p	e	p	e	p	e	p	e	p	e	p	e	p	e	p
10	Scen. 1	45	15	50	66	5	19	0	0	0	0	0	0	0	0	0	0
	Scen. 2	44	12	50	72	5	15	1	1	0	0	0	0	0	0	0	0
100	Scen. 1	1	0	75	42	20	50	4	8	0	0	0	0	0	0	0	0
	Scen. 2	1	0	72	8	26	67	1	20	0	4	0	1	0	0	0	0
200	Scen. 1	0	0	65	39	28	55	6	5	1	1	0	0	0	0	0	0
	Scen. 2	1	0	61	4	33	65	5	27	0	3	0	1	0	0	0	0
300	Scen. 1	0	0	70	37	25	58	3	5	2	0	0	0	0	0	0	0
	Scen. 2	0	0	65	4	28	64	6	25	1	7	0	0	0	0	0	0
400	Scen. 1	0	0	67	29	27	56	6	14	0	1	0	0	0	0	0	0
	Scen. 2	0	0	60	1	35	55	4	39	1	4	0	1	0	0	0	0
500	Scen. 1	0	0	93	22	7	64	0	13	0	1	0	0	0	0	0	0
	Scen. 2	0	0	53	1	46	57	1	35	0	7	0	0	0	0	0	0
1000	Scen. 1	0	0	69	28	23	66	8	5	0	0	0	1	0	0	0	0
	Scen. 2	0	0	88	0	12	51	0	39	0	9	0	0	0	0	0	1
1500	Scen. 1	0	0	91	20	7	76	1	4	1	0	0	0	0	0	0	0
	Scen. 2	0	0	66	1	33	45	1	41	0	13	0	0	0	0	0	0
2000	Scen. 1	0	0	68	69	22	26	8	5	2	0	0	0	0	0	0	0
	Scen. 2	0	0	3	0	56	1	41	70	0	25	0	3	0	1	0	0

Quality of the Solutions. From the experiments we observe that, in the Edge-payment model $\mathrm{SC}(S) = \mathrm{WC}(S)$ for all the tested instances S.

The next question is, how far could be the social cost of the Nash equilibrium from the social cost of the optimum? Since the minimum spanning tree is a

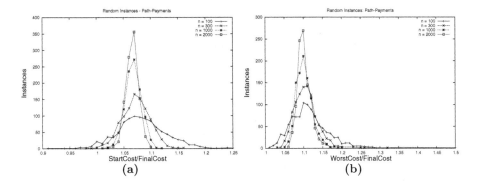

Fig. 2. Number of instances S with the same value $\texttt{SC}(S)/\texttt{FC}(S)$ (a) and $\texttt{WC}(S)/\texttt{FC}(S)$ (b) for random instances in the Path-payment model

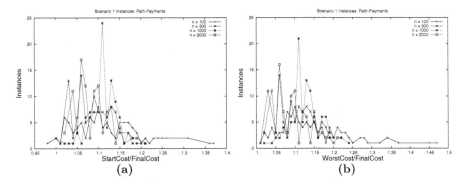

Fig. 3. Number of instances S with the same value $\texttt{SC}(S)/\texttt{FC}(S)$ (a) and $\texttt{WC}(S)/\texttt{FC}(S)$ (b) for mobility (scenario 1) instances in the Path-payment model

constant approximation of the optimum, we can compute the ratio between the cost of the minimum spanning tree (that is the starting solution) with the cost of the equilibrium. Notice that, due to the approximation property of the minimum spanning tree solution, this ratio is related with the cost-quality ratio.

In Fig. 2 it is shown the trend of the cost-quality ratio of the configurations generated by the $\texttt{findNE}$ algorithm for the Path-payment model. In particular, Fig. 2.(a) shows the number of instance S with the same ratio $\texttt{SC}(S)/\texttt{FC}(S)$ and Fig. 2.(b) shows the number of instance S with the same ratio $\texttt{WC}(S)/\texttt{FC}(S)$ for the Path-payment model.

For the Edge-payment model, the ratio $\texttt{SC}(S)/\texttt{FC}(S)$ has a similar trend to that in Fig. 2.(a). Notice that $\texttt{SC}(S) = \texttt{WC}(S)$ in the Edge-payment model.

We then conjecture that the Nash equilibrium created by the $\texttt{findNE}$ algorithm is a constant approximation of the optimal solution and the intermediate configurations are feasible solutions whose cost is an approximation of the optimal cost.

In Fig. 3 it is shown the number of instances S that have the same value $\texttt{SC}(S)/\texttt{FC}(S)$ and $\texttt{WC}(S)/\texttt{FC}(S)$ for mobility instances in the Path-payment model.

Also these results seem to support our conjecture that the Nash equilibria found by the findNE algorithm are constant approximations of optimal solutions. We obtain similar results for the other scenario and for the Edge-payment model.

4   Conclusions

In this paper we have studied the broadcast problem in the case of selfishly constructed ad-hoc networks. We used different non-cooperative game models depending on whether stations do not use payments, payments determine uniquely the existence of wireless links and agents can have profit. We have then considered two different payment policies, that is, Edge-payment and Path-payment, and we have proved the existence of a good (approximated) Nash equilibrium in the case of Path-payment and in the case of profit Edge-payment. Finally, we have given strong experimental evidence for the following conjecture (which is also the main problem left open by this paper): in the case of no-profit Edge-payment, there exists a polynomial time computable approximated Nash equilibrium that is an approximation of the optimal solution.

References

1. C. Ambühl. An optimal bound for the mst algorithm to compute energy efficient broadcast trees in wireless networks. In *Proc. of the 33rd International Colloquium on Automata, Languages and Programming (ICALP)*, 2005. To appear.
2. E. Anshelevich, A. Dasgupta, J. Kleinberg, É. Tardos, and T. Wexler. Near-optimal Network Design with Selfish Agents. In *STOC '03: Proceedings of the thirty-fifth annual ACM symposium on Theory of computing*, pages 511–520, New York, NY, USA, 2003. ACM Press.
3. E. Anshelevich, A. Dasgupta, J. Kleinberg, É. Tardos, T. Wexler, and T. Roughgarden. The Price of Stability for Network Design with Fair Cost Allocation. In *FOCS*, pages 295–304, 2004.
4. V. Bilò, M. Flammini, G. Melideo, and L. Moscardelli. On Nash Equilibria for Multicast Transmissions in Ad-Hoc Wireless Networks. In *Proc. of the 15th International Symposium on Algorithms and Computation (ISAAC)*, pages 172–183, 2004.
5. G. Călinescu, X.Y. Li, O. Frieder, and P.J. Wan. Minimum-Energy Broadcast Routing in Static Ad Hoc Wireless Networks. In *Proceedings of the 20th Annual Joint Conference of the IEEE Computer and Communications Societies (INFOCOM)*, pages 1162–1171, 2001.
6. A.E.F. Clementi, P. Crescenzi, P. Penna, G. Rossi, and P. Vocca. A Worst-case Analysis of an MST-based Heuristic to Construct Energy-Efficient Broadcast Trees in Wireless Networks. Technical Report 010, University of Rome "Tor Vergata", Math Department, 2001. Available at http://www.mat.uniroma2.it/~penna/papers/stacs01-TR.ps.gz (Extended version of [7]).
7. A.E.F. Clementi, P. Crescenzi, P. Penna, G. Rossi, and P. Vocca. On the Complexity of Computing Minimum Energy Consumption Broadcast Subgraphs. In *Proceedings of the 18th Annual Symposium on Theoretical Aspects of Computer Science (STACS)*, volume 2010 of *LNCS*, pages 121–131, 2001.

8. P. Crescenzi, M. Di Ianni, A. Lazzoni, P. Penna, G. Rossi, and P. Vocca. Equilibria for Broadcast Range Assignment Games in Ad-Hoc Networks. Technical Report, University of Rome "Tor Vergata", Math Department, 2005. Available at http://www.mat.uniroma2.it/~rossig/adhocnow2005extended.pdf.

9. S. Eidenbenz, V.S.A. Kumar, and S. Zust. Equilibria in Topology Control Games for Ad-Hoc Networks. In *Proceedings of the DIALM*, 2003.

10. A. Ephremides, G.D. Nguyen, and J.E. Wieselthier. On the Construction of Energy-Efficient Broadcast and Multicast Trees in Wireless Networks. In *Proceedings of the 19th Annual Joint Conference of the IEEE Computer and Communications Societies (INFOCOM)*, pages 585–594, 2000.

11. A. Jardosh, E. M. Belding-Royer, K.C. Almeroth, and S. Suri. Real world Environment Models for Mobile Ad hoc Networks. *IEEE Journal on Special Areas in Communications*, to appear.

12. R. Klasing, A. Navarra, A. Papadopoulos, and S. Perennes. Adaptive broadcast consumption (abc), a new heuristic and new bounds for the minimum energy broadcast routing problem. In *Proc. of the 3rd IFIP International Conference on Theoretical Computer Science (IFIP-TCS)*, volume 3042 of *LNCS*, pages 866–877, 2004.

13. D. Monderer and L.S. Shapley. Potential Game. *Games and Economic Behaviour*, 14:124–143, 1996.

14. C.L. Monma and S. Suri. Transitions in Geometric Minimum Spanning Trees. *Discrete & Computational Geometry*, 8:265–293, 1992.

15. M.J. Osborne and A. Rubinstein. *A Course in Game Theory*. MIT Press, 1994.

16. K. Pahlavan and A. Levesque. *Wireless Information Networks*. Wiley-Interscience, 1995.

17. P. Penna and C. Ventre. Energy-Efficient Broadcasting in Ad-Hoc Networks: Combining MSTs with Shortest-Path Trees. In *Proc. of the 1st ACM international workshop on Performance evaluation of wireless ad hoc, sensor, and ubiquitous networks (PE-WASUN)*, pages 61–68. ACM Press, 2004.

Efficient Mechanisms for Secure Inter-node and Aggregation Processing in Sensor Networks

Tassos Dimitriou

Athens Information Technology,
Markopoulo Ave., 19002, Athens, Greece
tdim@ait.edu.gr

Abstract. In this work we present a protocol for key establishment in wireless sensor networks. Our protocol is designed so that it supports security of data with various sensitivity levels. In particular, the protocol allows the establishment of a key that can be used for communication with the base station, pairwise keys that can be used to communicate with immediate neighbors and keys that allow for secure *in-network processing*. This last form of operation includes both secure aggregation and dissemination processing and is beneficial to sensor networks as it saves energy and increases network lifetime. Our proposed protocol is simple and scalable and exhibits resiliency against node capture and replication as keys are localized. Furthermore, our protocol allows for incremental addition of new nodes and revocation of compromised ones, while at the same time offers efficiency in terms of computation, communication and storage overhead.

1 Introduction

Sensor networks consist of hundred or thousands of inexpensive, low power devices equipped with one or more components for sensing, processing and exchanging information with other nodes and the environment. Potential applications include monitoring remote locations, target tracking, early fire detection, environmental monitoring, etc.

As sensor networks are usually deployed in hostile environments, many of these applications require that data must be exchanged in a secure and authenticated manner. Establishing secure communications between sensor nodes becomes a challenging task, given their limited processing power, storage, bandwidth and energy resources. Public-key algorithms, such as RSA are undesirable as they are computationally expensive. Thus, symmetric cryptosystems and hash functions constitute the basic tools for securing sensor network architectures [1].

The use of trust establishment protocols based on symmetric keys requires that the security of the network remains unaffected when some of the sensors become compromised by an adversary. Therefore it is necessary to maintain a balanced security level with respect to the constraints imposed by sensor nodes. Another issue that should affect the design and selection of a security protocol for sensor networks is the ability to allow for organization into clustered hierarchies.

V.R. Sirotiuk and E. Chávez (Eds.): ADHOC-NOW 2005, LNCS 3738, pp. 18–31, 2005.

In the clustered environment, data gathered by the sensors is processed within the network and only aggregated information is returned to the central location. In such a setting, certain nodes in the sensor network collect the raw information from the sensors, process it locally, and reply to the aggregate queries of the central server [2]. Such use of in-network processing allows for energy efficiency and increased lifetime of the network as the volume of communicated data is greatly reduced.

In this work we focus on the design of a simple and efficient key establishment protocol for sensor networks that supports security mechanisms for data with various sensitivity levels. Such separation allows efficient resource management that is essential for wireless sensor networks. In particular, our protocol supports the establishment of three types of keys for each sensor: a key that can be used for communication with the base station, pairwise keys that can be used to communicate with immediate neighbors and keys that allow for secure *in-network processing*.

The first two types of keys allow for the secure exchange of data and routing information with the base station and neighboring nodes, respectively. The third type was designed with both aggregation and dissemination processing in mind. Secure aggregation implies that data is forwarded from the sensors in a secure and authenticated way. Thus an adversary cannot issue false data into the network. Secure dissemination requires that lower level nodes are able to authenticate commands issued by their parents in the hierarchy. For example, in tracking applications the sensor network must be used in both modes: first to aggregate sensed data about the movement of the tracked object and then to disseminate commands to nearby sensors to enable further tracking. Most importantly, our protocol does not exclude the formation of clustering hierarchies as clusters are not necessarily confined to one hop neighbors.

Our protocol is simple and efficient since it does not involve the base station in establishing the security infrastructure. Furthermore, it offers resiliency against node capture as security breaches remain localized, and resiliency against replication as compromised nodes cannot be used to populate another part of the network. So, even if the keys of a node are exposed, an adversary can have access only to a small portion of the network centered around the compromised node. Finally, the protocol allows for incremental addition of new nodes.

2 Related Work

There are many protocols that attempt to establish pairwise keys between sensors. A typical example is the pebblenets architecture [3] where a global key is shared by *all* nodes. Having a network wide key is very efficient in terms of memory requirements and ease of use but it suffers from the obvious security disadvantage that compromise of a single node undermines the security of the entire network. Moving to the other extreme, one may try to establish pairwise keys between all pairs of nodes. Although this solves the network compromise problem mentioned above, scalability here is the main issue as sensors have only a few Kbytes of memory and cannot handle the requirements of such an architecture.

The observation that sensors can only directly communicate with their neighbors suggests that one should aim at securing the neighborhoods of such nodes. Random key pre-distribution schemes [4,5,6,7] offer a tradeoff between the level of security achieved and the memory storage required by each sensor to store the keys that have been *randomly* chosen from a key pool. In such schemes neighboring sensors can only communicate if they share one or more of these random keys. Although key establishment is usually easy, these schemes offer only "probabilistic" security as compromise of a single node may result in a breach of security in some other part of the network. Furthermore, these schemes have not been developed with the notion of aggregation processing in mind which makes establishing group keys less straightforward.

A protocol that uses similar techniques to ours is LEAP [8]. In this work the authors describe a methodology for establishing pairwise and other keys between sensor nodes with the help of a pre-deployed *master key* that eventually gets deleted from the memory of the sensors. However, even if the master key is deleted, the LEAP protocol can be attacked. More specifically an attacker may force a sensor node to compute pairwise keys with other (or *all*) nodes in the network. This is achieved by having the attacker broadcast a large number of HELLO messages (nothing prevents her from doing so). The recipient node, will compute all the pairwise secret keys according to the protocol. Then, once the neighbor discovery phase terminates, an attacker can compromise a sensor node and have in her procession keys that are shared between the compromised node and all other nodes in the network.

Two architectures that attempt to secure aggregation hierarchies but do not provide for different levels of keying material were also discussed in [9] and [10]. The authors in [9] propose a collection of mechanisms to address the requirements imposed by secure in-network processing. Their approach, however, makes heavy use of the base station as a means for introducing and delegating authorization of aggregators to group nodes and for establishing keys between the nodes and the aggregators. Furthermore, this scheme is not scalable as long lists of keys and topology information need to be transmitted.

In [10], the authors introduce gateway nodes, which function as the clusterheads in our proposal. The protocol, however, requires a large number of messages to be exchanged during the initialization phase. This is due to the fact that gateway nodes in their protocol are randomly assigned sensor node keys, which requires gateways to exchange large number of keys if a node happens to "land" in the wrong cluster.

In [11] a key management architecture is designed that provides two levels of security. The highest level is supported by supervised clusters, where the cluster head (i.e. the supervisor) is assumed to be tamper resistant and have high energy and computational resources. The lower level of security is supported be unsupervised clusters, where members share a common cluster key.

The use of different levels of security was studied in Slijepcevic *et al.* [12]. However, their model works under the assumption that sensor nodes are able to discover their exact location so that they can organize into cells and produce a location-based key. Moreover, the authors assume that sensor nodes are tamper resistant, otherwise the set of master keys can be revealed by compromising a single node and the whole network security collapses. Those assumptions are too demanding for sensor networks.

3   The Protocol

Our key establishment protocol is based on the use of symmetric keys and consists of sub-protocols that define how the various types of keys get generated as well as how keys are added, revoked or renewed during the lifetime of the network. As sensors have energy and computational constraints, it is necessary to maintain a balanced security level with respect to these constraints. The following notation will be used throughout the paper.

Notation	Meaning
ID_i or S_i	Identifier of sensor node i.
C_j	Identifier of j-th clusterhead.
M_1, M_2	Concatenation of messages M_1 and M_2.
$E_K(M)$	Encryption of message M using key K.
$MAC_K(M)$	Message Authentication Code (MAC) of message M using key K.
N_i	Nonce used by node i.
K_{ij}	Symmetric key shared by nodes i and j (i and j can be S_i, C_j.)

Each sensor node i has a unique ID_i that identifies it in the network and comes equipped with the following set of keys.

- K_i^B: This key is shared between node i and the base station B. It is not used in establishing the security infrastructure but only to encrypt data D that must be seen by the base station and reach it in a secure manner.
- K_i: This key uniquely identifies each node and is the result of the application of a secure one-way function F on a master key K_m and the node ID i, i.e.

$$K_i = F(K_m, i).$$

These keys will be used for specialized tasks like bootstrapping new nodes, associate new nodes to a cluster, etc. The one-wayness of F guarantees that if a node is compromised and K_i is known, an adversary cannot recover the master key K_m or the keys of other nodes.
- K_i^C: This group key will be used *only* by those nodes that will become clusterheads to secure information (or commands) that must be *disseminated* to nodes in the cluster. It will be convenient to assume that all such group keys result from the application of a secure pseudo-random function F on some master key K_m^C (cluster master key). This extra master key is not preloaded to sensors but it is known by the base station.

- K_m: A master key shared among all nodes, including the base station. This key will be used to secure information exchanged during the cluster key setup phase. Then it is *erased* from the memory of the sensor nodes.

The base station holds K_m along with the keys of each sensor node in the network. However, apart from the above keys that are preloaded to sensors during manufacture, there are also two types of symmetric keys that are constructed *dynamically* and allow information to be exchanged securely between a node and its neighbors as well as between a node and its clusterhead. The construction of such keys is the topic of Sections 3.1 and 3.2, respectively.

3.1 Secure Inter-node Communication

The use of pairwise keys allows a node to communicate securely with its *immediate* neighbors. The pairwise key-establishment phase takes place during neighborhood discovery. Each node broadcasts a hello message containing its ID and a nonce to avoid replay attacks, properly MACed as follows:

$$S_i \rightarrow Neighbors : ID_i, N_i, MAC_{K_m}(ID_i, N_i)$$

Upon receiving this message, sensor j uses the master key K_m to check the MAC. If the MAC verifies, node j adds node i in its list of neighbors. Otherwise it rejects it. If verification is successful, both nodes i and j compute their pairwise key K_{ij} as:

$$K_{ij} = F(K_m, i, j), \quad \text{where} \quad i < j.$$

Notice that the use of K_m in the messages above guarantees that an adversary cannot create faked associations as K_m is not known in advance and eventually will be deleted (Section 3.2). Finally, a good security practice is to use different keys for different cryptographic operations; this prevents potential interactions between the operations that might introduce weaknesses in a security protocol. Although not shown here, we use independent keys for the encryption and authentication operations, K_{encr} and K_{MAC} respectively, which are derived from a symmetric key K through another application of the pseudo-random function F, i.e. $K_{encr} = F(K, 0)$ and $K_{MAC} = F(K, 1)$. To simplify the notation, however, we usually omit including these keys in the security mechanisms described.

3.2 Secure Aggregation

In a clustered environment, the data gathered by the sensors is processed and aggregated *within* the network. In such a setting, certain nodes in the sensor network collect the raw information from the sensors, process it locally, and reply to the aggregate queries of the central server. This scheme results in energy efficiency as sensors are now communicating data over smaller distances.

We assume that the network is partitioned into distinct clusters and that each cluster is composed of an aggregator and a set of sensor nodes (distinct

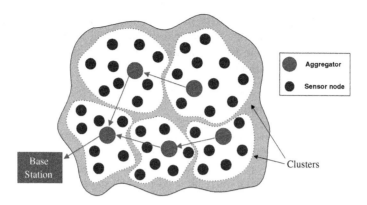

Fig. 1. A hierarchical sensor network. Arrows point to upper hierarchical levels of the tree which clusterheads may form. Although clusterheads are shown bigger than simple sensor nodes, it should not be inferred that they are more powerful.

from other sets), which gather information and transmit it to the aggregator of their cluster. The aggregator fuses the data from the different sensors, performs mission-related data processing, and advances it further to the hierarchy. In general, we assume a hierarchical model of aggregation processing where *clusterheads/aggregators* may form multiple levels of hierarchies encompassing any number of sensor nodes in the cluster (Figure 1).

To ensure privacy and integrity of messages sent *from* sensor nodes *to* their corresponding clusterheads a separate pairwise secret key between each sensor node and its clusterhead is required. Let S_i be a node that is attached to the cluster of clusterhead C_j and let K_{ij} the symmetric key that needs to be established between them. During the cluster formation process, each cluster node S_i transmits a hello message consisting of the node's ID and a nonce N_i, MACed using the master key K_m:

$$S_i \rightarrow C_j : ID_i, N_i, MAC_{K_m}(ID_i, N_i)$$

Upon receiving this message, C_j checks the MAC. If the MAC verifies, the sensor node is included in the cluster and the clusterhead stores all relevant information (such as ID_i, nonce identifier - to avoid replay attacks, etc.). Then both the sensor and the clusterhead compute the pairwise key K_{ij} by applying the pseudo-random function F:

$$K_{ij} = F(K_m, i, j), \quad \text{where} \quad i < j$$

After this process is completed for all the sensor nodes in the cluster, the master key is *deleted* from the memory of all sensor nodes. We have now established a secure channel between neighboring nodes as as well as between the clusterheads and their corresponding cluster nodes.

An implicit assumption here is that *the time required for the establishment of secure channels is smaller than the time needed by an adversary to compromise a node during deployment.* As security protocols for sensor networks should *not* be designed with the assumption of tamper resistance [13], we must assume that an adversary needs more time to compromise a node and discover the master key K_m (see also [8] for a similar assumption).

3.3   Secure Dissemination

A closely related form of in-network processing is *data dissemination*, in which control messages get disseminated from the central server to the clusterheads and eventually to the sensor nodes in the clusters. For example, in tracking applications the sensor network must be used in both modes: first to aggregate sensed data about the movement of the tracked object and then to disseminate commands to nearby sensors to enable further tracking.

Thus a mechanism is required so that commands can be propagated to *all* sensors within a cluster. One simple way to do this is to send a separate unicast message to each member of the group encrypted and authenticated using the key shared between the clusterhead and the specific node. However, this method is clearly inefficient as the same command must be transmitted unnecessarily many times resulting in wasting valuable energy resources.

A much simpler way is to create a *group* (or cluster key) and send it over a secure channel to each member of the cluster. It is this key that will be used subsequently for command dissemination. To propagate its group key K_i^C, the i-th clusterhead C_i uses the pairwise key K_{ij} it shares with each sensor S_j in its cluster to encrypt and authenticate the group key. The message sent to S_j is:

$$C_i \rightarrow S_j : \quad C_i, c, \sigma$$

where

$$c = E_{K_{encr}}(\text{``Group key''}, C_i, K_i^C) \quad \text{and} \quad \sigma = MAC_{K_{Mac}}(c).$$

First the clusterhead encrypts the group key using the key K_{encr} derived from K_{ij} and then creates a MAC σ of the resulting ciphertext c. Then it transmits the message. Observe here that if we make the reasonable assumption that nodes form hierarchies and are organized in a breadth first tree based on some routing protocol, then the group key can be distributed recursively using intermediate level clusterheads until all the sensor nodes at the leaves are reached.

Once the group key K_i^C is settled, the clusterhead can broadcast commands to all the nodes in the group encrypted and authenticated using K_i^C. However, although this approach defends against outside attacks in which the adversary does not hold any keys, inside attacks are possible after an adversary compromises a sensor node. In such a case, a malicious node may use the group key to send forged messages to other nodes in the same group.

To defend against the impersonation attack described above, we propose to use a one-way hash key chain $OWHC_i$ computed by each clusterhead C_i. Such a chain is a sequence of keys, $k_0, k_1, \ldots, k_{n-1}, k_n$ such that

$$\forall l, \quad 0 < l \le n, \quad k_{l-1} = F(k_l),$$

where as usual F is a secure pseudo-random function that is difficult to invert.

Each clusterhead generates a different one-way hash chain of length n and commits to the first key k_0 by transmitting this securely to all the sensor nodes during the group key generation phase. Whenever clusterhead C_i has a new command to disseminate to the nodes, it attaches to the command the *next* key from the hash chain. In particular, the l-th command contains the l-th commitment of the hash chain, encrypted and authenticated using keys derived from K_i^C as follows:

$$C_i \to Group: \quad C_i, c, \sigma$$

where

$$c = E_{K_i^C}(\text{``Some command''}, C_i, k_l) \quad \text{and} \quad \sigma = MAC_{K_i^C}(c).$$

A node receiving a command encrypted with the group key can verify its authenticity by checking whether the new commitment k_l generates the previous one through the application of F. When this is the case, it replaces the old commitment k_{l-1} with the new one in its memory and accepts the command as authentic. Otherwise it rejects it.

Note, however, that although the possibility of an impersonation attack is reduced, it is not completely eliminated by this scheme. If an adversary *jams* communications to sensor S_j then it can introduces new commands by "recycling" the unused commitments. One way to defend against this possibility is to assume that sensors are loosely synchronized [15] and commands are issued only at *regular* time intervals. Then a command issued by an clusterhead at time t will contain commitment k_t. If a sensor node does not receive anything within the next d slots and the last commitment was k_t, it will expect to see the commitment k_{t+d} that accounts for d missing commands. This way it cannot be fooled to authenticate unused commitments.

To avoid the computational burden imposed to sensor nodes by the previous scheme, a simpler defense is to have the sensor nodes *acknowledge* the issued command using the pairwise key K_{ij}. To decrease the traffic that will be generated in the cluster by such a scheme, sensor nodes may acknowledge messages according to some *probability* distribution. The acknowledgement probability p must be chosen so that a balance is achieved between increasing network traffic and detecting compromised nodes. As this solution can be part of an overall defensive strategy, we leave this part for future investigation.

Finally, when the hash key chain is exhausted the clusterhead creates a new chain $k_0', k_1', \ldots, k_n'$ and broadcast a "renew hash key chain" command which contains the new commitment k_0' authenticated with the last unused key of the old chain, essentially providing the connection between the two chains. In a similar manner the group key can be refreshed to defend against cryptanalysis and ensure forward secrecy of previously distributed messages. The clusterhead can broadcast a "refresh group key" command after which every node can refresh the group key through the application of the pseudo-random function F. When it does so, it erases the old key from its memory.

3.4   Eviction of Compromised Nodes

We now discuss a mechanism for evicting compromised nodes from the network. We assume the existence of a detection mechanism that informs the base station or the clusterheads about compromised nodes (see also the related work of [14] on detecting forged aggregation values).

If a node is compromised, the attacker cannot insert duplicates of that node in groups other than the group it originated from. This is the case, since no other clusterhead will be able to compute the pairwise key, which is essential for establishing the secure communication channel. Moreover, the replicated node will not be able to establish pairwise keys with nodes other than the original node's neighbors, hence security breaches remain *localized*. Therefore, it suffices to provide a mechanism for node revocations transmitted by the base station to be authenticated and eventually for clusterheads to revoke nodes within their groups.

As in the case of authenticating commands issued by the clusterheads, we will base the revocation scheme on the use of one-way hash key chains. To use this scheme, we follow a μTESLA-like approach [15] and assume that sensor nodes are loosely synchronized and each clusterhead is preloaded before deployment with the first key of the chain. The base station can then broadcast authenticated lists of compromised nodes to all clusterhead by disclosing the keys of the key chain one by one. When clusterheads receive such lists of compromised nodes, they verify the authenticity of the messages and then evict these nodes from the cluster by establishing a new group key with the rest of the nodes using the method described in Section 3.3. They can also notify group members about the identities of compromised nodes, so that individual nodes delete the corresponding pairwise keys from their memories. Eventually, only the nodes that are not in the revocation list will obtain the new group key while compromised nodes will not be able to communicate with the rest.

When a clusterhead has been compromised, the entire cluster is considered to be compromised as well since clusterheads hold pairwise keys with all the cluster nodes. In such a case, the parent nodes in the cluster hierarchy can establish new group keys with the rest of the sensors/clusterheads using the method described above.

3.5   Addition of New Nodes

This section address the problem of refreshing the network as sensors usually die of energy depletion. We assume that new sensors are arbitrary deployed. As they cannot be preassigned to a specific cluster, they must i) establish pairwise keys with the corresponding clusterhead and ii) acquire the keys of their neighbors for inter-node communication.

New sensor nodes come equipped with both master keys: K_m and K_m^C. Again we must assume that an adversary cannot compromise a new node during this phase. This is a valid assumption as the time to compromise a node and discover its keys is usually larger than the key establishment phase. When a new node S_k discovers that it is assigned to the j-th cluster, it can compute the group key using the formula

$$K_j^C = F(K_m^C, j).$$

In a similar manner it can compute the pairwise key K_{kj} it shares with the particular clusterhead using the expression

$$K_{kj} = F(K_m, k, j).$$

This is needed for data to be sent securely towards the clusterhead. However, as the clusterhead cannot compute K_{kj} as it does not have the master key K_m anymore, sensor S_k must transmit it securely to C_j. This is done as follows:

$$S_k : K_j = F(K_m, C_j), c = E_{K_j}(S_k, C_j, K_{kj}), \sigma = MAC_{K_j}(c)$$
$$S_k \rightarrow C_j : S_k, c, \sigma$$

First, the sensor node computes the unique key K_j held by the clusterhead and then transmits the pairwise key K_{kj} in a secure and authenticated manner. As K_j is known only to the clusterhead C_j, the clusterhead accepts the new node in its cluster and an adversary cannot populate the cluster with new nodes.

A similar procedure can be used by node S_k to establish pairwise keys with all its neighbors. First the new node broadcasts a node discovery request and then each node responds with its ID S_i authenticated using the key K_i. This is needed in order to prevent an adversary from realizing the following attack: the adversary sends faked messages containing various IDs. When the new node makes the association between the ID and the pairwise key and stores it in its memory, the adversary can later compromise the node thus having acquired pairwise keys with every node in the neighborhood (or even worse in the network). To prevent this type of impersonation attack the response sent by existing nodes is simply

$$ID_i, MAC_{K_i}(ID_i).$$

The new node, upon receiving these IDs, will first compute $K_i = F(K_m, i)$ and then verify the MAC. If everything is ok, it will construct the pairwise key K_{ki} and transmit it to each neighbor S_i encrypted each time with K_i, just like it was done with the clusterhead. When the phase is over, both master keys K_m and K_m^C are deleted from the memory of the new nodes.

4   Resource Requirements

In this section we analyze the computational, communication and storage requirements for our key establishment protocol. The individual key of each sensor node is pre-computed and does not involve any processing or transmission overhead. The cost for establishing a group key in our protocol is the same as updating the group key in the cluster, thus we only analyze the cost for establishing the group key. We should emphasize at this point that our protocol is independent of the underlying routing protocol used.

4.1   Computational/Communications Cost

In order for nodes to establish pairwise keys with their neighbors, every node has to broadcast his ID authenticated with K_m and then compute the key K_{ij} by applying the pseudo-random function F. Thus on average, a node has to perform $d + 1$ operations, where d is the average density of the network.

To join a cluster, a sensor node has to compute the message described in Section 3.2 which includes a MAC computation and then to compute a pairwise key. To process this message the clusterhead has to perform two operations. First, it needs to verify the message and then to compute the pairwise key by applying a secure pseudorandom function on the master key. Therefore, if we assume that a cluster includes D sensor nodes on average, the clusterhead has to compute D values from the pseudorandom function and perform D MAC verifications during the initialization phase.

In order to establish a group key, the clusterhead has to compute the message described in Section 3.3 for each sensor node in its cluster. This message includes an encryption and a MAC operation based on a secret key. Additionally the clusterhead has to compute the one way key chain. Again, assuming D sensor nodes in each cluster on average and an n level OWHC, the clusterhead needs to perform D encryptions, D MAC and n hash operations. For the same process, each sensor node has to perform a decryption and MAC verification operation.

Since both the initialization and group establishment phases take place *only* during network setup, we see that the computational/communications cost is really averaged out throughout the network's life. The only "expensive" operation is the computation of the hash key chain by the clusterhead. But since the clusterhead uses a commitment only when it has to issue a command, we see again that the computation cost per hash value is much smaller than the transmission cost of the transmitted command [1].

We have also experimented with the number of encryptions/decryptions required for a message originated from a sensor node to reach the base station as a function of the number of clusterheads in the network. This is needed since security mechanisms always impose an overhead to communications. Using the hierarchical structure of the network, however, this number remains small even for large network sizes (details omitted due to space restrictions).

4.2   Storage Requirements

Each sensor node has to store in memory the key it shares with the base station, its unique key K_i, the current group key, the pairwise keys it shares with the clusterhead and its neighbors, and the current value of the hash key chain. Assuming that the average density of the network is d and that each key is 10 bytes long (80 bits), each sensor has to store $10(d + 4)$ bytes of information. Each clusterhead needs to store the pairwise key shared with each sensor node in its cluster (D in total), at most n values of the OWHC and its individual key, plus the previous set of keys needed for communication with nodes higher in the hierarchy (for example with the Base Station). Assuming that each of the above values is 10 bytes long, each clusterhead has to store $10 * (D + d + n + 4)$ bytes of information.

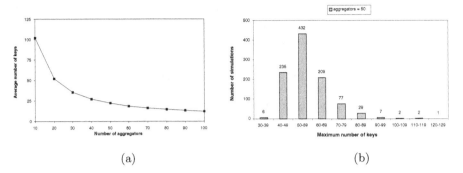

Fig. 2. (a) Average number of keys stored per clusterhead for a network of 1000 nodes and clusters of approximately the same size. (b) *Maximum* number of keys (excluding the hash chain key material) stored in clusterheads when the number of clusterheads is equal to 50 (approximately $D = d = 20$). The graph shows that in 432 out of 1000 simulations the number of keys is between 50 and 59. However, there always exist clusterheads that have twice as many keys (tail of distribution).

The values of d and D really depend on the density of the network and the particular clustering protocol used, while the value of n depends on the frequency the clusterhead issues commands. If we assume $D = d = 20$ and $n = 50$, each sensor node has to store 240 bytes of keying material while each clusterhead has to store 940 bytes in memory (see Figure 2). Since the value of n can be as small it takes (provided the hash chain is renewed regularly), the above analysis shows that memory requirements are not an issue in our scheme, since typical sensor nodes have significantly more available memory.

5   Security Analysis

We now analyze the security achieved by our scheme by discussing one by one some of the general attacks [16] that can be applied to routing protocols in order to take control of a small portion of the network or the entire part of it.

- *Spoofed, altered, replayed routing information and Selective forwarding.* In this kind of attack an adversary selectively forwards certain packets through some compromised node while drops the rest. Although such an attack is always possible once a node is compromised, its consequences are insignificant since nearby nodes can have access to the same information and detect these attacks. Then the compromised node and its keying material can be revoked.
- *Wormhole attacks.* In our protocol such an attack can only take place during the key establishment phase. But the authentication that takes place in this phase and its small duration makes this kind of attack impossible.
- *Sybil attacks.* Since every node shares a unique symmetric key with the trusted base station, a single node cannot present multiple identities. An

adversary may create clones of a compromised node and populate them into the same cluster or the node's neighboring clusters but this doesn't offer any advantages to the adversary with respect to the availability of the information to the base station.

- *Hello flood attacks.* In our protocol, nodes broadcast a HELLO message during the neighborhood discovery phase. Since, however, messages are authenticated this attack is not possible. (A necessary assumption for all key establishment protocols is of course that the duration of this phase is small so that an adversary cannot compromise a node and obtain the key K_m.)
- *Acknowledgment spoofing.* Since we don't rely on link layer acknowledgements this kind of attack is not possible in our protocol.

6   Conclusions

We have presented a key establishment protocol suitable for sensor network deployment. The protocol provides security against a large number of attacks and resiliency against node capture and replication mainly because of its authentication character. Our protocol allows nodes to share keys with the base station, their neighboring nodes, but most importantly it supports the notion of secure in-network processing.

The benefits of in-network processing include improved scalability since the aggregators themselves may be aggregated to form multi-level hierarchies, and prolonged lifetime since aggregation reduces the volume of data communicated through the network. In particular, we have designed the proposed mechanisms with both aggregation and dissemination in mind. Our proposal accommodates both secure aggregation so that data be forwarded securely to the base station, and secure dissemination so that commands issued by the base station and the clusterheads become authenticated by sensor nodes. This increases the versatility of the network as important applications such as target tracking can be efficiently implemented.

We have also demonstrated how to add new nodes to the network, a critical requirement, as sensors have limited energy and thus limited life expectancy. Additionally, given the existence of an intrusion detection mechanism, our protocol allows for the eviction of compromised nodes. Finally, the proposed protocol scales very efficiently as the key establishment mechanisms are very efficient in terms of computation, communication and storage requirements.

References

1. D. Carman, P. Kruus, and B.J.Matt, "Constraints and approaches for distributed sensor network security," Tech. Rep. 00-010, NAI Labs, June 2000.
2. C. Intanagonwiwat, R. Govindan, D. Estrin, J. Heidemann, and F. Silva, "Directed diffusion for wireless sensor networking," *ACM/IEEE Transactions on Networking*, vol. 11, February 2002.

3. S. Basagni, K. Herrin, D. Bruschi, and E. Rosti, "Secure pebblenet," in *Proceedings of the 2001 ACM International Symposium on Mobile Ad Hoc Networking & Computing, MobiHoc 2001*, pp. 156–163, October 2001.
4. L. Eschenauer and V. D. Gligor, "A key-management scheme for distributed sensor networks," in *Proceedings of the 9th ACM conference on Computer and communications security*, pp. 41–47, 2002.
5. H.Chan, A.Perrig, and D.Song, "Random key predistribution schemes for sensor networks," in *IEEE Symposium on Security and Privacy*, pp. 197–213, May 2003.
6. D. Liu and P. Ning, "Establishing pairwise keys in distributed sensor networks," in *Proceedings of the 10th ACM Conference on Computer and Communication Security*, pp. 52–61, October 2003.
7. W. Du, J. Deng, Y. S. Han, and P. K. Varshney, "A pairwise key pre-distribution scheme for wireless sensor networks," in *Proceedings of the 10th ACM conference on Computer and communication security*, pp. 42–51, October 2003.
8. S. Zhu, S. Setia, and S. Jajodia, "LEAP: efficient security mechanisms for large-scale distributed sensor networks," in *Proceedings of the 10th ACM Conference on Computer and Communication Security*, pp. 62–72, October 2003.
9. J. Deng, R. Han. S. Mishra, "Security Support for in-network Processing in Wireless Sensor Networks," in *ACM Workshop on Security of Adhoc Networks (SASN)*, 2003.
10. G. Jolly, M. Kuscu, P. Kokate and M. Younis, "A Low-Energy Key Management Protocol for Wireless Sensor Networks," in *Proc. 8th International Symposium on Computers and Communication*, 2003
11. Y. W. Law, R. Corin, S. Etalle, and P. H. Hartel, "A formally verified decentralized key management architecture for wireless sensor networks," in *4th IFIP TC6/WG6.8 Int. Conf on Personal Wireless Communications (PWC)*, vol. LNCS 2775, pp. 27–39, September 2003.
12. S. Slijepcevic, M. Potkonjak, V. Tsiatsis, S. Zimbeck, and M. Srivastava, "On communication security in wireless ad-hoc sensor networks," in *11th IEEE International Workshops on Enabling Technologies: Infrastructure for Collaborative Enterprises*, pp. 139–144, June 2002.
13. R. Anderson and M. Kuhn, "Tamper resistance – a cautionary note," in *Proceedings of the Second Usenix Workshop on Electronic Commerce*, pp. 1–11, November 1996.
14. B. Przydatek, D. Song and A. Perrig, "SIA: Secure Information Aggregation in Sensor Networks," *in ACM SensSys*, 2003.
15. A. Perrig, R. Szewczyk, V. Wen, D. Culler, and J. Tygar, "SPINS: Security Protocols for Sensor Networks," In *Proc.of Seventh Annual ACM International Conference on Mobile Computing and Networks(Mobicom 2001)*, 2001.
16. C. Karlof and D. Wagner, "Secure routing in wireless sensor networks: Attacks and countermeasures," *Elsevier's Ad Hoc Network Journal, Special Issue on Sensor Network Applications and Protocols*, vol. 1, pp. 293–315, September 2003.

Cluster-Based Framework in Vehicular Ad-Hoc Networks

Peng Fan[1], James G. Haran[2], John Dillenburg[2], and Peter C. Nelson[1]

[1] Artificial Intelligence Lab, University of Illinois at Chicago,
Department of Computer Science, Chicago, IL 60607, USA
{pfan, nelson}@cs.uic.edu
[2] University of Illinois at Chicago, Department of Computer Science,
851 S. Morgan (M/C 152), Room 1120 SEO, Chicago, IL 60607, USA
{jharan, dillenbu}@cs.uic.edu

Abstract. The application of Mobile Ad Hoc Network (MANET) technologies in the service of Intelligent Transportation Systems (ITS) has brought new challenges in maintaining communication clusters of network members for long time durations. Stable clustering methods reduce the overhead of communication relay in MANETs and provide for a more efficient hierarchical network topology. During creation of VANET clusters, each vehicle chooses a head vehicle to follow. The average number of cluster head changes per vehicle measures cluster stability in these simulations during the simulation. In this paper we analyze the effect of weighting two well-known clustering methods with the vehicle-specific position and velocity clustering logic to improve cluster stability over the simulation time.

1 Introduction

Vehicular Ad Hoc Networks (VANETs), an outgrowth of traditional Mobile Ad Hoc Networks (MANETs), provides the basic network communication framework for application to an Intelligent Transportation System (ITS). The U.S. Federal Communications Commission (FCC) has recently allocated the 5.85-5.925 GHz portion of the spectrum to inter-vehicle communication (IVC) and vehicle-to-roadside communication (VRC) under the umbrella of dedicated short-range communications (DSRC). This has fuelled significant interest in applications of DSRC to driver-vehicle safety applications, infotainment, and mobile Internet services for passengers.

Vehicles provide a robust infrastructure for the creation of highly mobile networks. In addition to providing a stable environment for the low cost and robust wireless communication devices typical of ad hoc networks, vehicles can easily be equipped with the storage, processing, and sensing devices necessary in any ITS implementation. A huge opportunity exists to leverage VANETs to enable a wide variety of service and societal applications.

VANETs have significant advantages over the traditional MANETs. Vehicles can easily provide the power required for wireless communication devices and will not be seriously affected by the addition of extra weight for antennas and additional hardware. Furthermore, it can be generally expected that vehicles will have an accurate

V.R. Sirotiuk and E. Chávez (Eds.): ADHOC-NOW 2005, LNCS 3738, pp. 32–42, 2005.

knowledge of their own geographical position, e.g. by means of Global Positioning Satellite (GPS). Thus, many of the issues making deployment and long term use of ad hoc networks problematic in other scenarios are not relevant in MANETs.

In addition, there is a wealth of desirable applications for ad-hoc communication between vehicles ranging from emergency warnings and distribution of traffic and road conditions to chatting and distributed games. As a consequence many vehicle manufacturers and their suppliers are actively supporting research on how to integrate mobile ad hoc networks into their products.

Vehicles in a VANET environment move within the constraints of traffic flow while communicating with each other via wireless links. Ah hoc networks use less specialized hardware for infrastructure support and leave the burden of network stability on the individual nodes within the network. Without routers, or other dedicated communication hardware, a possible method to optimize communication within the network is to develop a hierarchical clustering system within the network. This clustering system would identify certain lead or cluster head vehicles that act as the relay point of communication between vehicles local to that node and other vehicle clusters. To support the dynamic nature of the VANET environment, the vehicles clustering must be periodically updated to reflect topological changes and vehicle movements. Clustering within the network must be very fast to minimize time lost to clustering [16].

Association with and dissociation from clusters, as a result of the mobile nature of VANET nodes (vehicles) perturb the network and cluster selections. Cluster reconfiguration and cluster head changes are unavoidable. Therefore, a good VANET clustering algorithm should seek to regulate rather than eliminate cluster changes. This algorithm should also maintain cluster stability as much as possible during vehicle velocity and acceleration changes and/or traffic topology shifts. Otherwise, the overhead of cluster re-computation and the involved information exchange will result in high computational cost and negate the benefits of VANET communication. The ideal VANET cluster will maintain its cluster head and members over the longest possible time range. This concept will be explained and evaluated further later in this paper.

A significant amount of research focuses on optimal methods for clustering nodes in MANETs. VANETs, however, pose new challenges in cluster head selection and network stability. VANETs must follow a tighter set of constraints than MANETs, and therefore require specialized clustering algorithms. First, nodes or vehicles cannot randomly move within the physical space, but must instead follow constraints set in place by the real road network topology. Second, vehicle movements follow well-understood traffic movement patterns. Each vehicle is constrained by the movements of surrounding vehicles. Third, vehicles generally travel in a single direction and are constrained to travel within a two-dimensional movement. Given these movement restrictions and the knowledge of position, velocity, and acceleration common available to on-board vehicle systems it is possible to approach clustering more intelligently and possibly discover a better clustering methodology for VANET environments.

The constrained environmental conditions of VANETs warrant a constrained simulation environment. Many simulation tools and environments have been designed for MANET implementations. These tools, however, fail to adequately model the needs of a VANET network. Compared to the random movements modeled in MANET environments, VANET simulation movements must behave according to traffic patterns in terms of car-following, lane-changing, directional movement, velocity, and acceleration among others. Current MANET simulation environments cannot be considered suitable for VANET simulations even in the broadest sense. Therefore, simulation of the network environment is best performed with traffic micro-simulation tools. For the purpose of this study, simulation and traffic modeling was performed using a micro-simulation tool specially modified to perform randomized vehicle-based clustering under a number of algorithms and traffic constraints. This approach also allows further research on traffic statistics and flow improvements as a result of network communication. Further modifications to the environment were made to log vehicle cluster, position, velocity, and acceleration states during simulation activity.

This research concerns the augmentation of two well-known clustering algorithms with two additional traffic-specific algorithms to determine whether the clustering stability can be improved for VANETs. This works focuses on simulation of these algorithms in a constrained micro-simulation environment using a compound-weighting scheme devised at part of this work. Additionally, this document discusses a utility function implementation for determining the vehicle cluster head selection priority based on a multiple-metric weighting algorithm.

2 Backgrounds and Related Work

2.1 Clustering in MANETs

Communication network clustering organizes the network nodes into a hierarchical arrangement. Figure 1 provides and example of the organization of twelve nodes into three clusters. The basic communication capability between the twelve nodes is outlined as connections between the bottom tier of the hierarchy. These twelve basic nodes are then grouped into clusters using some algorithm. In the upper tier of Figure 1, the three cluster head nodes are displayed with connections between them representing the possible message paths under the cluster-constrained network [8,9, 14].

This clustered architecture reduces the communication relay points for each node to a small subset of the total network. Each cluster head aggregates local member topology and acts as a relay point for communication between its members and members of other clusters. This reduces the messages exchanged between individual network nodes and the overhead of information stored within those nodes [13].

Attention on clustering in MANETs has increased considerably as wireless technologies improve and MANET theories become practice [1,2,3]. Most of these approaches embrace the role of a cluster head that maintains the cluster and provides the entry point of that cluster into the broader network. Among several proposed cluster head selection algorithms the predominant approaches are the (i) Lowest-ID algorithm and (ii) Highest-Degree algorithm. Recent work has simulated the performance of these algorithms using random placement in a square grid with multi-directional node movement [2,4,7]. As previously stated, this does not translate well into the VANET environment.

This research does not consider network broadcasts requiring more than one hop in communication. This simplifies the overall communication and clustering strategy and reduces the overall bookkeeping necessary to maintain the clusters. This approach seeks to obtain optimal results by adding traffic-specific information to the clustering logic.

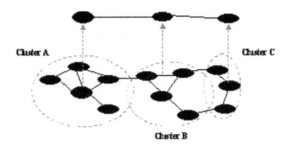

Fig. 1. Clustering within a 12-node MANET environment

2.2 Lowest-ID Algorithm

The Lowest-ID algorithm involves the selection of cluster heads by means of an absolute ordering of a fixed vehicle ID attribute. Cluster formation is performed using node-level election of cluster heads. During the clustering stage, each node within the network broadcasts its ID to all other reachable nodes. Each node, in turn, chooses as its cluster head the node with the lowest ID. This method has been discussed in great detail [4,5,6] in a number of works and is well known for its stability in general MANET applications. In each cluster, the node within range with the lowest ID becomes a cluster head and maintains the cluster membership information of all other nodes.

This simulation study in this research randomly assigns the ID values to each vehicle in the simulation. This approach approximates real-world situations in which the ID attribute relates to the MAC ID of the network hardware. The ID attribute of a vehicle is fixed for the lifetime of that vehicle. This property explains why repeated cluster head selection from a local set of vehicles tends to reselect previous cluster heads.

2.3 Highest-Degree Algorithm

This algorithm uses the degree of the nodes within the network to determine the cluster heads. The general idea that choosing high-degree nodes as cluster head candidates tends to uncover larger clusters. In MANET implementations, however, small movements in network nodes can often lead to a large number of degree changes throughout the network. This, understandably, has a detrimental effect on the stability of the clusters over time [4,8]. So cluster heads in Highest-Degree implementations are not likely to maintain cluster head status for long.

Table 1. Summary of the two algorithms

Algorithm	Strengths	Weaknesses
Lowest-ID	Fast and simple, Relatively stable clusters.	Small clusters, long cluster head duration.
Highest-Degree	Most connected nodes appropriately given higher priority.	Relatively unstable clusters.

Many additional clustering algorithms have been defined to meet special-case purposes. These algorithms are included in this research because they have constant time complexity and good scalability. For convenience, these algorithms have been summarized in Table 1.

The Lowest-ID clustering was generalized to a weight-based clustering technique, referred to as the DCA (Distributed Clustering Algorithm) in [2,8,15]. Our implementation will not consider network broadcasts requiring more than one hop in communication. This simplifies the overall communication and clustering strategy and reduces the overall bookkeeping necessary to maintain the clusters. This approach seeks to obtain optimal results under realistic traffic flow simulation.

3   Transportation-Specific Clustering Methodology

Review of current MANET research highlights the need for a transportation-specific review of clustering methodology and the discovery of traffic-optimized clustering schemes. This research chose to design a utility-based methodology for network cluster formation. In this approach, each vehicle implements some form of utility analysis of each proximally located possible cluster head. Periodically, each vehicle will broadcast general network information such as ID and current degree as well as vehicle-specific traffic statistics such as position and velocity. Upon receipt of this information, each vehicle chooses a cluster head by evaluating the utility of each potential head. The node with the highest utility is selected as the cluster head.

3.1   Utility Function

A utility-based approach to clustering requires the creation of a vehicle-specific agent model for periodic cluster formation. This model was implemented by augmenting each vehicle in a traffic micro-simulation platform to periodically determine and store cluster head information. The cluster head determination algorithm was implemented in a single weight method that produced a weight value for each vehicle with which the current vehicle can communicate. After implementation of this method, the Lowest-ID and Highest-Degree methods were implemented and tested [17]. This research applies compound clustering schemes based on the compound weighting of the Lowest-ID and Highest-Degree algorithms and the traffic specific Position and Closest Velocity to Average algorithms to find a more stable clustering method for use in VANET environments. The belief is that these traffic-specific algorithms will be better predictors of the common traffic situations that lead to cluster dissociation. Thus the new algorithms should augment the well-known and stable MANET clustering techniques to obtain a more stable technique for the constrained VANET problem.

As an important note on this investigation, an exhaustive investigation of vehicle parameters and parameter-specific cluster methods was not performed or intended. Many other vehicle state measurements exist and are equally predictors of traffic movement, but have been fixed for this experiment.

The two clustering methods used to attempt a stability improvement in the Highest-Degree and Lowest-ID algorithms are as follows:

a) Closest Position to Average: A vehicle attempts to choose as its cluster head in order of the absolute difference of candidate's position to the average position of all proximal vehicles.

b) Closest Velocity to Average: A vehicle attempts to choose as its cluster head in order of the absolute difference of candidate's velocity to the average velocity of all proximal vehicles.

These steps outline the procedure for implementation of this utility function:

1. Each vehicle determines the vehicles within range by polling the local broadcast region and tracking the candidate cluster head set C. All vehicles with broadcast range are considered candidate cluster heads.

2. Using candidate set C and the state information received by broadcast, each candidate is evaluated using the utility function.

3. The cluster head is chosen in decreasing order of utility. The petition for cluster membership is broadcast to the chosen vehicle. Should the chosen vehicle deny the request the vehicle with the next highest utility is selected and this step repeated.

A vehicle may deny the selection as cluster head if it has reached its maintainable limit of cluster members or if the vehicle has already chosen to join with another cluster head. Note, a vehicle may elect itself as its cluster head. Random selection of vehicles simulates asynchronous cluster formation at fixed time intervals.

3.2 Vehicular Considerations of Cluster Formation

Due to the dynamic nature of traffic flow, the member vehicles as well the cluster heads tend to move in semi-related motion throughout the roadway. This motion destabilizes the network clusters and warrants periodic cluster reformation. Reclustering may result in transition of nodes from one cluster to another, split of a cluster into more than one cluster, or convergence of multiple clusters into a single larger cluster. The frequency of cluster formation and cluster change is thus an important consideration in algorithm evaluation.

Equally important is the size of each cluster. Resource and relay algorithm performance considerations may limit the manageable size a cluster head's cluster. For simplicity this research used a common fixed upper bound on all vehicle's cluster size. The implication is that vehicles may reject nodes within range due to resource exhaustion.

The delicate balance between cluster size and coverage has major implications in network communication latency and throughput. Each vehicle communicates with vehicles in other clusters through the selected cluster heads. Care must be taken to ensure that the head selection algorithm does not have the unfortunate result of adding network transmission bottlenecks. Alternately, algorithms that yield too many cluster heads may result in a computationally expensive system. An important area of study is the selection of cluster algorithms that balance high throughput and lowest latency. The performance of the new algorithms must be measured relative to previously analyzed MANET algorithms. The objective of this research is to evaluate the number of cluster changes and the cluster size for simple combinations of the Highest-Degree and Lowest-ID algorithms with the Position and Closest Velocity to Average methods to determine whether traffic-specific augmentation can improve stability in VANET environments.

MANET research covers many compound or multi-dimensional clustering algorithms. In general, these methods are presented to overcome certain disadvantages of general MANET models such as power consumption, low mobility, or random multi-directional movement. These algorithms have not been modeled because their contributions to VANET implementations are not immediately apparent. Instead, general-purpose compound algorithms using traffic-specific information have been implemented to obtain a better overall clustering technique.

3.3 Weighted Clustering

This research utilized a weighting scheme incorporating traffic-related information and multiple clustering techniques. In this approach, the Lowest-ID and Highest-Degree clustering logic is augmented with traffic-specific logic. The goal of this research is to improve on the basic MANET clustering by applying domain-specific logic to aid in cluster head selection. This implementation uses a weighting scheme of 85% weighting to the Lowest-ID or Highest-Degree logic and 15% to the traffic-specific information of position or velocity. The desire is to improve upon the initial clustering logic to obtain better stability over the simulation time.

$$\text{Equation 1: Utility} = \text{Sum} (W_j * \text{Rank}_{j...})$$

The weighting algorithm calculates a weight for each potential cluster head using normalized results from each of the component algorithms. Equation 1 shows the method for aggregating weights of each of the cluster methods. Using the rank of each vehicle after clustering with the initial methods, the compound rank or utility is calculated using the method of Equation 1. After each vehicle's compound utility is determined, the vehicles are selected as cluster heads in order of highest utility.

4 Simulation Study

This study modified Traffic Simulation 3.0 [10], an Intelligent-Driver Model (IDM) [11] micro-simulation tool built to monitor traffic flow under various basic highway configurations. This environment simulates accelerations and braking decelerations of drivers (i.e. longitudinal dynamics), and uses the Minimized Overall Braking Induced by Lane changes (MOBIL) lane change model. All model parameters and the initial simulation source code are available at [10].

4.1 Implementation

The source code for the aforementioned simulation tool was modified to perform fixed interval cluster formation using six experimental algorithms (Lowest-ID, Highest-Degree, Lowest-ID augmented with Position, Lowest-ID augmented with Closest Velocity to Average, Highest-Degree augmented with Position, Highest-Degree augmented with Closest Velocity to Average) using a utility function based on our compound weighting scheme. To simplify this study, only simple two-dimensional compounding was analyzed.

To further constrain this simulation, the weights for each algorithm were fixed to use an 85% weight for the MANET algorithms and 15% weight for each of the two traffic algorithms studied.

4.2 Metrics

In addition to utility function and display changes, periodic state logging was implemented. This data provided the basis for the simulation result analysis and algorithm comparison. To measure the system performance, two metrics were identified: (i) the average cluster head change per step and (ii) the average cluster size. Metric (*ii*) alone does not accurately depict system performance, so the relative measurement (*ii*)/(*i*) was introduced to provide a reasonable comparison metric between the analyzed algorithms. A method is considered relatively better if it has either better stability using metric (*i*) or larger average cluster size.

5   Results Discussion

The simulation results represent the performance of each algorithm across various wireless transmission range values (0-300 meters) and maximum vehicle speed (40-140 kilometers/hour) with a fixed maximum cluster size of 50 vehicles. In addition, the simulation time duration was held constant across all tests. To minimize traffic flow variability between simulations and enable repeatable test results, the randomized features of the model were seeded with the same value at each simulation run.

Figure 2 summarizes the variation of the average number of cluster head changes with respect to the transmission range. It illustrates the performance of all six algorithms for a reasonably standard traffic flow environment with a fixed maximum speed of 100km/h. Notably, the algorithms based on the Lowest-ID clustering show rapid initial increase of cluster head changes as a result of transmission range increase. These algorithms quickly converge, however, in line with the uniform

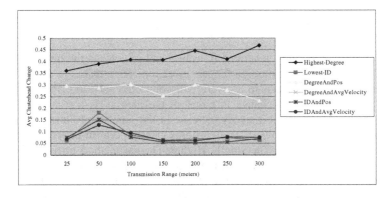

Fig. 2. Changes vs. Transmission Range

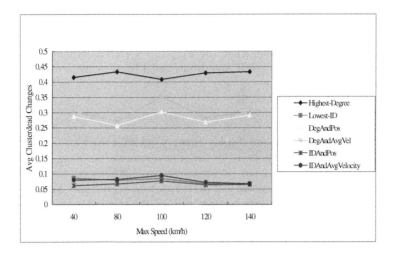

Fig. 3. Changes vs. Max Speed

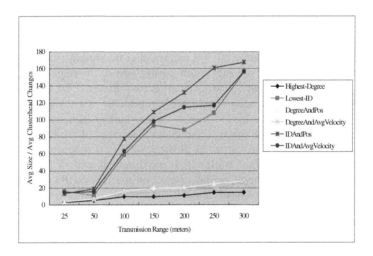

Fig. 4. Clustering Ratio vs. Transmission Range

distribution of the randomly generated ID values and vehicles in the Intelligent Driver Model [11]. For small transmission ranges, most vehicles remain out of each other's transmission range. This leads to a severely disconnected network. For the other algorithms, the likelihood of change in either of the metrics as a result of increased transmission range results in a steady increase in the number of clusters with transmission range. The three Lowest-ID algorithms clearly perform better than the three Highest-Degree algorithms. The simple Highest-Degree method proved to have the poorest performance. Clearly, the addition of traffic-specific information to the Highest-Degree clustering enabled more stable clusters. The three Lowest-ID algorithms show very similar performance. The traffic-specific logic did not have as great an effect on an already well-performing algorithm, but did help reduce the initial peak values seen with the simple method at a 50m range. Traffic patterns show that similarly located vehicles are more likely to share similar velocities. As a result, the algorithms weighted with Position and Closest Velocity to Average clustering logic tend to have a high correlation, but different overall performance.

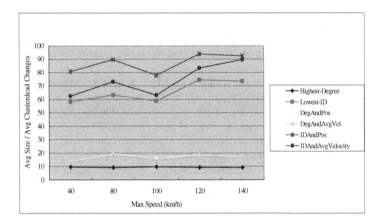

Fig. 5. Clustering Ratio vs. Max Speed

Figure 3 shows the effect of varying the maximum speed on the average number of cluster head changes with a fixed transmission range of 150m. Algorithm performance is consistent with those of Figure 2. Speed limits are only useful only in heavy-traffic situations. [10]

Figure 4 displays the performance of all algorithms over various transmission ranges. Higher curves indicate better overall performance. Figure 5 shows the overall performance across various speed limits.

6 Conclusion

The analysis performed in this research highlights the performance of the Lowest-ID clustering algorithms as optimal for the constrained MANET environment provides by VANETs. As in MANET studies, the Lowest-ID provides a stable cluster topology over long time durations due to its nature as an unbiased, uniformly distributed clustering methodology. The addition of traffic-specific clustering logic to this basic algorithm in a compound-weighted scheme did not provide any noticeable improvement. Some stability gain was realized, however, but the additional overhead may not justify the added complexity.

The study of the augmented Highest-Degree methods did, however, show some noticeable improvement when the traffic-specific logic was added. This result shows the potential for improvement of well-known and trusted MANET clustering algorithms when used in VANET environments and the advantage of applying domain knowledge to the specific MANET problem.

This study focused on a small subset of algorithms in an attempt to realize such a gain. Further research should be performed to realize further gains and determine optimum performance across the many possible compound-weighting scenarios.

References

1. Sivavakeesar, S and Pavlou, G, 2002. A Prediction-Based Clustering Algorithm to Achieve Quality of Service in Multihop Ad Hoc Networks, London Communication Symposium.
2. Basagni, S., 1999. Distributed Clustering for Ad Hoc Networks, Proceedings of the 1999 International Symposium on Parallel Architectures, Algorithms and Networks (I-SPAN'99), IEEE Computer Society, pp.310-315, Australia, Jun 23-25,1999.
3. Basagni, S., Chlamtac, I., Farago, A., 1997. A Generalized Clustering Algorithm for Peer-to-Peer Networks, Workshop on Algorithmic Aspects of Communication, satellite workshop of ICALP'97, invited paper, Bologna, Italy, Jul 11-12, 1997.
4. Gerla, M. and Tsai, J., 1995. Multicluster, Mobile, Multimedia Radio Network, Wireless Networks, 1(3) 1995, pp. 255-265.
5. Ephremides, A., Wieselthier, J.E., Baker, D.J., 1987. A Design Concept for Reliable Mobile Radio Networks with Frequency Hopping Signaling, Proceedings of the IEEE, Vol. 75, No. 1, January 1987, pp. 56_73.
6. Jiang, M., Li, J., and Tay, Y.C., 1999. Cluster Based Routing Protocol, IETF Draft, August 1999. Work in Progress.
7. Amis, A., and Prakash. R., 2000. Load-Balancing Clusters in Wireless Ad Hoc Networks. In Proceedings 3rd IEEE Symposium on Application-Specific Systems and Software Engineering Technology, pages 25–32 Mar. 24–25 2000.
8. Krishna, P., Vaidya, N.H., Chatterjee, M., Pradhan, D.K., 1997. A Cluster Based Approach for Routing in Ad Hoc Networks, ACM Computer Communications Review (CCR), 1997.
9. Ramanathan, R. and Steenstrup, M, 1998. Hierarchically-Organized Multihop Mobile Networks for Quality-of-Service Support, Mobile Networks and Applications, Vol. 3, No. 2, August 1998.
10. Traffic Simulation 3.0, Treiber, M , 2005 http://vwisb7.vkw.tu-dresden.de/~treiber
11. Treiber, M., Hennecke, A., and Helbing, D., 2000. Congested Traffic States in Empirical Observations and Microscopic Simulations, Physical Review E 62, 1805 2000.
12. MOBIL http://vwisb7.vkw.tu-dresden.de/ ~treiber/MicroApplet/MOBIL.html
13. Garg M. and Shyamasundar, R.K., 2004. A Distributed Clustering Framework in Mobile Ad Hoc Networks
14. Bettstetter, C., and Konig, S., 2002. On the Message and Time Complexity of a Distributed Mobility Adaptive Clustering Algorithm in Wireless Ad Hoc Networks, Proceedings of the Fourth European Wireless Conference.
15. Basu, P., Khan, N., and Little, T., 2001. A Mobility Based Metric for Clustering in Mobile Ad Hoc Networks.
16. Johansson, T. and Carr-Motyckova, L., 2004. Bandwidth-constrained Clustering in Ad Hoc Networks.
17. Haran, J., Fan, P., Nelson, P., and Dillenburg J., An Intelligent Vehicle Approach to Mobile Vehiclular Ad Hoc Networks, Proceedings of ICINCO 2005.

Randomized AB-Face-AB Routing Algorithms in Mobile Ad Hoc Networks

Thomas Fevens, Alaa Eddien Abdallah, and Badr Naciri Bennani

Department of Computer Science and Software Engineering,
Concordia University, Montréal, Québec, Canada, H3G 1M8
{fevens, ae_abdal, b_naciri}@cse.concordia.ca

Abstract. One common design for routing protocols in mobile ad hoc networks is to use positioning information. We combine the class of randomized position-based routing strategies called AB (Above-Below) algorithms with face routing to form AB:FACE2:AB routing algorithms, a new class of hybrid routing algorithms in mobile ad hoc networks. Our experiments on unit disk graphs, and their associated Yao sub-graphs and Gabriel sub-graphs, show that the delivery rates of the AB:FACE2:AB algorithms are significantly better than either class of routing algorithms alone when routing is subject to a threshold count beyond which the packet is dropped. The best delivery rates were obtained on the Yao sub-graph. With the appropriate choice of threshold, on non-planar graphs, the delivery rates are equivalent to those of face routing (with no threshold) while, on average, discovering paths to their destinations that are several times shorter.

Keywords: Mobile adhoc networks, randomized routing, position-based routing.

1 Introduction

A mobile ad hoc network (MANET) is a system of wireless autonomous hosts that can communicate with each other without having any fixed infrastructure. Each node in the network can communicate with all other nodes within its transmission range [1,2], which we will assume to be a fixed range r for all nodes. If two nodes are not able to communicate directly then a multi-hop routing protocol is needed for the nodes to send packets to each other. The absence of infrastructure in MANETs, together with possible dynamic topology changes and resource constraints, makes routing in these networks a challenging problem.

We are specifically interested in *position-based* routing (also known as online routing [3,4] or geographic routing [5]). In position-based routing protocols, a node forwards packets based on the location (coordinates in the plane) of itself, its neighbors, and the destination [6]. The position of the nodes can be obtained using GPS, for example. There are numerous ways to use position information in making routing decisions. In one class of algorithms, *progress-based* algorithms as categorized in [7], the algorithm forwards the packet in every step to exactly

V.R. Sirotiuk and E. Chávez (Eds.): ADHOC-NOW 2005, LNCS 3738, pp. 43–56, 2005.

one of its neighbors, which is chosen according to a specified heuristic, such as minimizing distance to the destination, or moving in a direction nearest to that to the destination.

In this paper we represent a MANET by a unit disk graph, where two nodes are connected if and only if their Euclidean distance is at most r. In GREEDY routing [8,9], a node forwards the packet to its neighbor which is closest to the destination. COMPASS or *directional* routing [10] moves the packet to a neighboring node such that the angle formed between the current node, next node, and destination is minimized. Both of these progress-based algorithms are known to fail to deliver the packet in certain situations.

To overcome the inability of progress-based routing to always deliver their packets, position information can be used to extract a *planar* sub-graph such that routing can be performed on the faces of this sub-graph, known as *face* routing or perimeter routing [10,3]. The advantage of this approach is that delivery of packets can always be guaranteed. The original face routing algorithm was called *Compass Routing II* in [10]. An optimization of this algorithm was given by [3] and called *Face-2*. These algorithms will be discussed further in §3. Other improvements to face routing have also been proposed. In [11], *AFR* (Adaptive Face Routing) is proposed where face routing is executed strictly within an area bounded by an adaptively sized ellipse, leading to an algorithm that is asymptotically optimal in that the cost of the algorithm steps are proportional to the square of the cost of the square of the optimal path. In [12], face routing is adapted to guarantee delivery on restricted classes of non-planar graphs.

One drawback of face routing is that, although delivery is guaranteed on a planar graph, the route discovered may be many times longer than the shortest path in the graph. To try to reduce the path length, [13] (also see [14]) and [15] use a hybrid routing algorithm that combines greedy routing with face routing. The idea is to use greedy routing until a local minimum (where all the neighbors of the current node c are further way from the destination node than c) is reached whereupon face routing is used. When the local minimum is bypassed, the algorithm switches back to greedy routing, and so on. This algorithm is termed *GFG* (Greedy-Face-Greedy) by [13] and *GPSR* (Greedy Perimeter Stateless Routing) by [15]. Other variations of this hybrid approach of combining progress based routing with face routing have appeared. *GOAFR* [16] combines greedy routing with a variation of *AFR* called *OAFR*. They show that their algorithm is worst case optimal and average case efficient.

In [17], the authors proposed a class of randomized algorithms called AB algorithms (Above-Below). Essentially, to decide on the next node to which the packet should be forwarded, the AB algorithms pick one neighbor of the current node *above* the line passing through the current node and the destination, and another neighbor *below* this line. The next node is chosen randomly from these two neighbors according to some probability distribution. The exact choice of the neighbors and the probability distribution determine the specific algorithm.

In this paper, we consider hybrid routing algorithms, called AB:FACE2:AB algorithms, that combine AB algorithms with face routing. The goal is to improve the delivery rates of randomized routing algorithms on non-planar graphs, such as the unit disk graph, while still discovering short paths to the destination. As will be shown, the performance of AB:FACE2:AB algorithms is on par with that of FACE2 routing on the unit disk graph but while choosing shorter paths.

Many routing strategies use a spanning sub-graph of the unit disk graph such that only the edges in the sub-graph are used for routing. These sub-graphs have fewer edges to maintain, and may provide useful properties such as planarity. We studied the performance of all the algorithms on the original unit disk graphs as well as Gabriel graphs [18] and Yao graphs [19] derived from the unit disk graphs. The rational for considering these sub-graphs is discussed at the end of §2. Our results show that the performance of nearly all algorithms in terms of the delivery rate and the length of the path discovered is best on the Yao sub-graphs.

The rest of the paper is organized as follows. The next section gives relevant definitions. Our routing strategies are discussed in §3. Then, §4 gives the empirical results of our simulations and provides an interpretation of the behavior of the algorithms. We conclude with a discussion of the results in §5.

2 Preliminaries

We assume that the set of n wireless nodes is represented as a point set S in the two-dimensional plane. Two nodes are connected by an edge if the Euclidean distance between them is at most r, the transmission range of the nodes. The resulting graph $UDG(S)$ is called a unit disk graph. For node u, we denote the set of its neighbors by $N(u)$. Given a unit disk graph $UDG(S)$ corresponding to a set of points S, and a pair (s, d) where $s, d \in S$, the problem of online position-based routing is to discover a path in $UDG(S)$ from s to d. At each point of the path, the decision of which node to go to next is based on the local position information of the current node c, $N(c)$, and d. Here, s is termed the source and d the destination. Frequently, we will also refer to the $\overline{cd}$ line passing through c and d. An algorithm is *deterministic* if, when at c, the next node is chosen deterministically from $N(c)$, and is *randomized* if any next step taken by a packet is chosen randomly from $N(c)$ (this description includes hybrid deterministic and randomized algorithms).

We are interested in the following performance measures for routing algorithms: the *delivery rate* which is the percentage of times that the algorithm succeeds in delivering its packet, and the *path dilation*[1], the average ratio of the length of the path returned by the algorithm to the length of the shortest path in the UDG. The path dilation is defined with respect to the shortest path sp in the UDG since, even when routing on a sub-graph of the UDG, sp is available to any routing algorithm and sp is equal in length to, or shorter than, any shortest path that may be discovered in a sub-graph. Here the length of the path is taken to mean the number of hops in the path.

---

[1] This term was suggested to us by Jaroslav Opatrny to differentiate it from *stretch factor* commonly associated with t-spanner definitions.

In this paper, we also consider the behavior of the routing algorithms on several sub-graphs of the unit disk graph. Define $P(G)$ as a t-spanner of G if the length of the shortest path between any two nodes in $P(G)$ is not more than t times longer than the shortest path connecting them in G, where t is the stretch factor. Let $G = UDG(S)$ be a unit disk graph. Define $d(x, y)$ to be the Euclidean distance between x and y. Denote the disk centered at the midpoint between the points u and v and with radius $d(u, v)/2$ by $disk(u, v)$. Then the *Gabriel Graph* of G, denoted $GG(G)$, is defined as follows [18]. Given any two adjacent nodes u and v in G, the edge (u, v) belongs to $GG(G)$ if and only if no other node $w \in G$ is located in $disk(u, v)$. It is known that $GG(G)$ is planar and connected if the underlying graph G is a connected unit disk graph [3], Also, $GG(G)$ is a $(4\pi\sqrt{2n - 4})/3$-spanner of G [20].

For a geometric graph G, a *Yao Graph* (also called a *Theta Graph* [19]) $YG_k(G)$ with an integer parameter $k \geq 6$ is defined as follows [21]. First, we will define a directed Yao graph, $DYG_k(G)$, for G. At each node u in G, k equally-separated rays originating at u define k cones. In each cone, only the directed edge (u, v) to the nearest neighbor v, if any, is part of $DYG_k(G)$. Ties are broken arbitrarily. For $G = UDG(S)$, the result is a sub-graph that may still have edges that cross. Let $YG_k(G)$ be the undirected graph obtained if the direction of each edge in $DYG_k(G)$ is ignored. The graph $YG_k(G)$ is a $1/(1 - 2sin(\pi/k))$-spanner of G [22].

Our rational for considering Gabriel sub-graphs is that they are planar graphs but their stretch factor is not bounded. Conversely, Yao graphs are not planar (although our experiments show that frequently they have very few crossing edges) and have bounded stretch factors.

3   Routing Algorithms

In what follows, we always assume that the current node is c, the next node is x, and the destination node is d. The algorithms consider in this paper differ in how x is chosen from among the set $N(c)$. The greedy and compass strategies have already been described in Section 1. In addition, we will use a deterministic progress-based routing algorithm, ELLIPSOID-BASED routing [23], a variation of the GREEDY routing, where the current node c forwards the packet to its neighbor n which minimizes $d(c, n) + d(n, d)$.

Next, we describe the class of algorithms called AB algorithms [17]. Each algorithm has two attributes, which is reflected in our naming convention: **AB(R,S)** where **R** is one of **C** (as in COMPASS), **G** (GREEDY), or **E** (ELLIPSOID-BASED), and **S** is one of **U**, **A**, or **D**. Each routing algorithm is based on initially determining two candidate neighbors, one neighbor of c from above the $\overline{cd}$ line, n_1, and, similarly, one neighbor of c below the $\overline{cd}$ line, n_2. Out of all the possible neighbors from above (below) the $\overline{cd}$ line, n_1 (n_2) is the one that would be chosen by the **R** protocol. Which of these two candidate neighbors is actually chosen depends on the symbol for **S**. If the symbol is **U**, then the next node x is chosen uniformly at random from n_1 and n_2. If the symbol is **A**, then the next node x is chosen from n_1 and n_2 with probability $\theta_2/(\theta_1 + \theta_2)$ and $\theta_1/(\theta_1 + \theta_2)$, respectively, where

$\theta_1 = \angle\, dcn_1$ and $\theta_2 = \angle\, n_2cd$. Finally, if the symbol is **D**, then the next node x is chosen from n_1 and n_2 with probability $dis_2/(dis_1 + dis_2)$ and $dis_1/(dis_1 + dis_2)$, respectively, where $dis_1 = d(n_1, d)$ and $dis_2 = d(n_2, d)$. If either of n_1 or n_2 is not defined, then the other neighbour is chosen by default.

As mentioned in the Introduction, message delivery on a planar graph, if the source and destination are connected, was first guaranteed with face routing [10,3]. The central idea of face routing is that of the exploration of the interior boundary of a face using the right hand rule analogous to exploring a maze by keeping one's right hand on a wall. Face routing walks around the perimeters of the faces of a planar graph, keeping track of the points where the boundary of a face intersects the $\overline{cd}$ line connecting the source and destination nodes. The primary difference between the original face routing algorithm [10] and *Face-2* [3] is as follows. In original face routing algorithm, the entire perimeter of a face is traversed and at the intersection point p of the $\overline{cd}$ line with the perimeter that is closest to the destination, routing switches to exploring the face sharing p. Thus the face routing progresses toward the destination along the $\overline{cd}$ line. In *Face-2*, this switch occurs at the first intersection point discovered that is closer than any previously discovered intersection points to the destination.

Here, we define our primary routing algorithms which combine the randomized AB algorithms with *Face-2* [3] (our implementation will be called FACE2). Since we have nine distinct AB algorithms, we have correspondingly nine hybrid algorithms which we will term as AB(R,S):FACE2:AB(R,S) algorithms where R is one of C, G, or E, and S is one of U, A, or D as in the description of the AB algorithms previously. Define a *progress halfplane* to be the half plane whose boundary passes through c, contains d, and direction of the normal on the boundary is in the same direction as the vector from c to d. Each algorithm starts as the particular AB(R,S) routing algorithm until the algorithm reaches a node within the graph, such that there are no neighbors in (or on the boundary of) the progress halfplane. At this node, the routing algorithm switches to the FACE2 routing algorithm. Then when FACE2 reaches a vertex where the protocol would begin exploring a different face, the routing algorithm switches back to the AB(R,S). Note that our characterization of the node at which the algorithm switches to FACE2 routing is slightly different than that of a local minimum — this difference is greater the closer a packet gets to the destination whereupon FACE2 routing is resorted to less often than would be the case with a local minimum. As shown in [17], using a uniform probability distribution tended to be out-performed by using a distribution based on angle or distance, so we will only consider AB algorithms using the latter two distributions.

Table 1. Classification of AB:FACE2:AB algorithms

Name of Algorithm	Selection of Neighbors	Probability Distribution
AB(C,A):FACE2:AB(C,A)	Compass	Angle
AB(C,D):FACE2:AB(C,D)	Compass	Distance
AB(G,A):FACE2:AB(G,A)	Greedy	Angle
AB(E,A):FACE2:AB(E,A)	Ellipsoid-based	Angle

Table 1 shows the classification of the subset of the AB(R,S):FACE2:AB(R,S) algorithms demonstrated in this paper, based on how the AB(R,S) components of each algorithm makes their initial choice of neighbors and the probability distribution used to choose the final neighbor.

4   Empirical Results

In the following, GREEDY, COMPASS, ELLIPSOID-BASED, GREEDY:FACE2: GREEDY (our implementation of GFG [13]) and FACE2 are deterministic routing algorithms, and all of the other algorithms are randomized. To evaluate the relative performance of these algorithms we will consider their packet delivery rates and path dilations. We first describe our simulation environment, and then describe and interpret our results, comparing our algorithms with previous work.

4.1   Simulation Environment

In the simulation experiments, we consider point sets that are uniformly distributed in the plane, as well as point sets that simulate clustering. The parameters like node density will determined experimentally as discussed below.

In the uniform distribution simulation experiments, a set S of n points (where $n \in \{75, 100, 125, 150\}$) is uniformly randomly generated on a square of 100 units by 100 units. For the transmission range of nodes, we use $r = 15$ units. After generating $G = UDG(S)$, we randomly choose a source node s and a destination node d from S. We determine if there exists any path from s to d in $UDG(S)$. If not, the graph is discarded; otherwise, all routing algorithms are applied in turn on G, $GG(G)$, and $YG_8(G)$.

In the clustered simulation experiments, $b = 50$ background points are uniformly randomly generated on a square of 100 units by 100 units. For the transmission range of nodes, again, we use $r = 15$ units. In addition to the fixed number of background points, three clusters, A, B, C, each with j cluster points where $j = 1 + 6 * i$, $i = 0, 1, 2, 3$, are added to the background points to give a set S of $n = b + 3j$ points. The clusters are uniformly randomly generated on a square of 20 units by 20 units and have their centers fixed such that the regions of two clusters A and B overlap (centered at $(11, 25)$ and $(25, 11)$) and the region of the third cluster C (centered at $(89, 89)$) is disjoint from the first two. Note: to facilitate comparison with the uniform distribution simulations results, within the region of a cluster, for the purposes of choosing the source and destination nodes, the additional cluster nodes are considered distinct from the background points. After generating $G = UDG(S)$, a source node s is randomly chosen from the cluster nodes of A and destination node d is randomly chosen from the cluster nodes of C. As with the uniform distribution simulations, we determine if there exists any path from s to d in $UDG(S)$. If not, the graph is discarded; otherwise, all routing algorithms are applied in turn on G, $GG(G)$, and $YG_8(G)$.

In Fig. 1 are shown histograms of the node degrees for the UDGs for the chosen simulation values of n. It is suggested in [24] to consider simulations with node density per unit disk of around 5, which would correspond to graphs

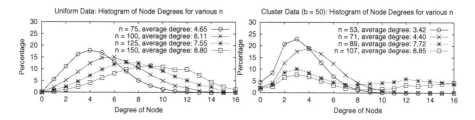

Fig. 1. The histograms of average node degrees in 10,000 generated UDGs for the indicated values of n

with average node degrees of around 4. For uniform data, graphs with $n = 75$ would most closely match this node density of interest, and for clustered data, graphs with $n = 71$ would be of particular interest for comparison. Nonetheless, we shall give experimental results for a variety of values of n to demonstrate relative behaviours with respect to n, but it may be noted in Fig. 1 that when n is larger than 100, that a substantial percentage of nodes have degrees larger than 10 which indicates highly connected graphs.

Clearly, an algorithm succeeds if a path to the destination is found. The progress-based deterministic algorithms and FACE2 (without threshold) are deemed to fail if they enter a loop, while GREEDY:FACE2:GREEDY, FACE2 (with threshold), and the randomized algorithms are considered to fail when the number of hops in the path computed so far exceeds a threshold. We will use the number of nodes, n, as this threshold (in the next section, we will explore the effect of varying the threshold value). To compute the packet delivery rate, this process is repeated with 100 random graphs and the percentage of successful deliveries determined. To compute an average packet delivery rate, the packet delivery rate is determined 100 times and an average taken. Additionally, out of the 100×100 runs used to compute the average packet delivery rate, the average path (hop count) dilation of successful paths is computed.

4.2   Discussion of Results

Detailed simulation results for all the routing algorithms, along with the associated standard deviations, are given in Tables 2 to 7 for unit disk graphs and their associated Gabriel graphs, and Yao graphs (with parameter 8). In particular, we are interested in the relative performance of the progress-based deterministic algorithms (GREEDY, COMPASS, and ELLIPSOID-BASED), the deterministic FACE2-based algorithms (GREEDY:FACE2:GREEDY, FACE2 with and without threshold), the randomized AB algorithms, and the new hybrid AB:FACE2:AB algorithms based on the same AB algorithms.

Performance of Algorithms in Unit Disk Graphs. We will now present the comparison between different groups of algorithms in terms of packet delivery rate and path dilation. As discussed earlier, for comparison purposes, we will

Table 2. Uniform data: Average packet delivery rate, D, and average path dilation, P, and associated standard deviations, σ, in UDG

Algorithms	$n = 75$				$n = 100$			
	D	σ	P	σ	D	σ	P	σ
AB(C,A):FACE2:AB(C,A)	98.33	1.19	2.44	3.09	96.48	2.04	2.46	3.20
AB(C,D):FACE2:AB(C,D)	97.34	1.58	2.87	3.56	94.67	2.27	3.07	4.21
AB(G,A):FACE2:AB(G,A)	98.00	1.43	2.87	3.26	95.38	2.23	2.80	3.23
AB(E,A):FACE2:AB(E,A)	97.65	1.57	2.96	3.74	95.08	2.22	2.99	3.50
GREEDY:FACE2:GREEDY	93.25	2.12	2.26	3.28	91.28	2.04	2.35	3.18
FACE2 (with threshold)	96.79	1.65	5.16	6.82	97.13	1.40	5.84	5.21
COMPASS	72.31	5.02	1.06	0.12	77.68	4.09	1.08	0.13
GREEDY	70.73	5.16	1.01	0.05	76.43	4.04	1.02	0.07
ELLIPSOID-BASED	56.09	5.21	1.08	0.14	59.16	5.41	1.11	0.16
AB(C,A)	87.74	3.66	2.22	2.99	90.04	2.54	1.91	2.09
AB(C,D)	92.16	2.68	2.53	2.74	93.99	2.52	2.20	1.96
AB(G,A)	86.17	3.80	2.17	2.95	88.54	2.77	1.83	2.04
AB(E,A)	82.78	4.44	2.50	3.70	84.75	3.55	2.13	2.48
FACE2 (without threshold)	98.60	1.01	5.64	7.78	99.23	0.57	6.99	5.93

focus on uniform distributions with $n = 75$ and clustered distributions with $n = 71$.

In UDGs, as presented in Tables 2 and 3, for both uniform and clustered distributions, the FACE2 (without threshold) and, closely in second place, the AB:FACE2:AB algorithms (with a threshold of n) outperform all the other randomized and deterministic algorithms in terms of the packet delivery rate. As we will see later in this section, by increasing the threshold, the delivery rate for the AB:FACE2:AB algorithms can be improved to match that of the FACE2 (without threshold) algorithm. Out of the AB:FACE2:AB algorithms, on the UDG, the algorithms based on the AB(C,A) and AB(G,A) have the best performance in terms of delivery rate. All algorithms tended to perform worse on clustered distributions than uniform distributions. Next in terms of delivery rate, and roughly equivalent in performance, are the FACE2 (with threshold) and GREEDY:FACE2:GREEDY algorithms, following by the randomized AB algorithms. Finally, all the progress-based deterministic algorithms have the worst delivery rates but the best path dilations. Although the best delivery rate is for FACE2 (without threshold), this algorithm has by far the worst path dilations. The results for GREEDY:FACE2:GREEDY (without threshold) is not included in the tables but our simulations show that the path dilations are about the same as those for AB:FACE2:AB algorithms given in Tables 2 and 3, but with lower delivery rates (as we will see in Fig. 2 when we consider the effect of threshold).

The main weakness in the AB algorithms is when the current node is on the edge of a face and there may be no, or a restricted choice of, nodes on one side of the $\overline{cd}$ line. At this point, the algorithm is essentially deterministic and inherits a greater possibility of looping while trying to discover a path. It is at these critical points within the graph that we find nodes that have no

Table 3. Clustered data: Average packet delivery rate, D, and average path dilation, P, and associated standard deviations, σ, in UDG, for $b = 50$

	$n = 71$				$n = 89$			
Algorithms	D	σ	P	σ	D	σ	P	σ
AB(C,A):FACE2:AB(C,A)	96.99	1.94	2.36	2.03	94.14	2.57	2.50	2.78
AB(C,D):FACE2:AB(C,D)	94.87	2.54	3.10	2.30	90.06	3.05	2.81	3.29
AB(G,A):FACE2:AB(G,A)	96.16	2.21	2.84	2.28	91.27	3.28	3.30	2.95
AB(E,A):FACE2:AB(E,A)	95.47	2.16	2.94	2.28	91.12	3.41	3.29	3.14
GREEDY:FACE2:GREEDY	93.23	2.46	2.17	1.66	90.50	3.65	2.44	2.54
FACE2 (with threshold)	91.58	2.64	4.01	2.84	85.10	3.80	4.45	3.59
COMPASS	46.66	5.21	1.09	0.09	46.85	4.86	1.10	0.09
GREEDY	44.06	5.29	1.03	0.05	45.02	4.73	1.03	0.05
ELLIPSOID-BASED	23.35	3.88	1.12	0.11	23.91	4.53	1.13	0.11
AB(C,A)	82.90	3.63	2.28	1.81	84.20	3.74	2.42	2.15
AB(C,D)	85.86	3.52	3.39	2.12	87.61	3.54	3.47	2.36
AB(G,A)	80.20	3.87	2.27	1.86	81.13	4.10	2.49	2.40
AB(E,A)	74.20	3.86	2.69	2.20	74.73	4.23	2.96	2.77
FACE2 (without threshold)	97.23	0.87	5.14	7.89	96.32	0.54	5.64	5.86

neighbors in the progress halfplane and our AB:FACE2:AB routing protocols switch to FACE2 routing to bypass these critical points and greatly reduce the possibility of looping. For uniformly distributed UDGs, this in turn leads to a greater percentage of path completions within the preset threshold for number of hops, with path dilations about 150% longer than optimal. Similar behavior is observed for clustered distributions. The only routing algorithm with a larger delivery rate is FACE2 (without threshold), but with path dilations that are 415–460% longer than optimal on average.

Effect of Type of Sub-Graph. The trend of the results in Tables 4 to 7 is similar within each type of graph studied. First, compare the performance of the algorithms between the different types of graphs in terms of delivery rates. In general, most algorithms have the best delivery rates in the Yao sub-graphs for both uniform and clustered distributions. The progress-based and AB algorithms, for uniform and clustered distributions, have their best performance on UDG graphs, their next best on Yao sub-graphs and the worst performance on Gabriel sub-graphs (occasionally much worst such as for progress-based algorithms on clustered distributions in Table 7). For the GREEDY:FACE2:GREEDY and FACE2 algorithms using a threshold, the trends are less pronounced. For uniform distributions, these algorithms have roughly the same best performance on all three graphs. For clustered distributions, the GREEDY:FACE2:GREEDY and FACE2 (with threshold) algorithms perform slightly better on Yao sub-graphs, followed by Gabriel sub-graphs and worst on the UDG. When no threshold is used for FACE2, as expected, the delivery rate is 100% for Gabriel graphs and more than 98% for all graphs and distributions except for the case of clustered distributions on UDG graphs where the delivery rate is slightly less at around 97%. Finally, for the AB:FACE2:AB algorithms, performance was best on Yao

Table 4. Uniform data: Average packet delivery rate, D, and average path dilation, P, and associated standard deviations, σ, in $YG_8(UDG)$

	$n = 75$				$n = 100$			
Algorithms	D	σ	P	σ	D	σ	P	σ
AB(C,A):FACE2:AB(C,A)	99.28	0.80	2.11	2.12	98.88	0.89	2.19	3.12
AB(C,D):FACE2:AB(C,D)	98.54	1.03	2.74	2.36	98.34	1.12	2.77	3.26
AB(G,A):FACE2:AB(G,A)	99.08	0.83	2.53	3.21	98.59	0.97	2.70	3.12
AB(E,A):FACE2:AB(E,A)	99.21	0.77	2.68	3.04	98.53	1.14	2.74	3.50
GREEDY:FACE2:GREEDY	92.19	1.82	2.39	2.57	95.80	1.88	2.12	2.47
FACE2 (with threshold)	98.44	1.24	4.51	5.55	99.61	0.72	4.40	5.96
COMPASS	70.57	5.07	1.04	0.09	76.24	3.79	1.06	0.11
GREEDY	69.94	5.12	1.02	0.06	75.28	4.13	1.03	0.06
ELLIPSOID-BASED	55.75	5.50	1.06	0.12	57.54	4.81	1.08	0.13
AB(C,A)	86.37	3.70	2.22	2.80	88.62	2.77	1.89	1.83
AB(C,D)	91.11	3.03	2.61	2.62	92.81	2.42	2.27	1.88
AB(G,A)	85.16	3.97	2.22	2.85	87.61	2.82	1.90	1.95
AB(E,A)	82.28	3.82	2.45	3.25	83.60	3.27	2.01	2.09
FACE2 (without threshold)	99.30	0.88	5.67	5.90	99.25	0.64	5.75	5.19

Table 5. Uniform data: Average packet delivery rate, D, and average path dilation, P, and associated standard deviations, σ, in $GG(UDG)$

	$n = 75$				$n = 100$			
Algorithms	D	σ	P	σ	D	σ	P	σ
AB(C,A):FACE2:AB(C,A)	98.47	1.28	2.75	2.91	97.71	1.48	2.46	2.54
AB(C,D):FACE2:AB(C,D)	97.49	1.61	3.26	3.21	96.38	1.84	3.11	3.17
AB(G,A):FACE2:AB(G,A)	97.93	1.46	3.13	3.24	97.06	1.62	2.85	3.04
AB(E,A):FACE2:AB(E,A)	98.13	1.26	3.14	3.20	96.80	1.57	2.96	3.13
GREEDY:FACE2:GREEDY	94.50	2.01	2.60	2.52	94.80	1.95	2.38	2.36
FACE2 (with threshold)	97.65	1.49	4.52	5.58	99.33	0.76	4.48	6.00
COMPASS	66.59	5.31	1.02	0.06	70.70	4.85	1.04	0.07
GREEDY	66.78	5.05	1.02	0.05	71.05	4.43	1.03	0.07
ELLIPSOID-BASED	54.00	5.27	1.03	0.07	55.71	5.23	1.04	0.08
AB(C,A)	83.42	4.01	2.37	2.59	83.83	3.48	2.03	1.69
AB(C,D)	88.86	3.09	2.83	2.51	88.84	2.92	2.54	1.87
AB(G,A)	82.85	4.40	2.36	2.63	83.90	3.75	2.06	1.78
AB(E,A)	80.70	4.13	2.44	2.85	81.58	3.77	2.08	1.86
FACE2 (without threshold)	100.00	0.00	4.76	5.37	100.00	0.00	4.86	4.44

sub-graphs followed by Gabriel sub-graphs and then the UDG graphs. We understand the better performance on the Yao sub-graph to be due to its constant stretch factor [22,25] as well as near planarity combined with preservation of the directionality of the neighbors of nodes. The constant stretch factor maintains the availability of short paths to the destination while the decreased number of edges in the subgraph reduce the number of possibly long detours discovered

Table 6. Clustered data: Average packet delivery rate, D, and average path dilation, P, and associated standard deviations, σ, in $YG_8(UDG)$, for $b = 50$

Algorithms	$n = 71$				$n = 89$			
	D	σ	P	σ	D	σ	P	σ
AB(C,A):FACE2:AB(C,A)	98.90	1.05	2.08	1.49	99.30	0.87	2.27	1.70
AB(C,D):FACE2:AB(C,D)	98.58	1.29	2.74	1.90	99.42	0.87	2.85	2.19
AB(G,A):FACE2:AB(G,A)	99.13	0.93	2.51	2.02	99.60	0.68	2.68	2.12
AB(E,A):FACE2:AB(E,A)	99.09	0.90	2.60	2.10	99.49	0.69	2.75	2.25
GREEDY:FACE2:GREEDY	93.85	2.49	1.92	1.04	93.64	2.54	2.23	1.08
FACE2 (with threshold)	97.08	1.78	3.52	2.34	97.83	1.35	3.78	2.55
COMPASS	45.44	5.69	1.07	0.08	45.78	4.56	1.07	0.08
GREEDY	43.81	5.45	1.03	0.05	44.18	4.17	1.03	0.05
ELLIPSOID-BASED	24.44	4.54	1.08	0.09	23.10	4.29	1.08	0.09
AB(C,A)	82.31	3.49	2.24	1.72	83.84	3.71	2.34	1.98
AB(C,D)	85.38	3.59	3.34	2.02	87.33	2.97	3.50	2.33
AB(G,A)	80.07	4.09	2.24	1.73	81.16	3.93	2.37	2.11
AB(E,A)	74.31	3.94	2.49	1.99	75.34	4.54	2.70	2.37
FACE2 (without threshold)	99.65	0.42	6.32	4.87	99.88	0.17	6.21	7.22

Table 7. Clustered data: Average packet delivery rate, D, and average path dilation, P, and associated standard deviations, σ, in $GG(UDG)$, for $b = 50$

Algorithms	$n = 71$				$n = 89$			
	D	σ	P	σ	D	σ	P	σ
AB(C,A):FACE2:AB(C,A)	97.83	1.58	2.29	1.51	98.33	1.12	2.42	1.60
AB(C,D):FACE2:AB(C,D)	95.92	1.85	2.77	2.04	98.00	1.48	3.07	2.01
AB(G,A):FACE2:AB(G,A)	97.53	1.53	2.69	1.74	98.80	1.04	2.82	1.81
AB(E,A):FACE2:AB(E,A)	97.33	1.60	2.81	1.75	98.79	1.13	2.84	1.99
GREEDY:FACE2:GREEDY	92.41	2.51	2.15	1.02	93.90	2.39	2.35	1.03
FACE2 (with threshold)	95.57	1.96	3.34	2.09	97.60	1.67	3.58	2.26
COMPASS	37.27	5.27	1.03	0.04	36.79	4.71	1.04	0.04
GREEDY	36.90	5.27	1.03	0.04	36.87	3.98	1.03	0.03
ELLIPSOID-BASED	21.07	4.37	1.04	0.04	20.59	3.80	1.04	0.04
AB(C,A)	76.32	4.61	2.16	1.45	76.01	4.41	2.24	1.66
AB(C,D)	78.70	4.41	3.28	1.74	80.13	4.42	3.42	1.97
AB(G,A)	75.71	4.80	2.14	1.42	76.37	4.20	2.26	1.64
AB(E,A)	70.49	4.91	2.22	1.53	70.93	3.97	2.34	1.75
FACE2 (without threshold)	100.00	0.00	6.53	6.51	100.00	0.00	6.47	5.73

by randomized routing. Locally, being nearly planar allows FACE2 to work well in by-passing critical nodes while maintaining neighbors in the eight directions of each node of the Yao sub-graph allows the AB algorithms to choose pairs of neighbors that lead to the progression of packets.

In terms of path dilation, the most prevalent trend is that the best path dilations for uniform distributions occur in the Yao sub-graph and UDG, and the worst in the Gabriel sub-graph. For clustered distributions, the Yao and Gabriel sub-graphs perform equivalently well, with the UDG being worst. Typically within a class of algorithms, the difference in path dilations between types of graphs is slight. The only algorithm which did not follow this trend is FACE2 (without threshold) where the path dilation is significantly smaller on the UDG for clustered distributions and on Gabriel sub-graphs for uniform distributions.

Effect of Threshold. Fig. 2 shows the effect of varying the threshold value (which was set to $2n$ in the above simulations) on the average delivery rate and average path dilation of the randomized AB:FACE2:AB and the deterministic FACE2 and GREEDY:FACE2:GREEDY (both with threshold) algorithms. The reason for using FACE2 for comparison is that it has the highest delivery rates, especially if using the Gabriel sub-graph where we can achieve 100% delivery but with path dilations greater than 4.7.

First, we note that by increasing the threshold to $2n$, the relative behaviour of the algorithms is established and the differences between algorithms are clear. Therefore, for the simulations with all the algorithms, we use a threshold of $2n$. Also, it may be noted that at this threshold, the average delivery rates of the AB:FACE2:AB algorithms increase to nearly 100% with an aver-

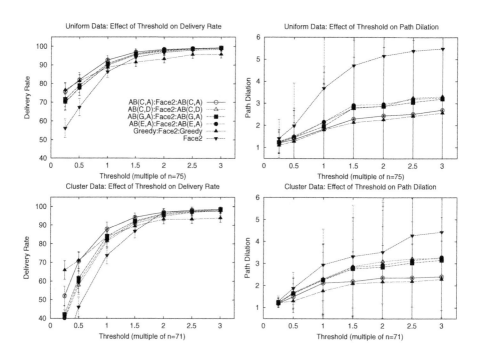

Fig. 2. The effect of changing the threshold used for uniform (upper row) and clustered data (lower row) on average delivery rate (left column) and average path dilation (right column). The error bars shown indicate standard deviation. The same legend applies to all four figures.

age stretch factor in the range of 2.1–2.5, comparable (and nearly matched by the AB(C,A):FACE2:AB(C,A) algorithm) to that of GREEDY:FACE2:GREEDY but with a significantly higher delivery rate. Meanwhile, as the threshold increases, the average delivery rate of FACE2 only slowly approaches that of the AB:FACE2:AB algorithms while suffering from an average path dilation greater than 4. Note that this behaviour occurs equivalently on both distributions.

5   Conclusions

In this paper, we proposed a new class of randomized position-based algorithms for routing in mobile ad hoc networks, called the AB:FACE2:AB algorithms, based on combining AB algorithms with the FACE2 algorithm. Our simulation results demonstrate that these hybrid algorithms, on non-planar graphs like the UDG, yield a definite improvement over all the other algorithms studied, when considered both in terms of the delivery rate and path dilation. When we consider the choice of $2n$ for a threshold for the AB:FACE2:AB algorithms, the delivery rate is equivalent to the performance of FACE2 (without threshold) algorithm but with average path dilations that are several times smaller. The best average path dilations are achieved by the progress-based deterministic algorithms, followed by GREEDY:FACE2:GREEDY, with the AB:FACE2:AB algorithms performing a close third with the latter still maintaining very high delivery rates. The best results were in terms of delivery rate and path dilations were seen when routing on the Yao sub-graph,

During our simulations, we assumed that the UDG was static. If the nodes were permitted to become mobile, we expect that, through the use of randomization, the proposed AB:FACE2:AB algorithms will continue to perform well on dynamic unit disk graphs. In essence, a dynamic graph is a version of a randomized graph and to be able to adapt to changes in topology a routing algorithm with a randomized component should fair better than a deterministic algorithm.

Acknowledgements. This work was partly funded by NSERC. B.N. Bennani was supported by an NSERC Undergraduate Student Research Award.

References

1. Basagni, S., Chlamtac, I., Syrotiuk, V., Woodward, B.: A distance routing effect algorithm for mobility (DREAM). In: Proc. of 4th ACM/IEEE Conference on Mobile Computing and Networking (Mobicom '98). (1998) 76–84
2. Barrière, L., Fraigniaud, P., Narayanan, L., Opatrny, J.: Robust position-based routing in wireless ad hoc networks with irregular transmission ranges. Wireless Communications and Mobile Computing Journal **3** (2003) 141–153
3. Bose, P., Morin, P., Stojmenovic, I., Urrutia, J.: Routing with guaranteed delivery in ad hoc wireless networks. Wireless Networks (2001) 609–616
4. Bose, P., Morin, P.: Online routing in triangulations. In: Proc. of 10th Annual Inter. Symposium on Algorithms and Computation (ISAAC '99). (1999) 113–122

5. Jain, R., Puri, A., Sengupta, R.: Geographical routing using partial information for wireless ad hoc networks. IEEE Personal Comm. Magazine **8** (2001) 48–57
6. Giordano, S., Stojmenovic, I., Blazevic, L.: Position based routing algorithms for ad hoc networks: A taxonomy. In Cheng, X., Huang, X., Du, D., eds.: Ad Hoc Wireless Networking. Kluwer (2003)
7. Mauve, M., Widmer, J., Hartenstein, H.: A survey of position-based routing in mobile ad-hoc networks. IEEE Network Magazine **15** (2001) 30–39
8. Finn, G.: Routing and addressing problems in large metropolitan-scale internetworks. Technical Report ISU/RR-87-180, USC ISI, Marina del Ray, CA (1987)
9. Stojmenovic, I., Lin, X.: Loop-free hybrid single-path/flooding routing algorithms with guaranteed delivery for wireless networks. IEEE Transactions on Parallel and Distributed Systems **12** (2001) 1023–1032
10. Kranakis, E., Singh, H., Urrutia, J.: Compass routing on geometric networks. In: Proc. of Canadian Conf. on Computational Geometry (CCCG '99). (1999) 51–54
11. Kuhn, F., Wattenhofer, R., Zollinger, A.: Asymptotically optimal geometric mobile ad-hoc routing. In: Proc. of the 6th International Workshop on Discrete Algorithms and Methods for Mobile Computing and Communications (DAILM '02). (2002)
12. Ansari, S., Narayanan, L., Opatrny, J.: A generalization of face routing to some non-planar networks. In: Proc. of Mobiquitous. (2005)
13. Datta, S., Stojmenovic, I., Wu, J.: Internal node and shortcut based routing with guaranteed delivery in wireless networks. Cluster Computing **5** (2002) 169–178
14. Stojmenovic, I., Datta, S.: Power and cost aware localized routing with guaranteed delivery in wireless networks. In: Proc. Seventh IEEE Symposium on Computers and Communications ISCC, Taormina, Sicily, Italia (2002) 31–36
15. Karp, B., Kung, H.: GPSR: greedy perimeter stateless routing for wireless networks. In: Proc. of 6th ACM Conference on Mobile Computing and Networking (Mobicom '00). (2000)
16. Kuhn, F., Wattenhofer, R., Zhang, Y., Zollinger, A.: Geometric ad-hoc routing: Of theory and practice. In: Proc. of Principles of Distributed Comp. 2003. (2003)
17. Fevens, T., Haque, I., Narayanan, L.: A class of randomized routing algorithms in mobile ad hoc networks. In: Proc. of 1st Algorithms for Wireless and Ad-hoc Networks (A-SWAN). (2004)
18. Gabriel, K., Sokal, R.: A new statistical approach to geographic variation analysis. Systematic Zoology **18** (1969) 259–278
19. Bose, P., Gudmundsson, J., Morin, P.: Ordered theta graphs. Computational Geometry: Theory and Applications **28** (2004) 11–18
20. Bose, P., Devroye, L., Evans, W., Kirkpatrick, D.: On the spanning ratio of gabriel graphs and beta-skeletons. In: Proceedings of the Latin American Theoretical Infocomatics (LATIN). (2002)
21. Yao, A.C.: On constructing minimum spanning trees in k-dimensional spaces and related problems. SIAM J. Computing **11** (1982) 721–736
22. Li, X.Y., Wan, P.J., Wang, Y.: Power efficient and sparse spanner for wireless ad hoc networks. In: Proc. of IEEE Int. Conf. on Computer Communications and Networks (ICCCN01). (2002) 564–567
23. Yamazaki, K., Sezaki, K.: The proposal of geographical routing protocols for location-aware services. Electronics and Communications in Japan **87** (2004)
24. Kuhn, F., Wattenhofer, R., Zollinger, A.: Ad-hoc networks beyond unit disk graphs. In: Proc. of the 2003 joint workshop on the found. of mobile comput. (2003) 69–78
25. Bose, P., Maheshwari, A., Narasimhan, G., Smid, M., Zeh, N.: Approximating geometric bottleneck shortest paths. Comp. Geom. Theory Appl. **29** (2004) 233–249

A Multiple Path Characterization
of Ad-Hoc Network Capacity

Peter Giese[1] and Ioanis Nikolaidis[2]

[1] TRLabs, 7th Floor, 9107 116 Street NW, Edmonton, AB, Canada T6G 2V4
pgiese@trlabs.ca
[2] Computing Science Department, University of Alberta,
Edmonton Alberta, Canada T6G 2E8
yannis@cs.ualberta.ca

Abstract. In this paper, we characterize the performance of wireless ad-hoc networks through a capacity model which is based on explicit modeling of the interference across multiple routing paths. A set of elementary equations were developed to describe single-channel interference based on radio transmissions within a unit disk (UD) under random mobility conditions. The distinguishing feature of the model is that it introduces micro-models for three elementary sources of interference: inter-path, intra-path, and common origin interference. Simulation experiments found that the proposed session-based model generates throughput results consistent with existing approaches for modeling single path (SP) throughput performance within a unit area disk. However, we go one step further to use the SP throughput equations as the basis for developing a set of equations that model multiple path (MP) throughput performance. Mobility simulations of MP routing indicate that inter-path interference plays a dominant role in defining throughput performance, while intra-path interference and common origin interference have different effects as multi-user session and nodal densities change.

1 Introduction

The capacity characterization of multi-hop mobile ad-hoc wireless networks has been the topic of both technological studies, e.g. from the point of view of Medium Access Control (MAC) models [1,2], as well as analytical studies using theoretical models [3,4 ,5]. In the case mobile ad-hoc networks (MANETs) capacity models have revealed poor scalability prospects [3] unless one is willing to sacrifice delay performance [6]. Designers of MANET protocols assume a unit disk (UD) model. In the UD model two mobile nodes (MNs) communicate if and only if the Euclidian distance between the nodes is at most one [7], representing the transmission radius. Several choices exist as to how traffic can be transmitted within the single-channel medium of a UD. Other work has shown that throughput can be increased if the UD model is replaced by more realistic physical layer models [8]. This work makes the most universal assumption of broadcasting radio transmissions within a UD and while one node is broadcasting all neighboring MNs within the UD cannot transmit. While it is clear that many MN sessions are bidirectional in nature, others are not (sensors

V.R. Sirotiuk and E. Chávez (Eds.): ADHOC-NOW 2005, LNCS 3738, pp. 57–70, 2005.

networks being a good example of the latter [9]). Hence, we adopt a unidirectional definition for traffic sessions, as it is the least constraining and can be easily extended.

A critical modeling consideration is how to capture the routing and forwarding decisions. Routing along the shortest (in terms of hops) path to the destination, are commonly used for MANET routing protocols [10,11]. Other than single path (SP) routing, there has been evidence that multiple path (MP) routing between the same source and destination could produce improved performance. Two main approaches [12,13] prevail in MP MANET routing. The first uses alternative paths as backup paths to improve availability when the primary paths fail. In the second, load-balancing schemes distribute traffic over multiple paths to improve delivery performance. Node-disjoint paths are highly desirable for multi-path routing. Nevertheless, it is difficult to argue that the use of multiple paths can counter the scalability problems revealed under the single path routing assumption. However, the scalability issues are evident in large node populations, and hence MP routing could still provide a relative (but limited) advantage in small node populations.

The modeling contribution of the paper is in characterizing the traffic carrying capacity over MP routing via the introduction of path-based micro-models. That is, it attempts to answer the question of *what is the carrying capacity* over MP routing. Specifically, a session model has been developed than enables the throughput of individual source-destination pairs to be determined under a wide range of SP and MP routing conditions. A salient feature of the model is that *it merges spatial and temporal aspects of single-channel interference into a set of inter-path and intra-path equations* that can be used to characterize throughput performance under an arbitrary arrangement of multi-user traffic conditions. The interference patterns we will consider extend beyond those encountered in shortest path routing. "Interference" is used here as a catchall term, indicating any form of inhibiting a node from transmitting. Interference thus covers both the counter-productive overhearing of other stations transmissions, but also the fraction of time that a node cannot transmit because it participates in forwarding traffic along a path. For example, in order for a node to relay traffic along a path, it has to be "interfered by" the upstream node of the path (to receive), but it also has to avoid transmitting when the downstream neighbor (as per the path) transmits. Thus, it can only transmit a third of the time in order to participate in forwarding along a particular unidirectional path. For a shared transmission medium bandwidth of W bits/sec the throughput would thus be $W/3$.

We note that if nodes in the shortest path were to move without breaking reachability to the destination, the path would not change even though the capacity over the given path could be reduced significantly. This is because existing MANET routing schemes do not recalculate paths until reachability of the destination is lost. Thus, situations like the one illustrated in Figure 1 where a path "folds onto itself" are not immediately detected because the path is still considered valid even though its carrying capacity changes. Our interference equations account for this case by defining the notion of *trapping of nodes and segments*. Finally, MP routing admits the possibility that the second and subsequent paths calculated are not shortest, and hence their contribution to the carrying capacity from source to destination could very well be less than $W/3$.

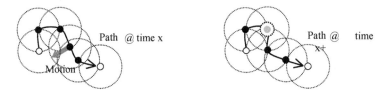

Fig. 1. Change of interference due to node mobility of a (originally) shortest path

The outline of the paper is as follows: In Section 2, we introduce the inter-path and intra-path interference micro-models and we discuss how they express the single path (SP) session model. We also introduce the common-origin interference which is a necessary ingredient to capture the MP session model. Section 3 reports the results of mobility simulations on the interactions between SP intra-path, SP inter-path and MP common-origin interference. Finally, Section 4 discusses the merits of the proposed SP and MP models and the implications of the experimental results.

2 Path Interference Micro-models

The set of interference models we consider are shown in Figure 2. We call them micro-models because they attempt to capture the fine-grained interaction of nodes participating in routing paths, rather than treating the nodes on a macroscopic (aggregate) basis. We opt for fine-grained models because they provide the means to quantify the interaction in the MP case.

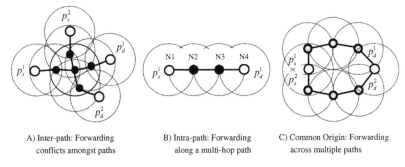

A) Inter-path: Forwarding   B) Intra-path: Forwarding   C) Common Origin: Forwarding
conflicts amongst paths   along a multi-hop path   across multiple paths

Fig. 2. A classification of interference models

In all micro-models, each source-destination pair is considered to be unidirectional. We let p^i represent the ith path and P represents the set of source-destination pairs present in the wireless network at any point in time. For example, in Figure 2-A, the set P consists of p^1 and p^2, whereas there is one MP consisting of p^1 and p^2 in Figure 2-C. Subscripts p_s, p_r, and p_d represent the source node, relay node(s), and destination node of a path, respectively. When viewed as variables, p_s^i, p_r^i, p_d^i define different sets of nodes unique to path i. To illustrate this consider Figure 2-B, in this figure $p_s^1 = \{N1\}$, $p_r^1 = \{N2, N3\}$, and $p_d^1 = \{N4\}$.

2.1 Inter-path Interference

We model throughput capacity of each node using the equation $C_n = W/g^n$, where g^n represents the set of transmitting nodes within the UD of node n. A session-oriented view allows us to define the set of transmitting nodes within the UD using a source-destination perspective. In this perspective, only source (p_s^i) and relay (p_r^i) nodes of any given path i can transmit. Determining g^n involves finding the total number of p_s^i and p_r^i nodes that reside within the UD of a given node. Let UD_n denote a set containing only the nth node in the capacity region, then $UD_n = \{n\}$. When stated in terms of sets of nodes belonging to source-destination pairs intersecting the transmission range of UD_n, the session-oriented equation for calculating nodal capacity is:

$$C_n = \frac{W}{\displaystyle\sum_{\forall i \in P} \left(\left\| p_s^i \cap UD_n \right\| + \left\| p_r^i \cap UD_n \right\| \right)} \ . \tag{1}$$

Equation (1) defines the maximum nodal throughput capacity due to interference caused by multiple unidirectional sessions. This equation is used to determine nodal throughput once the routes of all source-destination pairs are known. The maximum throughput that a single source-destination pair can achieve over path i is the minimum capacity of any node along the routed path between the pair, that is:

$$TH_{Inter-path}^i = \min_{\forall n \in P^i} (C_n) \tag{2}$$

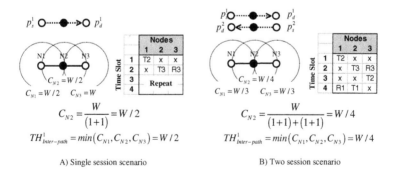

A) Single session scenario   B) Two session scenario

Fig. 3. Examples illustrating the application of inter-path equations

Application of the inter-path equations for a unidirectional session and two unidirectional sessions are illustrated in Figures 3-A and 3-B respectively. These illustrations indicate how the merging of path temporal relationships with unit-disk spatial relationships determines SP throughput.

In total, inter-path interference arises when distinct paths are no longer disjoint in any one of three ways: node, span, and unit-disk. Figure 4 illustrates the path topologies expressed by the inter-path equations.

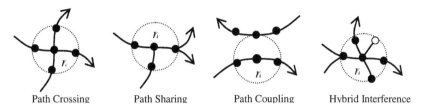

| Path Crossing | Path Sharing | Path Coupling | Hybrid Interference |

Fig. 4. Types of interference captured by inter-path equations

2.2 Intra-path Interference

Inter-path interference is not the only source of interference affecting session throughput. Inter-path equations do not take into consideration complex transmission schedules that can exist when SP paths re-enter the UD of a node more than once. SP throughput will be reduced if the transmission schedule of a node is restricted or "trapped" between the transmission schedules of other nodes belonging to the same path. What we will subsequently call *Intra-path interference* occurs when individual nodes belonging to the same SP become trapped within the UD of other nodes of the same path. By analyzing several different intra-path topologies it is found that per-session throughput varies as a function of the number of trapped nodes (t_n) and the number of trapped segments (t_{seg}) as shown in Figure 5. Intra-path equations have been developed to predict SP throughput under trapped transmission schedules.

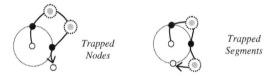

Fig. 5. Two trapped nodes within a single segment (left) and two trapped segments (right)

When path length is less than three hops no trapped transmission schedules can occur. Under this condition, the equation describing intra-path throughput for path i of hop length h is :

$$TH^i_{Intra-path} = \frac{W}{h} \text{ , } iff \text{ } 1 \le h < 3. \tag{3}$$

Analysis of transmission schedules under trapping conditions revealed that modular arithmetic, specifically modular three (mod 3), can be used to describe throughput performance as a function of t_n when the overall path length is greater than three hops. Maximum SP throughput corresponds to the minimum number of transmission time slots required to transport a payload the entire length of the path. Figure 6 shows how the minimum transmission schedules can vary depending on path length for one routing scenario.

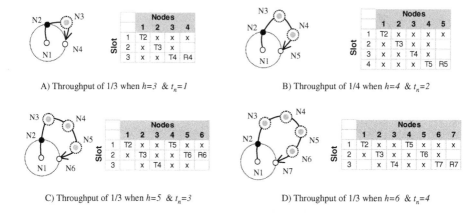

A) Throughput of 1/3 when $h=3$ & $t_n=1$

B) Throughput of 1/4 when $h=4$ & $t_n=2$

C) Throughput of 1/3 when $h=5$ & $t_n=3$

D) Throughput of 1/3 when $h=6$ & $t_n=4$

Fig. 6. Topologies involving source and destination transmission conflicts

A set of SP intra-path equations is proposed that takes into account different path topologies involving single-segment trapping. The intra-path equation for routing topologies where both end nodes p_s^i and p_d^i of path i are interfering with each other is described by:

$$TH_{Intra-path}^i = \begin{cases} \dfrac{W}{3} & \textit{iff } h \geq 3 \text{ \& } \mod(t_n+1,3) \neq 0 \\ \dfrac{W}{4} & \textit{iff } h \geq 3 \text{ \& } \mod(t_n+1,3) = 0 \end{cases} \quad (4)$$

A different equation is required when the routing topology of path i cause relay nodes p_r^i to be interfering with each other or relay nodes interfering with end nodes. Under this condition, the appropriate intra-path equation is described by equation (5).

$$TH_{Intra-path}^i = \begin{cases} \dfrac{W}{3} & \textit{iff } h > 3 \text{ \& } \mod(t_n,3) = 0 \\ \dfrac{W}{4} & \textit{iff } h > 3 \text{ \& } \mod(t_n,3) \neq 0 \end{cases} \quad (5)$$

The complete set of SP intra-path equations consists of (4) and (5) depending on the type of single-segment trapping present within a SP, and by equation (3) when no trapping occurs. An analysis of multi-segment trapping on end-to-end throughput was not conducted in this work because in-depth experiments need to be conducted to determine how often, if at all, trapping within one segment occurs let alone trapping within multi-segments. Moreover, for multi-segment trapping to occur path lengths must be long and when longer paths reside within the same UD overall throughput is ultimately limited by inter-path interference. The upper bound of SP throughput is defined by both inter-path and intra-path equations. Once both sets of equations have been calculated for a given unidirectional session the final SP throughput value (TH_{SP}^i) can be determined by the equation:

$$TH_{SP}^i = \min \left(TH_{Inter-path}^i, TH_{Intra-path}^i \right) \tag{6}$$

2.3 Common Origin Interference

MP throughput is influenced similarly as SP throughput, but also includes *common origin interference*. Multiple paths achieve maximum spatial reuse when the paths are disjoint. However, even under this condition MP throughput is limited to *W/2* because the overall transmission rate is limited by contention at the source node [13]. When multiple paths exist between a pair of MNs, forwarding of traffic can be modeled using a round-robin scheme that multiplexes along each path. Under this condition, a simple equation can be used to describe common origin interference that takes into account load sharing across the set of disjoint paths. Let k represent the number of disjoint paths that exist in MP. It is well known that MP throughput is:

$$TH_{MP} = \begin{cases} W & \text{iff } k = 1 \\ \dfrac{W}{2} & \text{iff } k > 1 \end{cases} \tag{7}$$

In this work, it was discovered that Equation (7) only applies when the path lengths of all disjoint paths belonging to the same MP are greater than or equal to 2. The existence of one single-hop path within the set of k paths affects overall MP throughput performance in a manner that is not predicted by Equation (7). It was observed that by defining k in terms of the sum of all single hop and non-single hop members within the same MP one could model all possible MP topologies and calculate valid MP throughput values. The equation for k is

$$k = \varepsilon + \delta, \tag{8}$$

where ε is the number of unit length (single hop) paths, and δ is the number non-unit (hops>1) paths.

There is one unique MP routing topology in which throughput not only depends on the set of $[\varepsilon, \delta]$ values that define k but also on the length of path δ. That unique topology occurs when $k=2$, such that $[\varepsilon, \delta] = [1,1]$. For this topology, a third variable n_δ is required to define MP throughput, where n_δ represents the number of nodes along a path of length δ. We let TH_{CO}^j represent the maximum throughput due to common origin interference in the jth MP and let the ith SP that is part of a MP be represented as $TH_{SP:MP}^i$. Two MP examples and the application of equation (8) are shown in Figure 7. These examples illustrate MP throughput due to common origin interference only. It is worth noting that the throughput performance of any individual SP within the MP group is significantly less than the total MP throughput value.

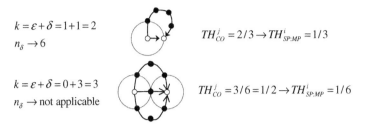

$$k = \varepsilon + \delta = 1 + 1 = 2$$
$$n_\delta \rightarrow 6$$

$$TH^j_{CO} = 2/3 \rightarrow TH^i_{SP:MP} = 1/3$$

$$k = \varepsilon + \delta = 0 + 3 = 3$$
$$n_\delta \rightarrow \text{not applicable}$$

$$TH^j_{CO} = 3/6 = 1/2 \rightarrow TH^i_{SP:MP} = 1/6$$

Fig. 7. Examples of $[\varepsilon, \delta]$ combinations and MP throughput values

The complete set of common origin throughput values for all possible combinations of $[\varepsilon, \delta]$ was derived and is summarized by equation (9). Determining the minimum transmission schedules of the source node determined the throughput values. All $[\varepsilon, \delta]$ sets of values directly apply to MP routing, with the exceptions of [0,1] and [1,0]. These two sets define SP routing topologies not MP routing topologies. All $\varepsilon > 1$ combinations represent specialized MP groups that possess multiple single hops. These specialized MP groups were not considered to be valid for the single-channel interference scenarios studied in this work, but could play an important role in the analysis of multi-channel wireless systems and/or wireless multicasting systems. Overall, the maximum MP throughput values can be determined by ε, δ and n_δ. Recall that path length does not play a role in defining MP throughput except in [1,1]. Namely, the equation derived capable of capturing MP is as follows:

$$TH^j_{CO} = \begin{cases} \dfrac{kW}{\varepsilon + 2\delta} & \text{iff } \varepsilon \neq 1 \,\&\, \delta \neq 1 \\[2mm] \dfrac{kW}{\varepsilon + 2\delta} & \text{iff } \varepsilon = 1, \ \delta = 1, \text{ and } \mathrm{mod}(n_\delta, 3) = 0 \\[2mm] \dfrac{kW}{\varepsilon + 3\delta} & \text{iff } \varepsilon = 1, \ \delta = 1, \text{ and } \mathrm{mod}(n_\delta, 3) \neq 0 \end{cases} \qquad (9)$$

Equation (9) is referred to as the common origin equation. The common origin equation defines the upper bound of MP throughput values based on the assumption that payload traffic is multiplexed equally across all k paths. By exploiting this multiplexing property, the maximum throughput of any individual SP within the MP group can be defined as

$$TH^i_{SP:MP} = \frac{TH^j_{CO}}{k} = \begin{cases} \dfrac{W}{\varepsilon + 2\delta} & \text{iff } \varepsilon \neq 1 \,\&\, \delta \neq 1 \\[2mm] \dfrac{W}{\varepsilon + 2\delta} & \text{iff } \varepsilon = 1, \ \delta = 1, \text{ and } \mathrm{mod}(n_\delta, 3) = 0 \\[2mm] \dfrac{W}{\varepsilon + 3\delta} & \text{iff } \varepsilon = 1, \ \delta = 1, \text{ and } \mathrm{mod}(n_\delta, 3) \neq 0 \end{cases} \qquad (10)$$

Equation (10) defines the maximum SP throughput that applies to all SPs (for all i) within the jth MP. As discussed in Section 2.2, the throughput of any individual SP is limited by inter-path and intra-path interference. Let TH^i_{SP} represent the SP throughput due to inter and intra-path interference, as defined by Equation 6, for the ith path in the MP containing K distinct SPs. Then MP throughput can be modeled on a per-session basis by using the equation

$$TH_{MP}^{j} = \sum_{\forall i \in K^{j}} \min\left(TH_{SP:MP}^{i}, TH_{SP}^{i}\right) \tag{11}$$

Equation (11) defines the upper bound of MP throughput performance in the presence of inter-path, intra-path and common origin interference for any multi-path topology. Like equation (6) in the SP case, equation (11) in the MP case is used to quantify end-to-end throughput for a unidirectional MP session. The MP equation only applies for disjoint SP routing within the MP group. MP throughput under non-disjoint routing conditions needs to be studied to ascertain whether or not interference caused by non-disjoint paths is already adequately dealt with by SP inter/intra-path equations or if additional equations are required.

3 Simulation Results

Mobility was simulated using a variant of the Random Direction Mobility Model [14]. A random starting direction and randomly generating a new direction for each MN after a fixed epoch time (10 cycles/units). Once the border of the region has been crossed a simple bounce back, without pause time, is applied to ensure that the MN remains within the simulation region. Simulation results that are not reported here for the sake of brevity, have validated the SP model by producing results in agreement with the previously proposed models by Gupta and Kumar and Gastpar and Vetterli's relay model. The simulation results presented here cover the topic of MP performance.

Fundamental to any MP design is the ability to efficiently perform multi-path forwarding across a set of disjoint paths. Path Algebra was used to perform shortest and disjoint routing path calculation between pairs of MNs in all simulations. The specific definition of a dioid structure used in this work is based on the definition of a semi-ring $\langle S, \oplus, \otimes, \phi, \varepsilon \rangle$ that lists elementary paths. Operations performed by the Path Algebra algorithm are based on the following linguistic-based mathematics [15]:

$$X \oplus Y = b(X \cup Y) \text{ and } X \otimes Y = \{\chi \circ \psi \mid \chi \in X \text{ and } \psi \in Y\} \tag{12}$$

where X and Y are set of languages associated with label edges of the graph. The theoretical foundation for Path Algebra is based on directed graphs so each edge can have up two labels (two if bidirectional). It is also assumed that wireless transmission within the UD is bidirectional.

The main advantage of the using Path Algebra is that finding the complete set of disjoint routes for all MN pairs is performed efficiently using sequences of matrix operations. The computational efficiency is highest when simulating a high degree of random motion. The SP and MP routes generated by this approach are all elementary is nature, meaning that routes are the shortest possible and disjoint. Using a centralized approach to determine all routes within a simulation cycle provides a means of characterizing MP throughput under idealized routing conditions.

In order to assess MP throughput performance over a wide range of routing scenarios, various MN densities were simulated over two rectangular areas. A small area with relatively high density of MNs was selected to assess throughput performance in a confined environment that produces short multi-hop path lengths. A second larger area with a medium density of MNs was selected to assess throughput performance in an open environment that produces longer multi-hop path lengths. The simulation parameters associated with each MP simulation area are shown in Figure 8.

Simulation time	900 cycles		Simulation time	900 cycles
Number of nodes	2 to 30		Number of nodes	2 to 60
Simulation area	$3*UD \times 3*UD$		Simulation area	$5*UD \times 5*UD$
MN range	$UD = 1$		MN range	$UD = 1$
Movement speed	$UD / 4$		Movement speed	$UD / 4$

A) Confined Area Simulation   B) Open Area Simulation

Fig. 8. Simulation parameters

In addition to varying MN densities, the number of active sessions was also varied to assess the affects of session densities on throughput performance. Three randomly generated session densities were simulated: 1) when only one unidirectional session exists for n MNs (denoted as S(1)), 2) when n unidirectional sessions exist for n MNs (denoted as S(n)), and 3) when $n(n-1)$ communication sessions exist for n MNs (denoted as S(all)). The set of SP and MP throughput simulation results for both the confined and open areas are presented in Figures 9-A and 9-B, respectively. By normalizing the value of W to unity, all throughput values are reported as $1/x$ fraction rather than W/x fraction. All reported results are based on averages taken over three different simulations, with each simulation using a different random seed value.

The simulation results revealed that overall per-session throughput is limited by the probability of connectivity occurring within a given simulation cycle. Under conditions of low node densities, the minimum being $n=2$, all throughput values are limited by the probability that a path exists for a given unidirectional path. From Figure 10 one can see that the probability of not finding a path is highest in the open area simulations. In both simulations the probability of routes existing for individual sessions increases as node density increases. In the open simulation there is a higher frequency of no-route-found events, especially at lower node densities. Thus, SP and MP throughput results for open area simulations are less than the confined area results. Also, in open areas the results are subject to more statistical variation due to the higher probability that no routes will be found.

No throughput values shown in Figure 9 reached a value of unity. This is an expected result as the upper limit of SP throughput is ultimately restricted by path length, where as the upper limit of MP throughput is restricted by common origin interference and path length. The plots shown in Figure 11 describe how path lengths increased as a function of node density. Open area simulations resulted in longer SP routes, up to 73% longer when measured in terms of the number of hops per path.

Also, longer MP routes occurred in the open area simulation, up to 44% longer when measured in terms of the sum of all hops across all paths divided by the number of paths. The impact of path length on throughput is most obvious when the session density is lowest and the nodal density is highest. Under these conditions, average path length combined with the number of times a path was found define an upper limit of SP and MP throughput performance.

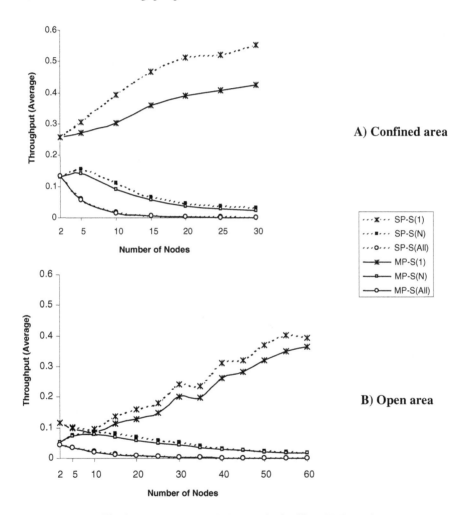

Fig. 9. Throughput simulation results for SP and MP routing

The simulation results indicate that session density has greater influence on throughput than nodal density. The simulation results shown in Figure 9 also indicate that MP throughput is always less than (or at best equal to) SP throughput. With high session densities it is expected that throughput of MP routing would be no better than SP routing because the capacity of relay nodes is diminished. The overall difference between SP interference (caused by high traffic loads) vs. interference caused by smaller traffic loads spread over several disjoint MP routes is negligible. Under low session density, it is the restriction of common origin interference that forces MP throughput to be less than SP throughput.

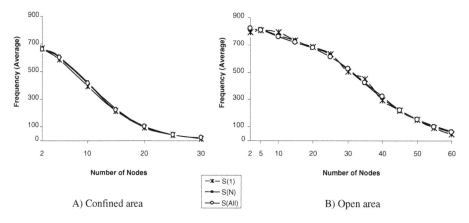

Fig. 10. No-route-found frequency of occurrence

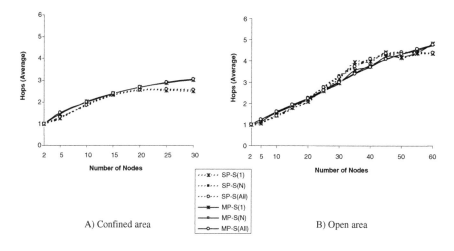

Fig. 11. SP & MP route lengths

A detailed analysis of the simulation results revealed that inter-path interference plays a dominant role in defining SP and MP throughput performance, while intra-path interference plays a minor role. The frequency plots illustrated in Figure 12 show that intra-path interference is most significant at higher node densities under MP routing conditions. The occurrence of trapping within SP routes was in the order of 1/100 less than trapping within MP routes and is not plotted. SP trapping exhibited near random behavior, with trapping events occurring more frequently as node density increased.

Simulation results produced by this work indicate that *MP routing does not enhance service delivery rate under conditions of elementary routing*. MP routing does not out-perform SP routing because of the dominance of inter-path and common origin interference. Even though MP routing does not enhance service delivery, it may still be useful for enhancing end-to-end service survivability.

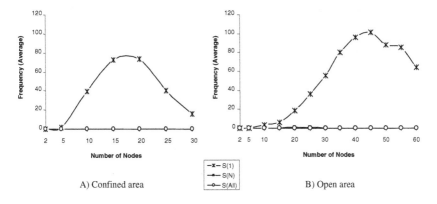

Fig. 12. Frequency of MP intra-path trapping

4   Conclusions

This paper demonstrated that SP and MP throughput performance could be modeled using a session approach that simulates single-channel interference in wireless networks. Throughput performance was successfully modeled by using a set of inter-path, intra-path and common origin interference equations. The proposed SP equations form a foundation for modeling the upper bound of throughput performance for unidirectional paths for a wide range of mobile traffic conditions. Experimental results, not reported here for the sake of brevity, have demonstrated that the SP session model produces results that are consistent with theoretical upper bounds when one selects optimum networking parameters.

It is shown in this paper, through MP simulation experiments involving random mobility, that load sharing over multiple disjoint routes results in lower overall per-session throughput when compared to sending all traffic over one single route. The difference is more pronounced in confined areas where multi-hop routes are shorter in length than in open areas. The simulation results also indicate that session density has a greater influence on throughput than nodal density, with both decreasing MP throughput performance as density values increase. MP experimental results indicate that inter-path interference plays a dominant role in defining throughput performance, while intra-path interference and common origin interference have different effects on capacity depending on session and nodal densities.

References

[1] S.Y.Wang, "Optimizing the packet forwarding throughput of multi-hop wireless chain networks," Computer Communications Vol. 26, pp 1515-1532, 2003

[2] G. Holland, and N. Vaidya, "Analysis of TCP Performance over Mobile Ad Hoc Networks," Wireless Networks, Vol. 8, pp275-288, 2002

[3] P. Gupta and P. R. Kumar, "The capacity of Wireless Networks," IEEE Trans. Inform. Theory, Vol 46, March 2000

[4] M. Gastpar and M. Vetterli, "On the Capacity of Wireless Networks: The Relay Case," IEEE Infocom, 2002

[5] S. Aeron and S. Venkatesh, "Scaling Laws and Operation of Wireless Ad-Hoc and Sensor Networks," IEEE Workshop on Statistical Signal Processing, pp 367 – 370, Oct. 2003

[6] M. Grossglauser and David Tse, "Moblity Increases the Capacity of Ad Hoc Wireless Networks," IEEE/ACM Transactions on Networking, Volume:10, Issue: 4, Aug. 2002

[7] B. N. Clark, C. J. Colbourn and D. S. Johnson, "Unit Disk Graphs," Discrete Mathematics, Vol 86, pp 165-177, 1990

[8] I. Stojmenovic, A Nayak, and J. Kuruvila, "Design Guidelines for Routing Protocols in Ad Hoc and Sensor Networks with a Realistic Physical Layer," IEEE Communications Magazine, March 2005

[9] J. N. Al-Karaki, A.E. Kamal, "A Taxonomy of Routing Techniques in Wireless Sensor Networks," in Sensor Networks Handbook, M. Ilyas and I. Mahgoub (Eds.), CRC Publishers, 2004.

[10] D. B. Johnson, D. A. Maltz, and J. Broch, "DSR: The Dynamic Source Routing Protocol for Multihop Wireless Ad Hoc Networks," in Ad Hoc Networking, C. E. Perkins (Ed.), pp. 139-172, Addison-Wesley, 2001.

[11] C. E. Perkins and E. M. Royer, "The Ad Hoc On-Demand Distance-Vector Protocol (AODV)," in Ad Hoc Networking, C.E. Perkins (Ed.), pp. 173-219, Addison-Wesley, 2001.

[12] Y. S. Liaw, A. Dadej and A. Jayasuriya, " Throughput Performance of Multiple Independent Paths in Wireless Multihop Network," IEEE International Conference on Communications, Volume: 7 , June 2004

[13] S. De and C. Qiao, "On throughput and Load Balancing of Multipath Routing in Wireless Networks," IEEE Wireless Communications and Networking Conference (WCNC), Vol. 3, pp 21-25 March 2004

[14] E. Royer, P.M. Melliar-Smith, and L. Moser, "An analysis of the optimum node density for ad hoc mobile networks," IEEE ICC Proceedings, 2001.

[15] B. Carre, "Graphs and Networks", Clarendon Press Oxford, 1979.

Increasing the Resource-Efficiency
of the CSMA/CA Protocol
in Directional Ad Hoc Networks

Matthias Grünewald, Feng Xu, and Ulrich Rückert

Department of Electrical Engineering,
System and Circuit Technology,
University of Paderborn / Heinz Nixdorf Institute,
Fürstenallee 11, Paderborn, Germany
{gruenewa, xu, rueckert}@hni.upb.de

Abstract. The use of directional communication can result in higher performance of ad hoc networks in terms of throughput and delay. To exploit these advantages, we propose a system architecture that applies k air interfaces on each node. Each interface is equipped with a directional antenna. However, the energy consumption of such a system would be too high intuitionally. Power management is required that switches off the air interfaces if they are not needed. Hence, we design a detailed energy model for the system and a MAC protocol based on CSMA/CA with extensions for parallel directional communication, radiation power control and air interface power management. We verify and evaluate our implementation in a simulation environment. Our results show that the proposed system can achieve an energy efficiency comparable to a single antenna system while increasing the throughput and time efficiency of the resulting ad hoc network by a factor of 2-3.

1 Introduction

Wireless connectivity is gaining increasing attention, due to its cable-free operation and mobility support. Wireless networks usually consist of base stations that service a number of mobile devices such as notebooks, PDAs and mobile phones. However, the operating cost of such networks is high and base stations may not be available everywhere. Ad hoc networks do not require base stations; the participating mobile devices discover themselves autonomously and form a dynamically changing multi-hop network. Traditionally, the applied radio technology employs an omni-directional antenna and a fixed radiation power. Hence, the radio signal is emitted in every direction at the largest possible range. All nodes that are in the same range need to suspend their own transmission in order to prevent collisions. Especially in crowded areas such as office buildings or public places, a high number of nodes have to share the same channel. This results in low data throughput, increased delays and higher energy consumption since the air interfaces need to stay on longer to finish their communication tasks.

V.R. Sirotiuk and E. Chávez (Eds.): ADHOC-NOW 2005, LNCS 3738, pp. 71–84, 2005.

A possible solution to this drawback is the use of directional antennas that send the radio signal in one direction only within a narrow beam. Compared to omni-directional antennas, directional antennas have a higher gain (cf. fig. 1), resulting in an increase of the range or in a decrease of the required radiation power if the same range is used. Hence, less space is blocked and the probability of parallel transmission is increased. By controlling the radiation power such that the signal is transmitted at the minimum range to reach the destination, the size of the blocked space can be further decreased.

In this paper, we propose a system architecture that applies k directional antennas and air interfaces on each node. To obtain the advantage of parallel directional communication and combat the shortcoming of high power consumption, a MAC protocol based on CSMA/CA with extensions for radiation power control and power management is further given. We study the performance gains of the proposed system and protocol extensions in terms of throughput, delay and battery power consumption by a network simulator. Our energy model includes a more realistic non-linear model for the power amplifier and also regards the consumed reception energy that is caused by collisions. Our results show that both the throughput and time-efficiency can be increased up to a factor of three in average, while an energy-efficiency can be achieved comparable to a single antenna system.

2   Related Work

A few researchers have already proposed extensions that enable the use of the carrier sense multiple access with collision avoidance (CSMA/CA) protocol with directional antennas. The early work of Nasipuri et al. [1] applies omni-directional exchange of Request-To-Send (RTS) and Clear-To-Send (CTS) control packets to establish directional links for data packets. Ko et al. [2] advocate the scheme of uni-directional RTS as well as omni-directional CTS to increase the network throughput further. However, these omni-directional or semi-directional exchanges of control packets limit the reuse probability of the space channel. Later, Takai et al. [3] have shown how virtual carrier sensing mechanism can be extended so that it can be used for both omni- and uni-directional transmission. Their approach can use uni-directional RTS and CTS packets, resulting in higher spatial reuse factor. Our approach is based on their extensions. To solve the problem of knowing in which direction a data packet needs to be sent, Korakis et al. have proposed in [4] to send RTS request in every direction first to find the destination. In our approach, a neighborhood discovery algorithm is used that avoids the need to send circular RTS for every data packet. In all of these previous schemes, it is assumed that the on-going directional transmission/reception on each node is exclusive. This property may reduce the spatial reuse factor and causes additional problems such as deafness as described in [5]. To our knowledge, only Lal et al. have studied the use of parallel receptions in a receiver-initiated polling MAC scheme in [6]. In this work, we propose that each node has the ability to perform parallel directional transmission/reception if it is allowed by our MAC protocol.

Until now, the energy efficiency is scarcely discussed in the area of directional MAC protocol design. Ramanathan [7] has analyzed the effect of link power control, which is widely implemented in military radios. The algorithm resulted in performance improvement on throughput and delay in their simulations. In [8], Nasipuri et al. have shown that their directional MAC protocol with power control (DMACP) reduces the average power up to 90% while the network throughput has nearly no degradation. Power control can decrease the energy consumption per packet as well as increase the throughput as less space is blocked. In this paper, we also analyze the influence of radiation power control for data and control packets, based on our previous work [9]. However, in all of these previous schemes (including our previous work), only the power for transmission (radiation power) is considered and the power consumed for receiving, carrier sensing and discarding (due to collision) is ignored. Actually, radiation power just accounts for small proportion of the whole battery power consumed for the transmission. Power management with omni-directional antennas has already been discussed in the literature. The IEEE 802.11 specification includes mechanisms to synchronize the nodes to wake up periodically in so-called ad hoc traffic indication message (ATIM) windows [10]. However, the clock synchronization algorithm requires a fully connected MANET. Tseng et al. have proposed power management algorithms that can also be used in multi-hop MANETs [11]. We are using a strategy comparable to the 802.11 approach.

3   System Architecture

In our studies, we assume a system that is based on multiple directional antennas. For each antenna, an individual air interface is used. Since an increased amount of computing power is necessary to process the medium access (MAC) protocol for each air interface in parallel, we have developed a multiprocessor system-on-chip (MPSoC) together with a suitable design methodology [12]. In most of the known literature, it is assumed that a smart antenna array is applied on the node. Smart antennas can be categorized into two types: steered beam and switched beam. The former has the advantage that the direction of the beam can be steered to any intended direction by the applied algorithm at the cost of more sophisticated and expensive signal processing than the latter [13]. However, Ramanathan et al. have shown that smart antennas with switched beams are nearly as good as steered beams if just spatial reuse is considered [7]. In our proposed system, the directional antennas are not combined to create the beam. Instead, the directional antenna pattern of each antenna is used. Therefore, for sending in a specific direction, only one antenna is used. By arranging the directional antennas properly, our proposed scheme can be compared to a switched beam smart antenna without the need for an additional signal processor that performs the beam-forming.

A major advantage of our scheme is the reduction of the *deafness problem*. To avoid excessive load on signal processing, smart antennas usually send or receive in one direction at one moment. Choudhury et al. have shown in [5] that this results in a deafness problem. If a node is beam-formed in one direction and a communication request is sent to the same node from another direction, the node can not hear the request because its antenna is not sensitive in that direction. In case of independent air interfaces, the node can theoretically receive and transmit in all directions in parallel and the deafness problem is reduced. However, suitable isolation is necessary to prevent the self-interference among the multiple directional antennas on the same node. In this paper, we assume that this problem is solved at the physical layer by signal processing or mechanical means. Another advantage of our multiple directional antennas system is that air interfaces can be switched off if they are not needed, to save energy. We show in this paper that the energy due to just sensing the medium is very high. For a k-antenna system, the energy efficiency is reduced up to the same factor k in the worst case. For a multiple directional antennas system, the energy consumption can be decreased below the one of a single antenna system by applying a suitable power management strategy. A smart antenna system can not achieve this advantage because it requires that the transceivers for all antenna elements are always powered on to generate the desired beam.

3.1   Directional Antenna Model

Our proposed communication system applies k identical directional antennas, each of which is designed so that its Half Power Beam Width (HPBW) is approximately $360°/k$. Furthermore, we assume that all of the nodes are located in the horizontal plane, and the antennas on each node are oriented so that their main beam directions distribute equally (circularly) around it in the horizontal plane. This divides the space around the node into k sectors (cf. fig. 1).

Via studying the existent directional antenna elements, we found that axial mode helical antenna (AMHA) holds very high directivity while its structure is quite simple. It has been theoretically well-analyzed since its invention on the late of 1940's [15], and can be practically realized due to its essence of simplicity. Thus, we apply the AMHA as the actual directional antenna model. To simulate the node equipped with 4, 6 and 8 air interfaces, we calculated three

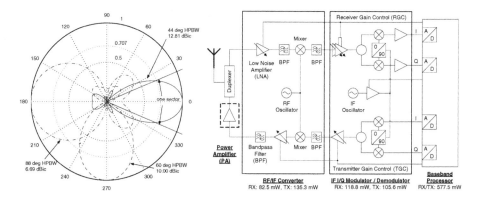

Fig. 1. Three types of normalized axial mode helical antenna radiation patterns (left) and general structure of the air interface (right) [14]

configurations of the AMHA model, which approximately hold HPBW with 90°, 60°, and 45°, respectively. Fig. 1 illustrates three normalized radiation patterns of AMHA applied in this paper. The directivity uses the unit dBic (with reference to an isotropic circular polarization antenna).

3.2 Energy Model of the Air Interface

The general structure of the air interface is shown in fig. 1 [14]. The baseband processor performs AD/DA, coding, decoding as well as other necessary digital signal processing to convert the bits of the data packets to a complex waveform and vice versa. The modulation part of the air interface converts between the complex and real representations of the signal by modulating/demodulating it with the carrier in the intermediate frequency (IF) band. The RF/IF-converter moves this signal from the IF band to the radio frequency (RF) band and vice versa. For receiving a radio signal, a low noise amplifier (LNA) and a series of cascaded amplifiers after the down-conversion are necessary to amplify the signal detected by the antenna. For sending, a power amplifier (PA) amplifies the generated radio signal to reach the necessary radiation power. The duplexer unit is necessary to switch between the receiving and sending mode of the antenna.

We have created a black-box energy model for the described components of the air interface. Feeney et al. have already shown that a linear approximation can be used to model the power consumption of an 802.11b based WLAN network card [16]. They have measured the complete energy consumption inclusive MAC protocol processing for different types of traffic (broadcast, point-to-point, overhearing traffic, ...). We want to create a model of the energy consumption of the air interface only, to include it in our simulation environment (cf. sec. 5). Hence, we can differentiate the modes of operation (receiving, sending and discarding packets as well as sensing the medium state) by collecting the required information in each simulated node. This has also the advantage that the (reception) energy consumption for unsuccessful attempts to aquire the channel and messages lost due to collisions can be considered. This is not possible if Feeney et al.'s approach is applied.

For the air interface, we assume three modes of operation: *sending* (TX) and *receiving* (RX) packets as well as *sensing* (SX) the state of the medium. The energy consumed if one packet is received is estimated by

$$E_{RX}(p) = P_{RX} \cdot p/B \qquad (1)$$

where p is the size of the packet in bits, B is the data rate at which the packet is received and P_{RX} is the power consumption of the air interface in receiving mode. In sensing mode, the air interface constantly listens if any signal power can be detected that is above the sensing threshold P_s. If a signal is detected, the air interface switches to receive mode. In modern communication systems, techniques such as direct sequence spread spectrum are applied to provide a number of communication channels within the same frequency band. Radio networks that operate in the same space and frequency band can coexist by using different channels. However, receivers that are sensing the medium have to perform a de-spreading procedure first before they can decide if a radio signal is above the sensing threshold in their assigned logical channel. Hence, the power consumption in sensing mode is nearly as high as in receiving mode, since all components in the receiving chain of the air interface have to be turned on. This effect has also been experimentally verified by Feeney et al. They have measured that the power consumption in sensing mode is only 12-18 % lower than in receiver mode [16]. We use the following equation to estimate sensing energy:

$$E_{SX}(\Delta t) = P_{SX} \cdot \Delta t \qquad (2)$$

where Δt is the duration the air interface is in sensing mode and P_{SX} is the power consumption of the sensing mode.

Since we also want to examine the effect of radiation power control on energy consumption, we have to include the battery consumption of the power amplifier. We use the following equation:

$$E_{TX}(p) = (P_{TX} + P_{PA}(P_t)) \cdot p/B \qquad (3)$$

where P_{TX} is the static power consumption (for powering the baseband processor and the up-converter). The function $P_{PA}(P_t)$ describes the dynamic power consumption of the power amplifier in dependence of the requested output radiation power P_t. For our simulations, we have sampled the power curve that can be found in the data sheet of Philip's 2.4 GHz SA2411 power amplifier [17] at 2 dBm steps and used a linear interpolation to obtain P_{PA}.

For the other components, we use power values provided in the data sheets of the Intersil's PRISM 11 Mbps 802.11b reference chip set [18] (cf. fig. 1). The values for P_{RX}, P_{SX} and P_{TX} are outlined in tab. 1. Please note that we assume a 15 % reduction in power consumption if the air interface is in sensing mode (in accordance with the results by Feeney et al) since the data sheet of the PRISM chip set does not differentiate between sensing and receiving mode. Additionally, the power consumption of the baseband processor is overestimated because the data sheet only gives the power consumption for both the baseband and the MAC processor.

4   Directional CSMA/CA with Power Management

The stack of our proposed protocol is shown in fig. 2. For each air interface, a physical (PHY), topology control (TC) and medium access control (MAC) layer is used. The PHY layers perform additional processing such as tagging packets that have the strongest reception power (cf. sec. 4.1). The local MAC layers perform the basic CSMA/CA protocol [10] and power management (PM) (cf. sec. 4.2). A global network layer (NET) performs the routing algorithm. The neighborhood discovery is done by a global Neighbor-MAC layer that sends broadcast HELLO packets with a period of T_{ND} in each sector. For this purpose, the local MAC layers inform the global one if they have received HELLO messages. The payload of the HELLO packets contain the neighbors that are already known

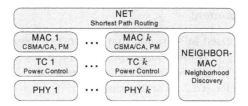

Fig. 2. The proposed protocol stack

by the sender. The global MAC layer checks if the sender knows this node. If it does not, it will send an HELLO ACK packet to the initiator to introduce itself.

The purpose of the TC layer is to maintain a list of neighbors for the sector it is responsible for. It also collects observed signal strengths for all known neighbors by collecting appropriate information from all received packets. With this information, it adapts the transmission power for outgoing packets such that a nominal reception power $P_{e,nom}$ is maintained at the destination. The neighbor list of the TC layer also allows to adapt the power for control packets such as RTS/CTS by adjusting their transmission power such that all known neighbors can be reached. However, the neighbor list has to be updated fast enough such that no hidden terminals are created. The NET and NEIGHBOR-MAC layer are periodically updated with the neighborhood information. With this information, the NET layer can decide in which sector the data packet needs to be routed. Details about the neighborhood discovery and power control mechanisms can be found in [9]. The new aspect of this paper is the analysis of battery energy consumption and power management.

4.1   Directional Communication

Takai et al. have developed a directional virtual carrier sensing [3] to allow transmissions in direction where no carrier has been sensed. It uses a NAV table whose entries mark specific angle ranges in which ongoing transfers have been sensed. A transmission can be started if the NAV table does not contain an entry for the desired direction. In our scheme, we are using k independent air interfaces. For every air interface, the CSMA/CA protocol is applied independently. Every instance of the MAC layer updates its NAV in parallel if it senses a carrier in its sector. Therefore, the NAV table mechanism is automatically obtained by the parallel working MAC layers. To handle the case that more than one receiver detects the same packet, the packet with the highest reception power is tagged by the PHY layer to indicate that it has been received in the main sector. Each MAC layer processes un-tagged packets as if they are not addressed to this node. Hence, the NAV will be updated in the other sectors and they are not starting a transmission. Therefore, the side lobes of the antennas in the other sectors can not interfere with the current transfer. Since the TC layers periodically updates the NET layer with the known neighbors, the NET layer has enough knowledge to decide in which sector to forward a packet. This can be compared to the direction-of-arrival (DOA) caching proposed by Takai et al.

4.2 Power Management

We have already discussed in section 3 that the energy required for just sensing the channel is very high. If directional transmission is used, the sensing energy becomes k times larger because k receivers are sensing the channels compared to one in the omni-directional case. For transmitting data from one node to another, only one sender and one receiver is required. Hence, if the sector (direction) of the data transfer has been determined, all receivers except the one in the reception sector can be switched off. However, there must exist time periods in which all receivers are on in order to start new transfers in other directions.

We are using a simple power management strategy comparable to the one found in the IEEE 802.11 specification [10]. It uses periodically occurring contact windows in which all receivers are powered on. Outside the contact windows, only the receivers that are required for the current communication task remain powered on, the other ones are switched off. During the contact windows, the nodes that have buffered data packets use STAY AWAKE control packets to request the destination node to keep the receiver that is responsible for the sector powered on. In order to implement this power management scheme, the clocks in the nodes have to be synchronized. This can be established by using distributed clock synchronization (e.g., IEEE 802.11 manages to synchronize the clocks of all nodes with an offset of a few μs).

The contact windows have a width of T_a seconds and a distance of T_d seconds (cf. fig. 3). The following function decides if the contact window is active:

$$cwin(t) = \begin{cases} 1 \text{ if} & t \geq \#cwin(t) \cdot (T_a + T_d) \wedge \\ & t \leq \#cwin(t) \cdot (T_a + T_d) + T_a \\ 0 \text{ else} \end{cases} \qquad (4)$$

with $\#cwin(t) = \lfloor t/(T_a + T_d) \rfloor$. Each instance of a MAC layer has an *awake neighbors list* (ANL) that contains all neighbors that are known to be awake outside the contact window. Each entry contains the address adr of the neighbor and the time point $t_s(adr)$ when it falls back to sleep.

If a packet has to be sent, a check is performed first if the destination is awake. The destination is awake if $cwin(t)$ is one or the ANL contains an entry for the destination for which $t < t_s(adr)$ is fulfilled. The transmission of the packet is deferred until the start of the next contact window if the destination is not awake. The transmission of broadcast packets is always deferred if the contact

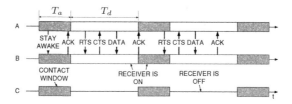

Fig. 3. Exchange of control and data packets if power management is applied

window is not active. This is due to the fact that broadcast requires that all neighbors have turned on their receivers. Please note that no STAY AWAKE packets are sent for broadcast messages. They are only delivered within contact windows.

To keep a neighbor awake outside the contact window, a STAY AWAKE packet is transmitted before the data packet if no valid entry is found in the ANL. The packet is treated like a data packet whose size is below the RTS threshold p_{RTS}. Hence, the usual medium access procedure is applied (involving retransmission of STAY AWAKE packets when the destination node does not answer, e. g. because it is busy). The destination node receives this packet during the next contact window. The MAC layer of the main sector in which the control packet is received will set an internal sleep time-point t_s to the end time of the upcoming #awake number of contact windows:

$$t_s = (\#cwin(t) + \#awake) \cdot (T_a + T_d) + T_a \qquad (5)$$

Outside the contact window, if the sleep time is not reached, the MAC layer keeps on its receiver. The sleep time is constantly updated when the MAC layer receives data packets. With this mechanism, the destination node will leave its receiver on as long as data packets are received. Hence, only one STAY AWAKE control packet need to be transmitted for a burst of data packets. The ATIM approach in 802.11 requires that each node sends an ATIM in every ATIM window if it has more data to sent. Therefore, our dynamic update of the sleep time reduces the overhead. If the initiating node receives the acknowledgment for the STAY AWAKE or data packet, it will update the corresponding entry for the destination in its ANL. The time point t_s is updated according to eq. 5.

Fig. 3 shows an example how our power management works. The number of awake periods is set to #awake = 1. Node C and B lie in one of A's sectors. Node A has buffered data for node B. Hence, it sends a STAY AWAKE request to node B during the next contact window. B acknowledges the request and keeps on its receiver in the sector that has received the request. Node A and B exchange two packets via the handshake sequence of the CSMA/CA protocol. Since node A has not sent any data packets in the third contact window, node B turns off its receiver at the end of the contact window. Node C is not involved in the communication, therefore it always switches off its receiver outside the contact windows.

The use of contact windows will introduce an extra initial delay before the transmission of a burst of data packets starts. In the worst case, the initiating node has to wait T_d seconds until the next contact window starts. Then it has to wait until the STAY AWAKE control packet has been successfully acknowledged by the receiver. This may take several contact windows if T_a is not large enough.

To find suitable values for T_a, T_d and #awake, we propose to set reasonable values for T_a and #awake first. The duration of the contact window T_a should be large enough such that every node can get an acknowledgment for its STAY AWAKE requests. To obtain approximately the same sensing energy compared to a single air interface system, we set the distance T_d between contact windows to $T_d = (k - 1) \cdot T_a$.

5 Experiments and Results

We have implemented the CSMA/CA protocol described in section 4 within our Packet Processing Library (PPL). The PPL allows a rapid implementation of network protocols. It has a hardware abstraction layer that allows to reuse the same implementation in different target environments, e. g., network simulators and real communication systems [12]. We use a network layer that has global network knowledge from the simulator to find the routes for delivering packets. Hence, no control packets for establishing the routes are necessary and the results just show the performance of the MAC layer. We use the network simulator SAHNE (Simulation of an Ad hoc Networking Environment) because it has been specifically designed for simulating directional networks [19].

Time is slotted in SAHNE, and packets are transmitted over several time slots depending on their size. The propagation of the radio signals is simulated according to the free space propagation model. To simulate collisions, SAHNE uses a signal-to-interference (SIR) ratio. If the SIR of the strongest-power-received-packet is below a threshold, the packet is discarded and the PPL is informed about the collision. SAHNE can simulate a single air interface system with an omni-directional antenna and a system with k air interfaces that uses a directional radiation pattern based on a helical antenna model (cf. sec. 3). We have implemented the presented energy model in the PPL. When a packet is received, sent, discarded or the medium is idle, the PHY layer is notified by SAHNE and the PPL will log the energy consumption according to eq. 1-3. The reception energy of discarded packets is counted as discarding energy (E_{DX}). In the same manner, various other metrics are computed (see tab. ?? for a summary).

5.1 Communication Scenario

Our communication scenario consists of N nodes that are randomly placed in a square area. The nodes do not move and can communicate with every other node via a single hop. Among these nodes, G traffic sources are randomly selected that send data packets to G randomly selected traffic destinations. Every traffic generator sends #p number of packets to its destination as fast as possible. The simulation stops if all packet generates are finished. Traffic sources can not act

Table 1. Parameter settings

Number of nodes	$N = \{20,\ 35,\ 50\}$	Time slot duration	$t_{slot} = 20\,\mu s$
Number of traffic generators	$G = \{1,\ 5,\ 10\}$	Sent packets per generator	#$p = 750$
Air interfaces per node	$k = \{1,\ 4,\ 6,\ 8\}$	Simulations per scenario	$W = 5$
Data packet size	$p_{data} = 578\,\text{bytes}$	Area width and height	$w = h = 443\,\text{m}$
Max. transmission range	$d = 627.61\,\text{m}$	Sensing/reception threshold	$P_s = P_r = -76\,\text{dBm}$
Nominal reception power	$P_{e,nom} = -75.9\,\text{dBm}$	Signal-to-interference ratio	$SIR_{min} = 10\,\text{dB}$
Radio frequency	$f = 2.4\,\text{GHz}$	Reception power	$P_{RX} = 0.78\,\text{W}$
Sensing power	$P_{SX} = 0.66\,\text{W}$	Interframe spacing	$t_{IFS} = 40\,\mu s$
Neighborhood discovery period	$T_{ND} = 0.5\,\text{s}$	Static transmission power	$P_{TX} = 0.82\,\text{W}$
RTS/CTS threshold	$p_{RTS} = 64\,\text{bytes}$	Min. contention window	$CWS_{min} = 32\,\text{slots}$
Max. contention window	$CWS_{max} = 1024\,\text{slots}$	Max. short and long retries	$RC_s = 4,\ RC_l = 7$
Awake periods	#$awake = 1$	Awake time	$T_a = 50\,\text{ms}$
Max. TX power	$P_{t,max} = \{20, 12.6, 6, 0.4\}\,\text{dBm}$		

as traffic destinations. All nodes perform the described neighborhood discovery (cf. sec. 4). Therefore, even if a node is not a traffic generator it generates additional load. The maximum transmission power $P_{t,max}$ has been adjusted such that every node can reach every other node independent of their orientation. For every scenario, simulations were performed with different settings for the number of air interfaces, power control (PC), and power management (PM). The performance metrics are first computed for each individual node and then accumulated over the total number of nodes in the simulation. Every setting was simulated W times on different scenarios that use different seed values for the random number generators. The performance measures shown here are the average of all performed number of simulations. The complete parameter set is summarized in table 1.

5.2   Results

In all our performed simulations, the MAC protocol achieved a packet delivery ratio (PDR) above 99 %. Hence, our protocol implementation has successfully managed to provide a reliable access to the medium. An exemplary energy distribution is shown in fig. 4. It can be seen that transmission energy E_{TX} only accounts for less than 3 % of the total energy consumption (also observed in the other scenarios). Hence, **power control and the higher gain of directional antennas may only have a weak impact on the total battery power consumption**. Therefore, we performed the remaining experiments without power control. The performance results of all considered parameters is shown in fig. 5. The following statements can be derived from the figure:

Throughput is increased up to 3 times compared to the omnidirectional system by applying directional communication. It also increases with increasing number of traffic generators. With directional communication, the more traffic generators are present, the higher is the probability that spatial reuse can be exploited.

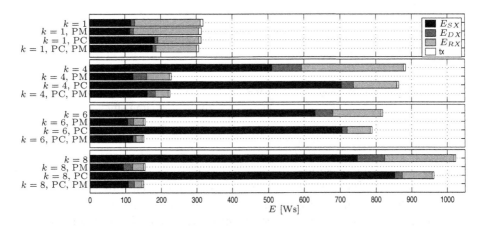

Fig. 4. Exemplary energy distribution ($N = 35$, $G = 10$)

Table 2. Used performance metrics

Throughput S	Accumulated number of bits received at the destination divided by the reception time of the last packet.
End-to-end delay D_{E2E}	Accumulated time the data packets have been in delivery.
Medium use time D_M	Accumulated time the data packets have been transmitted via the air.
End-to-end energy E_{E2E}	Accumulated energy required for sending and receiving data packets on the route to their destinations.
Total energy E	Energy spent for receiving, sending and discarding packets of any type and sensing the medium.
Packet delivery ratio PDR	The ratio of number of received packets at the destination to the number of sent packets at the source.

Time efficiency: $\eta_t = D_M / D_{E2E}$, Energy efficiency: $\eta_E = E_{E2E}/E$, Energy delay product: $EDP = E \cdot D_{E2E}$

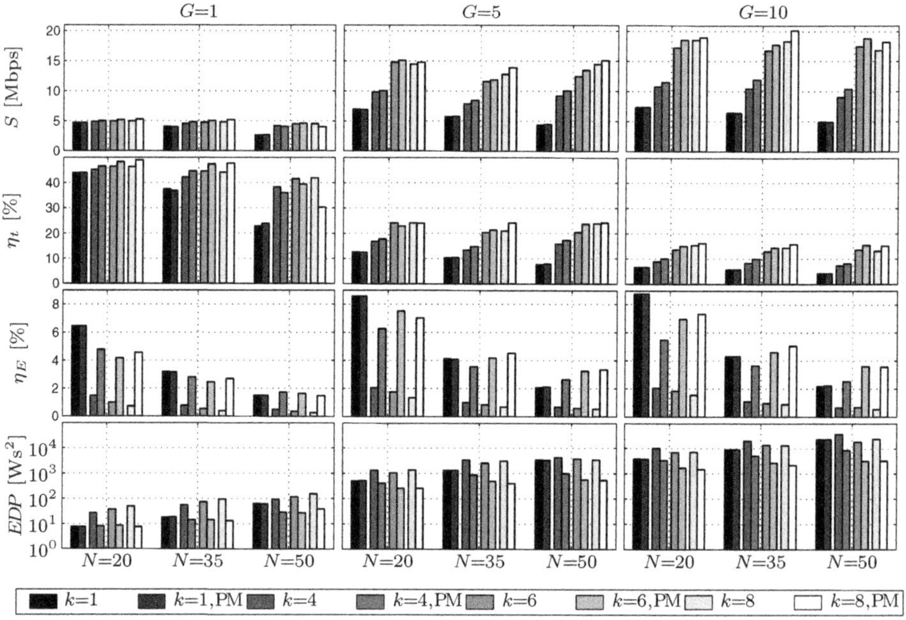

Fig. 5. Performance results for all considered number of nodes N, number of traffic generators G and number of air interfaces k

Power management does not degrade throughput and time efficiency. Moreover, in most cases throughput and time efficiency can even benefit from it due to a scheduling effect. However, the performance strongly depends on the considered communication task. If a constant bit-rate traffic is used with a low bit-rate such that the destination needs to be awaken for every packet, a performance degradation can occur. This can be changed by adapting the contact windows to the traffic, e. g., with the approach in [20].

Power management for directional communication achieves an energy efficiency comparable to omni-directional communication. In scenarios with high node density and a high number of traffic generators (here starting with $N = 35, G = 5$), the energy efficiency of directional transmission can even exceed that of omni-directional transmission.

The energy-delay-product (EDP) is always better if directional transmission with power management is applied. The EDP shows if the trade-off between increasing the energy consumption for obtaining more performance pays off. If no power management is applied, the EDP for directional transmission is always higher compared to omni-directional transmission. Therefore, we have to invest over-proportional more energy for reducing the end-to-end delay. With power management, the EDP is below the one of a single antenna system. Hence, the trade-off between more energy for less delay is better than that of the single antenna system.

6 Conclusion

We have proposed a communication system that uses k air interfaces, each equipped with an directional antenna. To enable communication in k directions in parallel on each node, we have designed a suitable MAC protocol based on CSMA/CA. To decrease energy consumption, a power management scheme is applied. To verify and analyze the performance of our network protocol, we have used our protocol implementation within a network simulator and compared important performance measures such as throughput, delay and battery energy consumption. Tne used energy model differentiates energy consumption not only for transmitting, but also for receiving and carrier sensing. Our simulation results show that carrier sensing energy dominates the total energy consumption. Power management can successfully reduce the energy required for sensing the medium. With power management, the directional communication system can obtain an energy efficiency comparable to a single antenna system with an increase of throughput and time efficiency by a factor of up to 3. However, it is still open how this system compares to one that uses k omni-directional radios, each able to establish an independent communication channel.

Acknowledgments

This work was supported by the DFG-Sonderforschungsbereich 376 "Massive Parallelität: Algorithmen, Entwurfsmethoden, Anwendungen".

References

1. Nasipuri, A., Ye, S., Hiromoto, R.E.: A MAC Protocol for Mobile Ad Hoc Networks Using Directional Antennas. In: Proc. of IEEE WCNC, Chicago, IL (2000)
2. Ko, Y.B., Shankarkumar, V., Vaidya, N.H.: Medium Access Control Protocols Using Directional Antennas in Ad Hoc Networks. In: Proc. of INFOCOM. (2000)
3. Takai, M., Martin, J., Bagrodia, R., Ren, A.: Directional Virtual Carrier Sensing for Directional Antennas in Mobile Ad Hoc Networks. In: Proceedings of ACM MobiHoc. (2002)

4. Korakis, T., Jakllari, G., Tassiulas, L.: A MAC protocol for full exploitation of Directional Antennas in Ad-hoc Wireless Networks. In: Proc. of MobiHoc, Annapolis, USA (2003)

5. Choudhury, R.R., Yang, X., Ramanathan, R., Vaidya, N.H.: Using Directional Antennas for Medium Access Control in Ad Hoc Networks. In: Proceedings of MobiCom, Atlanta, Georgia, USA (2002)

6. Lal, D. and Toshniwal, R. and Radhakrishnan, R. and Agrawal, D. P. and Caffery, J.: A Novel MAC Layer Protocol for Space Division Multiple Access in Wireless Ad Hoc Networks. In: Proc. of ICCCN, Miami (2002)

7. Ramanathan, R.: On the Performance of Ad Hoc Networks with Beamforming Antennas. In: Proceedings of MobiHOC, Long Beach, CA, USA (2001)

8. Nasipuri, A., Li, K., Sappidi, U.R.: Power Consumption and Throughput in Mobile Ad Hoc Networks using Directional Antennas. In: Proc. of ICCCN, Miami (2002)

9. Grünewald, M., Xu, F., Rückert, U.: Power Control in Directional Mobile Ad Hoc Networks. In: ITG Fachtagung 'Ambient Intelligence', Berlin, Germany (2004)

10. LAN/MAN Standards Committee of the IEEE Computer Society: ANSI/IEEE Std 802.11 – Wireless LAN Medium Access Control (MAC) and Physical Layer (PHY) Specifications. (1999)

11. Tseng, Y.C., Hsu, C.S., Ten-Yueng, H.: Power-saving protocols for IEEE 802.11-based multi-hop ad hoc networks. Computer Networks **43** (2003) 317–337

12. Grünewald, M., Niemann, J.C., Porrmann, M., Rückert, U.: A framework for design space exploration of resource efficient network processing on multiprocessor socs. In: Proceedings of the 3rd Workshop on Network Processors & Applications, Madrid, Spain (2004) 87–101

13. Lehne, P.H., Pettersen, M.: An Overview of Smart Antenna Technology for Mobile Communications Systems. IEEE Communications Surveys **2** (1999) 2–13

14. Feng, M., Shen, S.C., Caruth, D., Huang, J.J.: Device Technologies for RF Front-End Circuits in Next-Generation Wireless Communications. Proceedings of the IEEE **92** (2004) 354–375

15. Kraus, J.D.: Antennas. McGRAW-HILL (1950)

16. Feeney, L.M., Nilsson, M.: Investigating the energy consumption of a wireless network interface in an ad hoc networking environment. In: Proceedings of IEEE INFOCOM, Anchorage, Alaska (2001)

17. Philips: SA2411: +20dBm single chip linear amplifier for WLAN. (2003)

18. Intersil Americas Inc.: Intersil PRISM 2.5 chip set: baseband processor ISL3873B, I/Q modulator/demodulator HFA3783, RF/IF converter ISL3685. (2000-2002)

19. Rührup, S., Schindelhauer, C., Volbert, K., Grünewald, M.: Performance of distributed algorithms for topology control in wireless networks. In: Proceedings of the International Parallel and Distributed Processing Symposium, Nice, France (2003)

20. Zheng, R., Kravets, R.: On-demand Power Management in Ad Hoc Networks. In: Proceedings of INFOCOM, San Francisco, California, USA (2003)

Location Tracking in Mobile Ad Hoc Networks Using Particle Filters

Rui Huang and Gergely V. Záruba

Computer Science and Engineering Department,
The University of Texas at Arlington,
416 Yates, 300NH, Arlington, TX 76019
`rxh1725@omega.uta.edu, zaruba@uta.edu`

Abstract. Mobile ad hoc networks (MANET) are dynamic networks formed on-the-fly as mobile nodes move in and out of each others' transmission ranges. In general, the mobile ad hoc networking model makes no assumption that nodes know their own locations. However, recent research shows that location-awareness can be beneficial to fundamental tasks such as routing and energy-conservation. On the other hand, the cost and limited energy resources associated with common, low-cost mobile nodes prohibits them from carrying relatively expensive and power-hungry location-sensing devices such as GPS. This paper proposes a mechanism that allows non-GPS-equipped nodes in the network to derive their approximated locations from a limited number of GPS-equipped nodes. In our method, all nodes periodically broadcast their estimated location, in term of a compressed particle filter distribution. Non-GPS nodes estimate the distance to their neighbors by measuring the received signal strength of incoming messages. A particle filter is then used to estimate the approximated location from the sequence of distance estimates. Simulation studies show that our solution is capable of producing good estimates equal or better than the existing localization methods such as APS-Euclidean.

1 Introduction

Studies have shown that innovative algorithms can aid mobile ad hoc network (MANET) protocols if the nodes in the network are capable of obtaining their own as well as other nodes' location information. For instance, algorithms such as LAR [7], GRID [10], and GOAFR+ [9] rely on the location information to provide more stable routes during unicast route discovery. The location information is also applied to geocast (multicast based on geographic information) [6] for algorithms such as LBM [8], GeoGRID [11] and PBM [13]. To minimize the power consumption, The GAF algorithm [20] uses the location information to effectively modify the network density by turning off certain nodes at particular instances.

The algorithms listed earlier all rely on the availability of reasonably accurate location information. This assumption is valid for networks in which some location sensing devices, such as GPS receivers, are available at all nodes. However,

V.R. Sirotiuk and E. Chávez (Eds.): ADHOC-NOW 2005, LNCS 3738, pp. 85–98, 2005.

in reality this is rarely the case; although GPS receivers are increasingly cheaper to produce and becoming more widely available, they are still relatively expensive and power-hungry, and it is too general to assume that they will be available to every node in ad hoc networks. For this reason different algorithms have been proposed to derive approximated locations of *all* nodes based on the relaxed assumption that direct location sensing devices (such as GPS) are available to only a *subset* of the nodes.

This paper presents a solution to the location tracking problem based on particle filters. Given an ad hoc network with limited number of location-aware nodes, our solution estimates the locations of all other nodes by measuring the received signal strength indication (RSSI) from neighbors. For each node, the estimated location is viewed as a probabilistic distribution maintained by a particle filter. Unlike other location tracking methods, our solution has low overhead because it is purely based on local broadcasting and does not require flooding of the location information over the entire network. Simulation studies show that even without flooding, our solution can still generate good estimates comparable to other existing methods, given that the network is well connected and the percentage of anchors is not extremely low. In addition when connectivity is low and the percentage of anchors is small, our algorithm is still able to derive location information which is not the case with most of the other approaches.

1.1 Related Work

Generally speaking, there are two categories of distributed localization methods depending on whether sensory data are used. The methods that do not use sensory data are simpler but tend to perform poorly especially when anchor ratio is low or the network is sparse. The methods that do use sensory data generally perform better but tend to be significantly more complex. The performance in the latter case is also largely affected by the noise introduced to the sensory data which tends to aggregate rapidly as sensory data is propagated through the network.

The Centroid method [2] provides the most straight-forward solution that does not use sensory data. Assuming that a non-anchor node is capable of receiving the location information from multiple anchor nodes, the Centroid method derives the location of a non-anchor node as the average of its neighboring anchor nodes' locations. The method is simple and efficient, but it requires the anchor nodes to redundantly cover large areas for an acceptable performance. The APIT method [4] estimates the node location by isolating the area using various triangles formed by anchor nodes. The location of the node is narrowed down by analyzing overlapping triangles to determine whether the node is contained within the triangles. Both the Centroid method and the APIT method require the transmission range of anchors to be much greater than non-anchors (by an order of magnitude [4]) in order for nodes to obtain reasonable location estimates.

The DV-Hop method [17] allows the location information from anchor nodes to propagate through multiple hops. The locations of anchors are periodically flooded throughout the network much like the routing packets in a distance vector routing protocol. The locations of non-anchor nodes are derived geometrically by performing trilateration of the distance estimates from at least three anchor nodes. Here the distance estimates are obtained by multiplying the number of hops to the anchor node to a predefined average-distance-per-hop value. The DV-Hop method does not require a greater transmission range of anchors, and it works well even when the ratio between anchor and non-anchor nodes is low. However, the message complexity is rather high due to the flooding of the location information. Furthermore, because the average-distance-per-hop is an estimated value over the entire network, the accuracy of the location estimation suffers when the nodes are not uniformly placed over the network.

Other, significant location tracking methods make use of additional sensors. In [12], the location, velocity and acceleration of mobile nodes are estimated by measuring the received signal strength indicator (RSSI) from multiple base stations in a cellular network. The measured power levels are fed into a Kalman filter to smooth out (filter) the erratic readings and thus be able to derive the distance. Since base station locations are assumed to be well-known in a cellular network, mobile nodes can use them as reference points for location estimation. In [16], the authors assume that non-anchor nodes are equipped with devices that measure the incoming signal directions. The directional information allows the receivers to obtain the angle of arrival (AoA) of the signal thus allowing more accurate location estimates than the pure DV-Hop method. The DV-Distance method [17] is similar to the DV-Hop method but uses the estimated distance instead of the hop count during trilateration. In [19] after obtaining the initial location estimates from the DV method, the nodes obtain the estimated locations from the neighbors via local broadcast. The RSSI readings also provide the distance estimates from the neighbors. Using the distance estimates along with the estimated locations from the neighbors, the nodes can refine their initial location estimates via trilateration.

2   Localization Using Particle Filters

"Geometrically speaking," in order to find the location of a node in a 2-dimensional space, the distances and locations of at least three anchors need to be known (as each of these anchors define a circle where the target node could be). In a network where the percentage of anchors is low, the major challenge is to obtain the distances and locations of anchors when the node is several hops away from the anchors. Previous works resolve this problem by either 1) assuming a greater transmission range of anchors [2,4] (thus, anchors are always 1-hop away), or 2) broadcasting the anchor locations hop-by-hop over the entire network [17,16,19]. The assumption made in the first solution requires the network to be heterogeneous in the node types (in which anchors' radios are considered different than those of non-anchors) and requires homogeneity (uniformity) for anchor nodes' location over the area. The flooding of the location packets in the second solution requires extra overhead. This overhead can be especially heavy when nodes are mobile, where location packets need to be re-broadcasted repeatedly by nodes.

Recognizing various shortcomings of previous approaches, we propose a different location tracking method that is based on Bayesian filters using Monte Carlo sampling (also known as particle filters) introduced in [3]. Our method can be considered as a probabilistic approach in which the estimated location of each node is regarded as a probability distribution captured by samples, thus the term particles. The distribution of particles (the probability distribution of a node's location over the area) is continuously updated as the node receives location estimates from its neighbors along with the distance estimates from RSSI reading. Essentially, the nodes estimate their own locations by interchanging the location distributions with their neighbors. Our method has several advantages over existing methods. First of all, our method does not require a greater range for anchors, which allows it to work in homogeneous networks. Secondly, our method employs a simple computation and communication model which relies solely on local broadcast (broadcast to neighbors only). This allows our method to be naturally integrated in periodical Hello messages used by mobile nodes in ad hoc networks to declare their existence. Comparing to existing methods such as APS, our method generally converges with less message overhead. Finally, while previous works do not provide simulation result for mobile scenarios, we demonstrate via simulation that our method can be effectively used in mobile ad hoc networks.

2.1   Classic Monte Carlo Sampling-Based Bayesian Filtering

This section describes the theoretical background behind Bayesian filtering and how it can be applied to location estimation using RSSI. Let us envision a grid system superimposed over the entire tracking area, and let the state s_t be the location of the node to be tracked in the grid system at the time t . Our goal is to estimate the posterior probability distribution, $p(s_t|d_1, \ldots, d_t)$, of potential states - s_t, using the RSSI measurements, $d_1, \ldots, d_t$. The calculation of the distribution is performed recursively using a Bayes filter:

$$p(s_t|d_1, \ldots, d_t) = \frac{p(d_t|s_t) \cdot p(s_t|d_1, \ldots, d_{t-1})}{p(d_t|d_1, \ldots, d_{t-1})}$$

Assuming that the Markov assumption holds, i.e., $p(s_t|s_{t-1}, \ldots, s_0, d_{t-1}, \ldots, d_1)$ $= p(s_t|s_{t-1})$, the above equation can be transformed into the recursive form:

$$p(s_t|d_1, \ldots, d_t) = \frac{p(d_t|s_t) \cdot \int p(s_t|s_{t-1}) \cdot p(s_{t-1}|d_1, \ldots, d_{t-1}) ds_{t-1}}{p(d_t|d_1, \ldots, d_{t-1})},$$

where $p(d_t|d_1, \ldots, d_{t-1})$ is a normalization constant. In the case of the localization of a mobile node from RSSI measurements, the Markov assumption requires that the state contains all available information that could assist in predicting the next state and thus, an estimate of the non-random motion parameters of the nodes is required as part of the state description. Starting with an initial, prior probability distribution, $p(s_0)$, a system model, $p(s_t|s_{t-1})$, representing the motion of the mobile node (the mobility model), and the measurement model,

$p(d|s)$, it is then possible to drive new estimates of the probability distribution over time, integrating one new measurement at a time. Each recursive update of the filter can be broken into two stages:

Prediction: Use the system model to predict the state distribution based on previous readings

$$p(s_t|d_1, \ldots, d_{t-1}) = \int p(s_t|s_{t-1}) \cdot p(s_{t-1}|d_1, \ldots, d_{t-1}) ds_{t-1}$$

Update: Use the measurement model to update the estimate

$$p(s_t|d_1, \ldots, d_t) = \frac{p(d_t|s_t)}{p(d_t|d_1, \ldots, d_{t-1})} p(s_t|d_1, \ldots, d_{t-1})$$

To address the complexity of the integration step and the problem of representing and updating a probability function defined on a continuous state space (which therefore has an infinite number of states), the approach presented here uses a sequential Monte Carlo filter to perform Bayesian filtering on a sample representation. The distribution is represented by a set of weighted random samples and all filtering steps are performed using Monte Carlo sampling operations. Since we have no prior knowledge of the state we are in, the initial sample distribution, $p_N(s_0)$, is represented by a set of uniformly distributed samples with equal weights, $\{(s_0^{(i)}, w_0^{(i)})|i \in [1, N], w_0^{(i)} = 1/N\}$ and the filtering steps are performed as follows:

Prediction: For each sample, $(s_{t-1}^{(i)}, w_{t-1}^{(i)})$, in the sample set, randomly generate a replacement sample according to the system (mobility) model $p(s_t|s_{t-1})$. This results in a new set of samples corresponding to $p(s_t|d_1, \ldots, d_t)$:

$$\{(\tilde{s}_t^{(i)}, w_t^{(i)})|i \in [1, N], w_t^{(i)} = 1/N\}$$

Update: For each sample, $(\tilde{s}_t^{(i)}, w_t^{(i)})$, set the importance weight to the measurement probability of the actual measurement, $\tilde{w}_t^{(i)} = p(d_t|\tilde{s}_t^{(i)})$. Normalize the weights such that $\sum_i \eta \cdot \tilde{w}_t^{(i)} = 1.0$, and draw N random samples for the sample set $\{(\tilde{s}_t^{(i)}, \eta \cdot w_t^{(i)})|i \in [1, N]\}$ according to the normalized weight distribution. Set the weights of the new samples to $1/N$, resulting in a new set of samples $\{(s_t^{(i)}, w_t^{(i)})|i \in [1, N], w_t^{(i)} = 1/N\}$ corresponding to the posterior distribution $p(s_t|d_1, \ldots, d_t)$.

2.2 Modified Particle Filtering for Location Estimations

The classical Monte Carlo method is often implemented using particle filters. To apply the filter to the location tracking problem a system model and a measurement model must be provided. We use a simple random placement model as our system model (please note that this is the mobility model used in the filter which is different from the mobility model used in the simulations to enable node movement). The model assumes that at any point in time the node moves with a random velocity drawn from a Normal distribution with a mean of $0m/s$ and a fixed standard deviation σ. No information about the environment is included in this model, and as a consequence, the filter permits the estimates to move along arbitrary paths. Thus, our system model is simply $p(s_t|s_{t-1}) = N(0, \sigma)$, where N is a Normal distribution. Note that while such system model should work well in stationary networks, it's not best suited for mobile networks. In reality, mobile nodes follow a certain kind of movement profile instead of random motion. The system model should closely resemble the current movement profile of the node. However, since it's difficult to obtain a reliable movement profile when the location is unknown, the assumption of random movement is probably the best we can do at this stage.

The measurement data are obtained by observing the periodical location data broadcast from neighbors. To minimize the impact of the measurement error, we apply a simple Kalman filter to the RSSI sensor readings [5] before feeding the measurement data to the particle filter. When a node u receives broadcast location data from node v, the broadcast data consist of the unique identifier of v, and the probability distribution, X_v, of the location estimate of v at time t. The X_v distribution is a compressed version of the actual particle distribution at v. The detail method of compressing and decompressing the particle distribution is the topic of the next section. For now, let us assume that X_v contains a set of sample particles that represents v's location. Along with the RSSI reading of the broadcast, $RSSI_v$, the complete measurement metrics d_t is therefore $(id, X_v, RSSI_v)$.

After the measurement from the neighbor v is collected, the particle filter at node u is updated. In the classic particle filtering, particles are re-sampled based on weights, which are in turn assigned based on the measurement. More weights are assigned to the particle values that are more consistent with the measurement reading. After re-sampling, the particle distribution becomes more consistent with the current measurement. In our situation we have a unique scenario where the measurement itself consists of a particle distribution, X_v. Furthermore, both X_u and X_v are *imprecise*. Our task during the update step is to modify the particle distribution X_u so that it becomes more consistent with $RSSI_v$ while taking into account the inherent impreciseness of X_u and X_v. First, we obtain a distance estimate from the inverse of the signal propagation model P:

$$D^{(RSSI)} = P'(RSSI_v)$$

Note that P can be arbitrary as long as it depends on the distance from the sender to the receiver. Noise can be added to the model, but we disregard it when calculating the inverse and let it be filtered out by the particle filtering (note, that in the simulations noise is indeed added to the RSSI measurements).

For each particle x_u in X_u, we randomly select a particle x_v in X_v and calculate their distance $D^{(x_u,x_v)}$. We then measure the difference between $D^{(x_u,x_v)}$ and $D^{(RSSI)}$, and select a new location for re-sampling based on the difference as well as the variances of the particle distribution X_v and X_u. For instance, before the update step x_u and x_v are located at point A and B, respectively.

Thus, $D^{(x_u,x_v)} = |AB|$. Let A' be the location of x_u based on the RSSI reading on the same line, i.e., $D^{(RSSI)} = |A'B|$. Intuitively, if the location estimate given by the distribution X_v is accurate and the actual location for node v is indeed at x_v, then the new location for particle x_u should be at point A'. Conversely, if the location estimate of the distribution X_u is accurate, the new location for x_u should stay at A. Therefore, we select the new location based on the perceived accuracy, i.e., the variances, of the distribution X_u and X_v. Let the variance of a distribution X be $var(X)$. We select the new location of x_u, x_u', along the line $|AA'|$ such that

$$\frac{|Ax_u'|}{|x_u'A|} = \frac{var(X_u)}{var(X_v)}$$

A new particle is then randomly re-sampled by a Normal distribution centered at x_u' with the variance being the average of the variances of X_u and X_v. We consider the variances of both X_u and X_v during re-sampling because the spread of both distributions affects the spread of the updated distribution X_u'.

Comparing to the re-sampling method of classic particle filters, our method is different in that we do not use a weight based re-sampling method. Instead, we re-sample by comparing the two distributions together against the measurement reading. But, the concept is the same as we are updating the distribution to fit the measurement readings. Our re-sampling method has a number of advantages over the traditional method. First, our method does not re-sample directly from the original particle location using a weight based Gaussian distribution. Instead, it re-samples from a more accurate location influenced by neighbor's distribution. Thus, our method requires less amount of random probing and converges more quickly. Secondly, since our method requires less amount of random probing, a significantly smaller number of particles are required. With less particles, the particle filter update procedure computes more efficiently.

2.3   Compressing and Decompressing Particle Filter Distributions

The previous section makes the assumption that the complete location distribution is received from the neighbor. Since the complete distribution consists of a large number of particles with their location data, doing so is obviously not very practical due to the limited bandwidth of ad hoc networks. Therefore, we propose a simple yet effective mechanism to compress this information.

Given a particle distribution X, we locate the most likely value, $\hat{x}$, as the particle in the distribution that has the minimum overall distance between itself and other particles, i.e., $\hat{x} = \arg\min_{x \in X} (\sum_{y \in Y} |x - y|)$. In other words, $\hat{x}$ is the most representative particle of the entire distribution. From $\hat{x}$, we count the number of particles n within the predefined range r. We then calculate the variance, σ^2 within those n particles. Thus, we obtain a quadruple $(\hat{x}, r, n, \sigma^2)$. From there, we remove the n particles in the previous quadruple from the distribution and repeat the process of finding the most likely value, a larger range (explained later) and the variance. By continuing the same process until all particles have been covered, we obtain a sequences of quadruples that approximates the original particle distribution (each of these quadruples thus will represent modes of the particle density distribution). When the quadruples are received by the receivers, a decompressing algorithm runs to reproduce the distribution by randomly generating particles based on the expected value, range, particle number and variance for each quadruple.

Our experiment has shown that the compression method reduces the amount of data exchange by nearly 90 percent without a significant increase to the location estimation error (results are omitted in this paper due to space limitations).

3   Simulation Results

We have conducted a number of experiments to validate the effectiveness of our particle filter based solution. Our experiments attempt to duplicate real world scenarios as closely as possible. In our simulations we assume a network in which all nodes have an identical transmission power, with a certain percentage of nodes (simulation parameter) being anchor nodes. For a network of fixed size, the connectivity of the network depends (almost solely) on the transmission range. When a node is located within the transmission range of another node, we assume that it is capable of receiving signal from the sender when noise is not present. The received signal strength depends on the distance to the sender as well as a signal propagation model and a noise model.

The signal propagation model is given by $P = c \cdot d^{-2}$, in which the power of the received signal P is inversely proportional to the second power of the distance d (while c is a constant including the unit distance received power and the inverse square of the carrier frequency - among others). When the received signal power P is below a threshold P_{min}, it is considered too weak to be captured by the receiver thus the link breaks. Note the c and P_{min} selection does not affect the overall simulation results, as long as the same values are used consistently in the observation model of the filters. In fact, the same can be said about all other signal propagation models - all we require is a model that represents the receiving power as a function of distance, and we let the filter to filter out the noise. For the particle filter itself, we use a total number of 200 particles at each node.

We use an isotropic type of network of 100 randomly placed nodes with an average degree of 7.6. Noise is added to the signal strength calculated via the signal propagation model as a percentage of the calculated signal strength. For instance, a 10 percent noise means that the received signal strength may vary within a plus-minus 10 percent range of the calculated signal strength (uniformly distributed). Note that our network configuration and noise model is identical to that of the isotropic topology in [15], so that we can effectively compare our method with APS.

3.1   Filter Convergence

Figure 1 shows how the estimation error converges as more measurement readings are processed in a static network. We are interested in how long and how many messages it takes for the error to reach an acceptable level from which it only reduces marginally. We added a noise level of 50 percent to the measurement readings. The estimation error is calculated as the difference between the most likely value given by the particle distribution and the actual location. The difference is then measured in term of the ratio against the maximum transmission range. Thus, an estimation error of 1.0 means that difference between the expected value and actual location equals to the maximum transmission range. The data is collected of enough simulation runs to claim a 90 percent confidence that the error is less than 10 percent; the error ratio is the average of all non-GPS nodes (i.e., the perfect "estimates" of anchor nodes are not biasing the results).

Two obvious facts can be observed from Figure 1: i) networks with higher GPS ratio produce better estimations and ii) estimation error reduces quicker with higher GPS ratio. Both of those observations can be explained by the fact that GPS ratio determines how fast and how accurate location information can be propagated through the network. With a higher GPS ratio, non-GPS nodes will be able to obtain the necessary location information faster because non-GPS nodes are closer (i.e., less number of hops) to GPS nodes. Also, since measurement error is aggregated at each hop, the location information will be more accurate with higher GPS ratio.

Figure 1 also shows that the estimation error converges to the minimum between 2 to 5 seconds depending on the GPS ratio. Considering that the location broadcast occurs every 0.5 seconds, it takes about 4 to 10 rounds of broadcasts for the error to reach the minimum. Since the average degree of the network is 7.5 with a total of 100 nodes, each round of broadcast is equivalent to 750 messages. Therefore, it takes about 3000 to 7500 messages to minimize the error depending on the GPS ratio. Note that when even in the worst case where the GPS ratio is low, error converges very quickly and is close the minimum after 2 seconds. The results are at least as good as those of APS, where the "DV-distance" method uses 6500 messages (when GPS ratio is 0.1) to 9000 messages (when GPS ratio is 0.9), and the "Euclidean" method takes from 3000 to 8500 messages (results

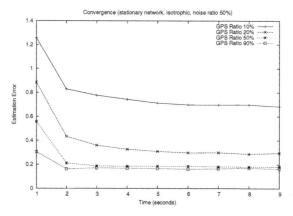

Fig. 1. Filter convergence (stationary network, 50% noise)

taken from Figures 7 and 11 of [15]). Note that our method converges quicker and takes less number of messages when the GPS ratio is high.

3.2 Minimum Estimation Error

Figure 2 compares the estimation error of varying GPS ratio and measurement noise in the network. (The DV-Hop and Euclidean curves are obtained from figures in [15]. The Euclidean curves are normalized based on the coverage.) Again, the simulation scenario is duplicated from that of the isotropic topology in APS [15]. Our results show that the estimation error continues to decline with higher GPS ratio compared to APS, making it performing more like the "Euclidean" method in APS. Like the "Euclidean" method, our method outperforms DV-Hop and DV-Distance when the GPS ratio is higher. The drawback of the "Euclidean" method is that, when the GPS ratio is low and the noise ratio is high, its coverage area (i.e., the number of nodes able to obtain the location estimates) is very limited comparing to the DV-Hop and DV-Distance, and thus it brings down the overall estimation error. As shown in Figure 2, our method improves over the "Euclidean" method especially in the difficult cases of low GPS ratio and high noise ratio. At a very low GPS ratio (i.e., 0.1), DV-Hop still performs better because it does not depend on any sensor readings and thus it is not affected by noises, although this is an unfair comparison due to the relaxed assumptions in our method.

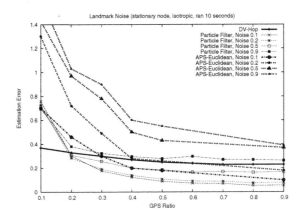

Fig. 2. Estimation error of varying GPS ratio and measurement noise

3.3 Connectivity

Simulation results in previous work are based on a rather dense network with an average degree of 7.67. Similar networks were used in [15], and thus allow a more sensible comparison. Fig. 3 shows the estimation error of our particle filter based localization method in more sparse networks. Here, we vary the network connectivity by changing the transmission range while maintaining the network size (100 nodes). As expected more error is introduced in sparser networks. Roughly speaking, the error halves when the network connectivity doubles. In theory a node needs to receive signal readings from a minimum of three neighbors in order to pinpoint its location. Thus, a network with degree of at least three will be needed to localize all its nodes. With our localization method, respectable estimations are obtained even with very sparse networks of degrees less than three (no other approaches are able to derive estimates for such situations).

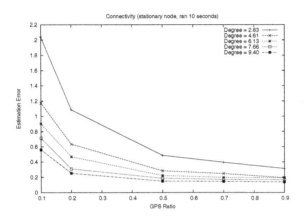

Fig. 3. Effect of network connectivity on estimation error

3.4   Results on Mobile Networks

Previous work on MANET localization generally do not contain extensive simulation and analysis when the network is indeed mobile (as the definition of MANETs imply). This section discusses simulation results on running the particle filter localization method on mobile networks. Again, we use a network with a population of 100 nodes and average degree of 7.5. We use the epoch-based mobility model of [14] to simulate node movement, which is widely accepted as a good mobility model for ad hoc networks - more realistic than, e.g., simple Brownian motion models. The entire movement path of the node is defined by a sequence of "epochs," i.e., $(e_1, e_2, \cdots, e_n)$. The duration of each epoch is I.I.D. exponentially distributed with a mean of $1/\lambda$. Within each epoch nodes move with a constant velocity vector. At the end of each epoch, nodes randomly select a new velocity vector. The direction of the movement is I.I.D. uniform between 0 and 2π. The absolute value of the velocity is I.I.D. normal with a mean μ of and a variance of σ^2. Our simulation uses a fixed mean and variance such that $\mu = \sigma$. The result is obtained by varying μ and σ from 1m/s to 20m/s. The expected amount of time a node maintains its current velocity is set to 5 seconds, i.e., $\lambda = 5$.

Figure 4 shows the filter convergence on mobile networks with measurement noise level set to 50%. Comparing to the results of stationary networks in Figure 1, the random movement of the nodes causes the estimation error to swing.

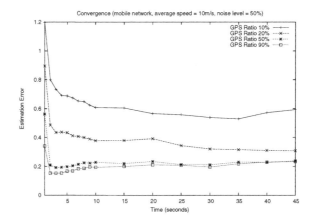

Fig. 4. Filter Convergence with Mobile Networks (10m/s speed, 50% noise)

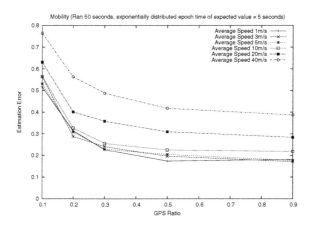

Fig. 5. Estimation error of varying GPS ratio and network mobility

However, the error variances are not very high once the nodes determine their initial locations after the first couple of seconds. This indicates that the filter is able to adapt to the node movement well enough to maintain its overall estimation accuracy. Figure 5 shows the average estimation error of mobile networks. The error does increases gracefully as the speed increases. Considering that neighbors exchange location information every 0.5s, in a network with an average nodal speed of 40m/s nodes move an average of 20 meters per observation; yet our method is capable of producing usable location estimates.

4   Conclusions

This paper described a novel solution to the location tracking problem for mobile ad hoc networks that uses a Monte Carlo sampling-based Bayesian filtering (i.e., particle filtering) method. The estimated location for nodes is regarded as a probability distribution represented by a collection of sample points. The location information from the anchors is propagated through the network via local broadcasting of the location estimates. When a node receives the location estimates from neighbors, it updates its location distribution using the particle filtering method. Simulation study has shown that the particle filter solution is capable of producing good estimates equal or better than the existing localization methods such as APS-Euclidean. Our solution also performs quite well when the network connectivity is low. Study has also shown that the solution is resilient to network topology change, making it suitable for ad hoc networks with significant mobility.

Our particle filter based localization method currently uses RSSI as the sole measurement. However, because our method is based on a rather generic algorithm of probabilistic filters, it can be easily extended to incorporate other measurement types such as angle of arrival (AoA). To do so, only the filter update step needs to be changed in order to meaningfully update the filter according to the properties of the new measurement, but the basic algorithm remains the same. In fact, it is easy to implement our method with multiple types of measurements coexisting in the network. The same particle filter method can be used in a network where an arbitrary portion of nodes are capable of measuring RSSI, another part of the nodes are capable of measuring AoA, and some are capable of measuring both. This makes our method truly versatile and ideal for such heterogeneous networks.

References

1. p. Bahl, and V.N. Pamanbhan, "RADAR: An in-Building RF-Based User Location and Tracking System" Proceedings of the IEEE INFOCOM'00, March 2000.
2. N. Bulusu, J. Heidemann, and D. Estrin, "GPS-less low cost outdoor localization for very small devices," IEEE Personal Communications Magazine, vol. 7, no. 5, pp. 28–24, October 2000.
3. N. Gordon, "Bayesian Methods for Tracking," Ph.D. thesis, University of London, 1993.
4. T. He, C. Huang, B. M. Blum, J. A. Stankovic, and T. F. Abdelzaher, "Range-Free Localization Schemes in Large Scale Sensor Networks," CS-TR2003 -06, ACM MobiCom 2003.
5. R. Huang, G.V. Záruba, and M. Huber, "Link Longevity Kalman-Estimator for Ad Hoc Networks," Proceedings of the VTC2003, IEEE 54th Vehicular Technology Conference, 2003.
6. X. Jiang and T. Camp, "Review of geocasting protocols for a mobile ad hoc network," Proceedings of the Grace Hopper Celebration (GHC), 2002.
7. Y. Ko and N. H. Vaidya, "Location-Aided Routing (LAR) Mobile Ad Hoc Networks," MOBICOM '98, Dallas, TX, 1998.
8. Y. Ko, and N. Vaidya, "Geocasting in Mobile Ad Hoc Networks: Location-Based Multicast Algorithms," IEEE Workshop on Mobile Computing Systems and Applications (WMCSA), 1999.

9. F. Kuhn, R. Wattenhofer, Y. Zhang, and A. Zollinger, "Geometric Ad Hoc Routing of Theory and Practice," PODC 2003, pp. 63-72, 2003.
10. W.-H. Liao, Y.-C. Tseng, and J.-P. Sheu, "GRID: A Fully Location-Aware Routing Protocol for Mobile Ad Hoc Networks," Telecommunication Systems, vol. 18(1) pp. 37–60, 2001.
11. W.-H. Liao, Y.-C. Tseng, K.-L. Lo, and J.-P. Sheu, "Geogrid: A Geocasting Protocol for Mobile Ad Hoc Networks Based on Grid," Journal of Internet Technology, vol. 1(2) pp. 23–32, 2000.
12. T. Liu, P. Bahl, and I. Chlamtac, "A hierarchical position-prediction algorithm for efficient management of resources in cellular networks," Proceedings of the IEEE GLOBECOM'97, Phoenix, Arizona, November 1997.
13. M. Mauve, H. Fuler, J. Widmer, and T. Lang, "Position-Based Multicast Routing for Mobile Ad-Hoc Networks," Technical Report TR-03-004, Department of Computer Science, University of Mannheim, 2003.
14. A. B. McDonald, and T. Znati, "A mobility-based framework for adaptive clustering in wireless ad-hoc networks," IEEE Journal on Selected Areas in Communication Special Issue on Wireless Ad-Hoc Networks, vol. 17, no. 8, August 1999.
15. D. Niculescu, and B. Nath, "Ad hoc positioning system (APS)," Proceedings of the IEEE GLOBECOM'01, San Antonio, 2001.
16. D. Niculescu, and B. Nath, "Ad hoc positioning system (APS) using AoA," Proceedings of the IEEE INFOCOM, San Francisco, 2003.
17. D. Niculescu, and B. Nath, "DV Based Positioning in Ad Hoc Networks," Journal of Telecommunication Systems, 2003.
18. P.J. Nordlund, F. Gunnarsson, and F. Gustafsson, "Particle filters for positioning in wireless networks," Proceedings of the XI. European Signal Processing Conference (EUSIPCO), 2001.
19. C. Savarese, J. Rabay and K. Langendoen, "Robust Positioning Algorithms for Distributed Ad-Hoc Wireless Sensor Networks," USENIX Technical Annual Conference, Monterey, CA, June 2002.
20. Y. Xu, J. Heidemann, and D. Estrin, "Geography-informed Energy Conservation for Ad Hoc Routing," Proc. seventh Annual ACM/IEEE International Conference on Mobile Computing and Networking (MobiCom), pp. 70–84, 2001.

Biology-Inspired Distributed Consensus in Massively-Deployed Sensor Networks

Kennie H. Jones[1], Kenneth N. Lodding[1], Stephan Olariu[2], Larry Wilson[2], and Chunsheng Xin[3]

[1] NASA Langley Research Center, Hampton, VA 23681
{k.h.jones, kenneth.n.lodding }@nasa.gov
[2] Old Dominion University, Norfolk, VA 23529
{olariu, wilson}cs.odu.edu
[3] Norfolk State University, Norfolk, VA 23504
cxin@nsu.edu

Abstract. Promises of ubiquitous control of the physical environment by large-scale wireless sensor networks open avenues for new applications that are expected to redefine the way we live and work. Most of recent research has concentrated on developing techniques for performing relatively simple tasks in small-scale sensor networks assuming some form of centralized control. The main contribution of this work is to propose a new way of looking at large-scale sensor networks, motivated by lessons learned from the way biological ecosystems are organized. Indeed, we believe that techniques used in small-scale sensor networks are not likely to scale to large networks; that such large-scale networks must be viewed as an ecosystem in which the sensors/effectors are organisms whose autonomous actions, based on local information, combine in a communal way to produce global results. As an example of a useful function, we demonstrate that fully distributed consensus can be attained in a scalable fashion in massively deployed sensor networks where individual motes operate based on local information, making local decisions that are aggregated across the network to achieve globally-meaningful effects.

1 Introduction

The use of sensors and actuators in industrial applications ranging from nuclear power plants to traffic control, to perimeter surveillance, to metallurgy and precision engineering goes back more than a century [9]. These sensors were generally bulky devices wired to a central control unit whose role was to aggregate, process, and act upon individual data collected by the sensors. Wireless sensor network research, as we know it today, can be traced back to the DARPA-sponsored *SmartDust* program [5]. The vision of SmartDust was to make machines with self-contained sensing, computing, transmitting, and powering capabilities so small and inexpensive that they could be released into the environment in massive numbers. These devices have come to be called *motes* and serve as nodes in a sensor network. As the motes are severely energy-constrained, they cannot transmit over long distances, restricting interaction to motes in their immediate neighborhood.

V.R. Sirotiuk and E. Chávez (Eds.): ADHOC-NOW 2005, LNCS 3738, pp. 99–112, 2005.
© Springer-Verlag Berlin Heidelberg 2005

Since building massively-deployed sensor networks is prohibitively expensive under current technology, in the past few years we have witnessed the deployment of small-scale sensor networks in support of a growing array of applications ranging from smart kindergarten, to smart learning environments, to habitat monitoring, to environment monitoring, to greenhouse and vineyard experiments, to forest fire detection, and to helping the elderly and the disabled [e.g., see 1,2,3,4,17,18]. These prototypes provide solid evidence of the usefulness of sensor networks and suggest that the future will be populated by pervasive sensor networks that will redefine the way we live and work [9]. It is, thus, expected that in the near future, in addition to the examples above, a myriad of other applications including battlefield command and control, disaster management and emergency response, will involve sensor networks as a key mission-critical component [9].

An examination of current prototypes reveals both successes and limitations to the promises of sensor networks. In spring 2002, researchers from UC Berkeley implemented a wireless sensor network on Great Duck Island, Maine. The initial application was to monitor the microclimates of nesting burrows in the Leach's Storm Petrel, and by disseminating the data worldwide, to enable researchers anywhere to non-intrusively monitor sensitive wildlife habitats. The sensor motes formed a multi-hop network to pass messages in an energy efficient manner back to a laptop base station. The data was intermediately stored at the base station and eventually passed by satellite to servers in Berkeley, CA, where it was distributed via the Internet to any interested viewer. The sensors measured temperature, humidity, barometric pressure, and mid-range infrared by periodically sensing and relaying the sensed data to the base station. The largest deployment had 190 nodes with the most distant placement over 1000 feet from the nearest base station.

Other implementations demonstrate similar capabilities. In the SmartDust demonstration at the Marine Corps Air/Ground Combat Center, Twentynine Palms, CA [21], six sensor motes were dropped from an Unmanned Aerial Vehicle (UAV) along a road. These motes self organized by synchronizing their clocks and forming a multi-hop network, and magnetically detected velocity of passing vehicles. In Aug. 2001, researchers from UCB and the Intel Berkeley Research Lab demonstrated a self-organizing wireless sensor network to those attending the kickoff keynote of the Intel Developers Forum. In another demonstration, quarter-sized motes hidden under 800 chairs in the presentation hall were simultaneously initiated forming as "... *the biggest ad hoc network ever to be demonstrated*" [6]. The FireBug system in [2,3] is a network of GPS-enabled, wireless thermal sensors that self-organizes into clusters, allowing user interaction for controlling the network to display real time changes in the network.

These early implementations represent pioneering work that demonstrates the enormous promise of the sensor network concept. They share a number of similarities. Most of the motes used are similar in capability to those used in the UCB-Intel large-scale demonstration [6]. Each network has homogeneous motes although the motes may be multifunctional having multiple sensing capabilities and serving other functions such as routers, cluster leaders, etc. PCs serve as base stations for interfacing with the outside world. However, all are passive, sensing systems (i.e., no effectors) observing the environment and reporting these observations to a central authority where decisions will be made often by human observers.

However, sensors for measuring the parameters in this implementations and the ability to feed their measurements into a computer are not new. What was novel about these approaches is the small size of the sensors and their wireless networking connection. This allowed easy and inexpensive installation of these motes directly into the environment. Using wireless communication, motes do not require an expensive communication infrastructure and expense does not increase exponentially with the number of motes as in wired networks (i.e., more motes would need more wire). Most importantly, both the small size of the motes and the wireless network enabled an installation that was un-intrusive compared with what was previously possible. An obstructive installation required by larger equipment may interfere with the ecosystem in a biological application. In other applications, the introduction of wiring infrastructure can delay projects due to intrusions into other infrastructure (i.e. may require additional approval or may interfere with design), and therefore be unacceptable.

These demonstrations used between 6 and 800 motes, thus, they do not approach the high fidelity information architecture advertised by proponents of sensor networks. Will the techniques used in small-scale sensor networks scale to massive numbers? We believe not.

In these designs, behavior is predetermined, its results collected, and otherwise managed by a central authority. Motes appear to be working together, though only in simple ways: multi-hop networks are formed and clocks are synchronized. An examination of the method used for clock synchronization in the Twentynine Palms demonstration [21] reveals that it closely models the behavior of biological singletons such as bacteria. Each mote periodically broadcasts its current time. Each mote that receives the broadcast updates its time if the broadcast value is greater than its own. Though effective, are the motes really working together? Or are they singletons following simple rules: they periodically broadcast their time and they listen for the broadcast of packets containing time values. Do they know or care that other motes exist? From the individual application of these simple rules, a simple behavior for the group *emerges*: their times are synchronized. However, neither the individual or group behavior can be described as cognitive.

Table 1 summarizes the promises made by visionaries compared with the realities presented in most current sensor network research. Current sensor networks are for the most part modeled after conventional computing networks under centralized control and involve a small number of motes. Motes simply observe and report with the majority of work done by a central processor to make conclusion and direct action. In many suggested techniques, motes must consistently remain in receive mode except when transmitting (i.e., an inefficient consumption of energy) or depend upon a method such as TDMA to limit power consumption. Neither method scales to massive sized networks. It is, therefore, not clear that they provide a credible approximation of the massive deployment envisioned by the proponents of sensor networks [e.g., see 1,5,6]. Rather than adapting conventional techniques of centralized computer control, new techniques dependent on local cooperation among network nodes will lead to self-sustaining communities of machines with emergent behavior that autonomously operate and adapt to changes in the environment. This evolution so parallels the development of life on Earth that living systems are likely to provide realistic models for sensor network design.

Table 1. Vision versus reality

Visionaries Promise	Reality
Massive deployment	Most implementations under 50 motes
Unlimited capabilities	Motes only observe and report. Cooperation only under centralized control for routing, data aggregation, etc.
Distributed control	Most implementations use centralized control
Energy efficiency	Implementation either have all motes listening when not transmitting, or depend on TDMA which requires strict, global synchronization
New paradigm	Most techniques are adaptations of existing technologies

1.1 Our Contributions

The main contribution of this work is to propose to look at large-scale sensor networks in a novel way. This was motivated by our intuition that in order to scale to massive deployment, sensor networks can benefit from lessons learned from the way biological ecosystems are organized. Indeed, in the presence of a massive deployment, sensor networks must behave as a community of organisms, where individual motes operate based on local information, making local decisions that are aggregated across the network to achieve globally-meaningful effects. Along this line of thought, our work centers on three fundamental characteristics of future sensor networks that are not demonstrated by the current implementations:

- We are developing techniques that eliminate bottlenecks even if the number of motes in a sensor network is dramatically increased.
- We demonstrate motes that do more than sense a value and report that value. Our motes are designed to be asynchronously autonomous, learning from their environment to make better decisions.
- We design sensor networks that do not require centralized control.

This differs greatly from the current direction of sensor network research. Rather that applying conventional techniques requiring centralized control and the establishment and maintenance of complicated infrastructure, our motes, upon deployment, will begin to operate autonomously without the establishment of any infrastructure. Each mote will truly operate and cooperate as if it and its immediate neighbors are they only motes in existence.

We are aware that the examples detailed here will not solve all problems facing the development of sensor networks. Rather, our solutions demonstrate a different approach to solving these problems that we believe can be employed in many situations without requiring expensive centralized control and infrastructure which are prone to single points of failure.

We will also show that memory of past experience and local cooperation among neighbors can lead to beneficial adaptive behavior. We do not claim that our examples here demonstrate much learning or intelligence. But again, we are arguing for a change in direction where local learning and cooperation is paramount in design leading to systems that adapt to environmental change without centralized direction.

The two important metrics of concern are time required to do work and the energy expended to do that work. If work must be done serially, then the time required to do that work will scale linearly (or worse) with the size of the network. The total energy consumption of the network is expected to increase with the size of the network, but a desirable property for scalability is for the energy consumption per mote to remain stable regardless of the size of the network. We will show that both desired characteristics are found in our examples: we will show that both the time required for our task and the energy required per mote to accomplish the task stabilizes are the size of the network grows. We believe consideration of these properties is paramount in the design of a scalable sensor network.

2   An Ecological Model for Sensor Networks

We think of motes as *organisms* within a community. At *birth* (i.e., at deployment time) the motes are endowed with *genetic material*, containing, among others, an initial *state* and *rules* by which they interact with the environment. The state and the rules may change as the motes interact with the environment, reflecting their dynamic adaptation to conditions in their neighborhood. Additionally, the motes may *remember* and *record* their interaction with the environment by storing information in their limited on-board memory. Memory and its use to change state or rules are considered learning. Changing state conditions or rules based on learning demonstrates some level of cognition.

Although genetic algorithms are a popular algorithmic paradigm, they rely in a crucial way on extremely fast computational speed to evaluate many random mutations of some genetic specification. While most of these new combinations will prove useless, or worse, harmful to the objective, the search is for the small percentage of mutations that prove beneficial. Consequently, while genetic algorithms may be useful for small-scale sensor networks, they are useless to large-scale systems. Indeed, the limited computational power of the motes would make the use of genetic algorithms prohibitively expensive. Furthermore, we view a mote as an individual organism. Just as with living organisms, successful changes in behavior or other capabilities must be based on experience and learning. Random changes would be highly likely to result in death (i.e., failure) of the mote and catastrophe for the network. Afterwards, there would be no chance to try another mutation.

One of the main goals of this work is to demonstrate how learning and cognition can facilitate adaptability of a sensor network without centralized control. In particular, we are interested in using these attributes to enable *local decisions* based on *local information* that effect *global* results. Limiting decisions to localities is important for reasons of scalability and autonomy. Local decisions allow distributed control. In turn, distributed control through local decisions provides a natural redundancy affording fault tolerance – as some motes fail or exhaust their energy budget others will continue to make decisions.

We propose to use *cellular automata* (CA, for short) to model large-scale sensor networks operating as organisms in an ecosystem. A cellular automaton represents, in most ways, a distribution of sensor motes throughout a geographic region. As illustrated in Fig. 1, each internal cell is surrounded by eight neighbor cells. Border cells have three or five neighbors. Neighbor cells represent those motes that can receive a transmission from a cell. Thus, the regularity displayed by the grid represents a logical indication of physical proximity: a cell adjacent to another is within transmission distance of that cell; it does not have to be a consistent distance from the cell (i.e., it represents a random deployment of physical placement). For simplicity of calculation and display, throughout this work we assume that each sensor has exactly eight neighbors. Visibly, the set of neighbors need not be limited to the eight adjacent cells. Increasing the number of cells that can receive transmissions from the selected cell is specified by the neighbor *radius*. A radius of two would include in addition to the eight adjacent cells, the 16 cells adjacent to these neighbors. One apparent limitation of this model is that the number of neighbors is fixed for a given radius; however, this can be changed by disabling some of the neighbors. This can be done randomly to simulate a random deployment. Our simulator also allows cells to fail by not responding to a request and not taking action on their own.

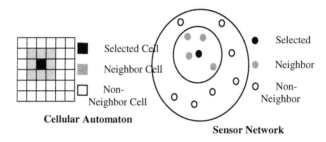

Fig. 1. Illustrating the neighborhood of a cell

It is important to note that the simplification of calculation and display we achieve by using a constant number of neighbors and assuming no failures does not oversimplify our simulation such that our results do not apply to a random deployment with motes that may fail. We will show that the time required and the energy consumed per mote to accomplish our tasks will remain consistent regardless of the size of the network. We will further show that the tasks are accomplished using only local information to attain a global goal. *None* of these properties would change if we used a random number of neighbors or introduced failed motes.

3   Aggregating Sensed Data: The Majority Rule

There are, essentially, two ways in which data sensed by the motes can be aggregated. In a centrally-controlled network, data aggregation is a two-stage process: in the first stage the motes forward the collected data to the sink. Some of the data may be fused en route but the final responsibility for aggregation rests within the sink. In the second stage, the aggregated result is broadcast back to the network. Though straightforward, this method does not scale very well [1]. By contrast, in a truly distributed system, as is the case in an autonomous sensor network, the motes cannot rely on the sink to perform the aggregation. Instead, the aggregation must be performed *in-situ* by the sensors themselves. An often studied problem is that of *majority rule* or *density* where each mote in a community must decide upon a binary value (e.g., on or off). Its decision is based on the requirement to align with the major frequency: if most are on then the mote will turn on. More generally, it is often the case that a community must reach a *consensus*, that is, an agreement on a common state when individuals are in different states from some domain of states. The challenge is how a mote determines the major frequency. In a centralized approach, all motes report their value to the sink, the sink determines the major frequency, and the sink reports back that value to all motes. In a distributed approach, each mote must determine for itself the major frequency. One of the key contributions of this work is to show that majority rule can be determined in a scalable fashion in massively deployed sensor networks. We prefer a distributed approach where a sensor network is modeled after a community of organisms. The motes work asynchronously and autonomously using only local information to make local decisions, yet a correct decision is made for the community at large. Yet, they cooperate with their neighbors sharing information to facilitate better decisions by all. A mote may decide work is done and become inactive only to be reactivated by a neighbor if more work is needed.

4 Using Cellular Automata to Calculate Majority Rule

Mitchell, *et al.*, [12] report on the application of genetic algorithms to attempt to discover rules to solve the problem they describe as *density classification*. In the density classification problem the goal is to find a rule set that allows the CA to use only local information to determine whether or not its initial configuration contains a majority of 1s (high density) or 0s (low density). The authors also refer to this as the "$\rho_c = \frac{1}{2}$" task, ρ indicating the density of 1s in the CA and "$\frac{1}{2}$" being the "critical" density of interest. The authors write "Designing an algorithm to perform the $\rho_c = \frac{1}{2}$ task is trivial for a system with a central controller or central storage of some kind, such as a standard computer with a counter register or a neural network in which all input units are connected to a central hidden unit. However, the task is nontrivial for a small radius ($r << N$) CA, since a small-radius CA relies only on local interactions" [12, p. 4]. The authors note that other researchers have argued that "*no finite-radius, finite-state CA with periodic boundary conditions can perform this task perfectly across all lattice sizes.*" [12, p. 4]. A periodic boundary CA is one in which the opposite end cells are repeated and wrapped for calculations.

In the experiments reported by [12], the genetic algorithms developed were rather unsophisticated, generally employing the strategy that the absence or existence of large blocks of 1s is a good predictor of the CA density, since statistically a high density CA is more likely to have adjacent blocks of 1s than a low density CA. This is referred to as a block-expanding strategy. Lacking from this strategy is any global communication or information passing. The algorithm performs poorly depending upon the CA size (large N), and the initial distribution pattern.

A more successful algorithm evolved by [12] successively uses local information and interaction to produce the desired state. Basically, the algorithm "successively classifies "local" densities with a locality range that increases with time. In regions where there is some ambiguity, a "signal" is propagated." The purpose of the signal is to flag that the analysis of this local area must be made at a later time and with a large radius.

The performance of the various evolved algorithms never reaches certainty. The following table of data concerning these experiments is abstracted from [13]: The data show the percent of correct classifications produced by the rule on an unbiased (i.e., density is close to 0.5) 1-D CA of size N.

Table 2. Percentage of correct majority predictions

CA Rule	N = 149	N = 599	N = 999
Majority	0.000	0.000	0.000
Expand 1-blocks	0.652	0.515	0.503
Particle-based-1	0.697	0.580	0.522
Particle-based-2	0.742	0.718	0.701
Particle-based-3	0.769	0.725	0.714
GKL	0.816	0.766	0.757

While the purpose of the experiments was to evaluate the ability of genetic algorithms to evolve strategies for solving the density problem, it provides a good indication of the difficulty in performing the actual task, with no algorithm being able to consistently solve the problem. Notice that the simpler algorithms (local majority and expand the local block) have extremely poor performance, no better than flipping a fair coin!

Epstein [14] investigated a similar problem, the evolution of norms and the conformance of individuals to the developing norm. He developed an agent-based model "Best Reply to Adaptive Sample Evidence", which demonstrates intriguing parallels to the CA density problem just presented (see [14] for details).

5   Calculating Majority Rule Within a Sensor Network

In the following, we first examine the issues of using a centralized approach for majority rule, and then propose our autonomous and distributed approach. As our cellular automaton is defined such that each cell can only transmit to its immediate neighbors, then in the centralized approach, data collected by the sink from each cell must pass through half the span of the grid on average. It is possible to minimize the number of transactions by en-route aggregation or *piggybacking* values in messages as they are routed to the sink. Nevertheless, this requires overhead for complex coordination and infrastructure establishment (e.g., constructing spanning trees). In the second stage (broadcasting majority rule), to conserve mote energy, a suggested approach is for the sink to broadcast a return message to all motes. But this method does not scale well as the power of the sink's transmission would have to grow with the size of the sensor network distribution and the transmission from the sink must be directly receivable by all motes (i.e., there can be no blockage). We assume a cost of flooding the computed average to all cells, which increases with the size of the grid. Besides transmission costs, there are additional problems with the centralized approach. Routing tables to reach the sink must be discovered and maintained.

Disruption in these routes must be handled to assure messages arrive at the sink. Regardless of how this is done, this method is open to single points of failure.

It was recently noticed by many researchers [e.g., 11,15,16,19,20] that centrally-controlled sensor networks are prone to uneven energy depletion leading to the creation of *energy holes* in the vicinity of sinks. This uneven energy depletion creates an *energy hole* around the sink, severely curtailing network longevity. Specifically, Wadaa *et al.* [11] and Olariu and Stojmenovic [15] showed that by the time the motes close to the sink have expended all their energy, other motes in the sensor network still have up to 94% of their original energy budget.

In our cellular automaton model, the cells closest to the sink must relay messages from every cell in the grid and their energy budget will decrease rapidly relative to cells further away. Our simulation results reveal that the cells closest to the sink (i.e., active cells) consume most energy:

- 50% most active cells use 87 % of the transactions.
- 25% most active cells use 70 % of the transactions.
- 10% most active cells use 52 % of the transactions.
- 1% most active cells use 40 % of the transactions.

Our objective is for the sensor network to calculate a majority rule across the network without any mote or any central authority having global knowledge of all mote values. The problem is for each mote to obtain and maintain the global average by iteratively using only data that is available locally.

We now give an informal description of our algorithm. Assume that each mote maintains a binary switch (e.g. [0,1], [on,off], [sense,sleep], etc.). Upon deployment, the motes are endowed with genetic material (see [9,10] for details): an activity *status*, a *transmission time* within a specified *time period* and an independent *clock*. The interested reader is referred to [7,11] for more details and other applications of the genetic material.

When the algorithm begins, each mote starts its own clock. The time period is divided into one or more slots (a parameter to the simulation). The transmission time assigned to a mote is one of these time slots within a time period (e.g., one mote may transmit at slot 3, another at slot 18, but all motes will transmit some time during a time period). Because the transmission time for each mote is a random number, there is no guarantee that two or more motes will never be transmitting at the same time. If two motes are at a neighbor radius greater than 3, they may execute the algorithm simultaneously, as the results of their calculations are independent of each other. If the radius is less than 3, the result of calculations is *order-dependent*. In this case, simultaneous transmissions will cause collisions, thus a MAC layer protocol is assumed to decide cellular execution order. To simulate this control, all motes (i.e. cells) transmitting at the same time slot are executed in random order. The order is randomized anew each time period. Thus, the transmission time is fixed at deployment (being part of the genetic material), but the execution order in each time slot is a "function of the environment" and may change at different times.

Setting the time period to a larger value increases the probability that each cell will transmit independently. This provides the most repeatable simulation, as a simulation with a given color distribution and a given transmission time assignment will execute exactly the same each time, provided none transmit simultaneously.

However, this does increase the execution time of the simulation, as fewer cells execute in each transmission time. Reducing the time period speeds the simulation but introduces more randomness for each time slot. Setting the time period to one guarantees that all cells will execute in that time slot, though in new random order for each period. The results are very similar regardless of the length of the time period; specifying a longer time period just requires more time during which nothing happens. Even with MAC layer negotiations, by this method each cell is acting *asynchronously* and *autonomously*. Because the time and cost of reaching a global decision is a function of initial switch value distribution, transmission time assignment, and execution order in each time slot, we ran multiple executions varying all three parameters and averaged the results.

When a cell reaches its transmission time within the time period, it is *selected* for action. If its status is inactive, it does nothing. Otherwise, if its status is active, it begins a series of *transactions*. A transaction is either a request for information from a neighbor (e.g., what is your switch value?) or a specification given to a neighbor (e.g., set your switch to off). Transactions are significant because they require radio transmissions, typically the most costly activity of a mote. If all neighbors are in agreement on the switch value, it sets its status to inactive and will not participate again until it is reactivated by one of its neighbors. An inactive status is significant in that no transactions occur when a selected cell is inactive. If any neighbor's switch value is not equal to the selected cell's value, the selected cell determines the major frequency for the neighborhood and sets itself and all neighbors to this value. During this process, if any neighbor is inactive, it is reactivated. The simulation continues until all cells are set to inactive. Thus, the simulation ends using only local information; no global control is required. Inactive cells are identified by the color red. Interesting color patterns are displayed as the simulation nears completion: the number of red cells varies but increases until all are red. As these local neighbor cells are cooperating with each other, acting on local environmental information, and remembering information from one action that will affect a future action, we argue that this system demonstrates simple cognition.

Fig. 2 depicts an initial distribution in a cellular automaton with a 30 x 30 grid of cells showing a random distribution of switch values, represented by the colors black and white. Fig. 3 illustrates the color change after the first time period (i.e. all cells have been selected once). Notice that at the end of the first iteration, the cells are grouped in large blocks of the same color. Fig. 4 shows the color change at the end of four iterations. Soon thereafter, all cells are the same color indicating the major frequency is determined. For larger grids, interesting patterns display as groups of colors appear to move around the grid while cells on group boundaries compete for majority value.

Fig. 5 shows that the time required for consensus escalates quickly for smaller grid sizes but after 300x300, the acceleration slows and then stabilizes for massive grid sizes. Fig. 6 confirms that the same is true for the required number of transactions, an indicator of energy expenditure.

A greater advantage of our decentralized approach is in the distribution of energy expenditure. The "funnel effect" of multi-hop routing required for the centralized approach described above will deplete the energy of cells much faster when their distance to the sink is shorter. In our decentralized approach, the workload is not only evenly distributed, but the workload required of an individual mote actually decreases as the grid size increases, as shown in Fig. 7. Each time a cell is selected, it initiates transactions resulting in energy expense. Thus, the fewer the selections, the smaller the energy consumed. As Fig. 7 shows, the cost in energy per cell does not escalate for larger grids.

Fig. 2. Initial distribution of colors for majority rule

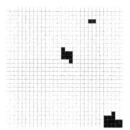

Fig. 3. Distribution of colors after the first time period

Fig. 4. Distribution of colors after 4 time periods

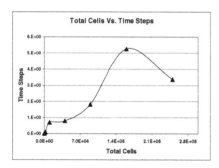

Fig. 5. Time required for majority rule

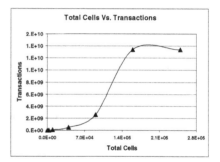

Fig. 6. Transactions required for majority rule

For some applications, it is not necessary for every mote to agree; it may be sufficient for most motes to be set to the majority value. As depicted in Fig. 2–4, this algorithm has as useful property for such cases, as more cells change to the majority value very rapidly. Fig. 8 shows that about 70% of the cells have the majority value with very few time steps and, furthermore, this time requirement stabilizes for larger grid sizes. Fig. 9 shows that while the number of transactions continues to increase for a final solution, the cost of 70% stabilizes for larger grid sizes. We call this *good enough computing*, and where applicable, it can substantially reduce the cost in time and energy of the calculation.

If carried to completion, this algorithm will always result in the community agreeing on a single switch value. However it suffers from the same problems described by Mitchell, et al. and Epstein. In some cases, particularly when there are large blocks within the initial distribution of the same color and especially if the distribution is close to equal, the final configuration will have a majority frequency opposite from the initial configuration, because the selection of cells for action is random and it is possible that initially the cells with minority color near the border of the color boundary are selected, which would turn cells of majority color in that neighborhood to minority color, thus incorrectly flip-flopping the minority-minority color. We solve this problem by using averaging to incrementally calculate the majority frequency. Instead of keeping an integer value for the switch, we maintain a real value. All cells are initially assigned genetic material for the switch value of 0.0 or 1.0. The algorithm proceeds as before, but now, instead of the selected cell calculating a binary frequency for the neighborhood, it calculates the average of all neighbors and itself. This real value is stored as the switch value. Whenever the binary value of the switch is required this value is assessed as (Value $> =0.5$) $= 1.0$ or (Value < 0.5)=0.0.

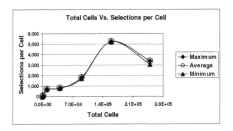

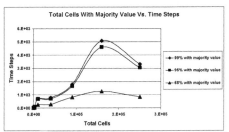

Fig. 7. Minimum, average, and maximum number of cell selections

Fig. 8. Stabilizing time requirement for percentage of cells with majority value

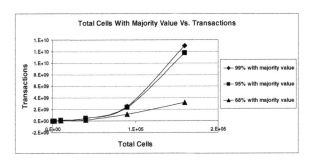

Fig. 9. Stabilizing transaction requirement for percentage of cells with majority value

This works because of an interesting property of our algorithm of incrementally averaging neighborhoods. Although we never break the rule that no cell can ask for any information outside of his neighborhood, the converged average is always equal to the average of the initial distribution. Furthermore, the average value of the entire community following each cell selection is also equal to the initial average of the community. Re-examining the process reveals the reasons for this state of affairs. A group of nine cells of different colors contributes to the average of the total grid. When these cells are averaged and set to the same average value for the group, they contribute exactly the same to the average of the entire grid as they did with differing values. If the initial distribution has a majority of 1s, then the initial average will be greater than or equal to 0.5 and if the majority is of 0s, the initial average will less than 0.5. Because the average remains constant throughout the process, the majority at the end will be the same as the majority at the beginning.

The averaging technique shares many of the same properties as the major frequency calculation described above. Fig. 10 shows a similar requirement for cell selection to calculate the average. Thus, the energy requirement per cell stabilizes for larger grids. Fig. 11 shows that, like the total transactions required for a solution, solutions with a large percentage of cells within the tolerance asymptotically slope towards zero as the grid size increases. Most importantly, solutions up to 95% with the major frequency value reach that asymptote quickly with few transactions regardless of grid size.

Using real values may seem like "cheating" compared with the unsuccessful method of Mitchel and Epstein. However, it is not our concern to solve the academic problem. We are only concerned with a solution that will work for sensor networks.

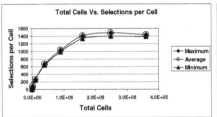

Fig. 10. Minimum, average, and maximum number of cell selections for averaging

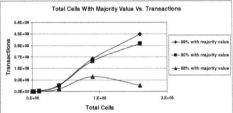

Fig. 11. Percentage of cells with majority value

6   Concluding Remarks

In this work we have demonstrated a function completed by a "social" sensor network: autonomous motes functioning asynchronously cooperative to achieve a common goal. The function is carried out without centralized control and without any mote needing to know all information known within the society. We have also shown that the goal can be closely approached with few costs in time and resources compared with the much more costly final answer.

References

1. I. F. Akyildiz, W. Su, Y. Sankarasubramanian, and E. Cayirci, Wireless sensor networks: A survey, *Computer Networks*, 38(4), 2002, 393-422.
2. D. M. Doolin, S. D. Glaser and N. Sitar, Software Architecture for GPS-enabled Wildfire Sensorboard. *TinyOS Technology Exchange*, February 26, 2004, University of California, Berkeley CA.

3. D. M. Doolin and N. Sitar, Wireless sensors for wildfire monitoring *Proc. SPIE Symposium on Smart Structures and Materials* (NDE 2005), San Diego, California, March 6-10, 2005

4. D. Estrin, R. Govindan, J. Heidemann and S. Kumar, Next century challenges: Scalable coordination in sensor networks, *Proc. MOBICOM*, Seattle, WA, August 1999.

5. J. M. Kahn, R. H. Katz, and K. S. J. Pister, Next century challenges: Mobile support for Smart Dust, *Proc. ACM MOBICOM*, Seattle, WA, August 1999, 271--278.

6. 6. D. Lammers, Embedded projects take a share of Intel's research dollars, *EE Times*, August 28, 2001. Retrieved April 5, 2004, from http://today.cs.berkeley.edu/800demo/eetimes.html

7. S. Olariu, A. Wadaa, L. Wilson and M. Eltoweissy, Wireless sensor networks: leveraging the virtual infrastructure, *IEEE Network*, 18(4), 204, 51-56.

8. S. Olariu and Q. Xu, A simple self-organization protocol for massively deployed sensor networks, *Computer Communications*, to appear, 2005.

9. P. Saffo, Sensors, the next wave of innovation, *Communications of the ACM*, 40(2), 1997, 93-97.

10. K. Sohrabi, J. Gao, V. Ailawadhi and G. Pottie, Protocols for self-organization of a wireless sensor network, *IEEE Personal Communications*, 7(5), 2000, 16-27.

11. A.Wadaa, S. Olariu, L. Wilson, M. Eltoweissy and K. Jones, Training a wireless sensor network, *Mobile Networks and Applications*, 10, 2005, 151-167.

12. M. Mitchel, J. Crutchfield. and R. Das, Computer science application: Evolving cellular automata to perform computations, in T. Bäck, D., Fogel and Z. Michaelewics, (Eds.), *Handbook of Evolutionary Computation*, Oxford University Press, 1997.

13. M. Mitchell, *An Introduction to Genetic Algorithms,* The MIT Press, 1999.

14. J. Epstein, Learning to be thoughtless: Social norms and individual computation, Center on Social and Economic Dynamics Working Paper No. 6, revised January 2000.

15. S. Olariu and I. Stojmenovic, Design guidelines for maximizing lifetime and avoiding energy holes in sensor networks with uniform distribution and uniform reporting, a manuscript, 2005.

16. V. Mhatre, C. Rosenberg, D. Kofman, R. Mazumdar, and N. Shroff, A minimum cost heterogeneous sensor network with a lifetime constraint, *IEEE Transactions on Mobile Computing,* 4(1), 2005, 4-15.

17. J. Polastre, R. Szewcyk, A. Mainwaring, D. Culler and J. Anderson, Analysis of wireless sensor networks for habitat monitoring, in *Wireless Sensor Networks*, Raghavendra, Sivalingam, and Znati, Eds., Kluwer Academic, 2004, 399-423.

18. M. Srivastava, R. Muntz and M. Potkonjak, Smart Kindergarten: Sensor-based wireless networks for smart developmental problem-solving environments, *Proc. ACM MOBICOM*, Rome, Italy, July 2001.

19. J. Li and P. Mohapatra, Mitigating the energy hole problem in many to one sensor networks, a manuscript, 2005.

20. J. Lian, K. Naik and G. B. Agnew, Data capacity improvement of wireless sensor networks using non-uniform sensor distribution, *International Journal of Distributed Sensor Networks,* to appear, 2005.

21. UCB/MLB 29 Palms UAV-Dropped Sensor Network Demo (2001). University of California, Berkeley. Retrieved April 5, 2004, from http://robotics.eecs.berkeley.edu/~pister/29Palms0103

A Key Management Scheme for Commodity Sensor Networks

Yong Ho Kim[1,*], Mu Hyun Kim[1], Dong Hoon Lee[1], and Changwook Kim[2]

[1] Center for Information Security Technologies (CIST), Korea University,
Seoul, Korea
{optim, mhkim, donghlee}@korea.ac.kr
[2] School of Computer Science, University of Oklahoma, Norman,
OK 73019, USA
ckim@ou.edu

Abstract. To guarantee secure communication in wireless sensor networks, secret keys should be securely established between sensor nodes. Recently, a simple key distribution scheme has been proposed for pairwise key establishment in sensor networks by Anderson, Chan, and Perrig. They defined a practical attack model for non-critical commodity sensor networks. Unfortunately, the scheme is vulnerable under their attack model. In this paper, we describe the vulnerability in their scheme and propose a modified one. Our scheme is secure under their attack model and the security of our scheme is proved. Furthermore, our scheme does not require additional communication overhead nor additional infrastructure to load potential keys into sensor nodes.

Keywords: security, key management, wireless sensor networks

1 Introduction

Wireless sensor networks are well recognized as a new paradigm for future communication. Sensor networks consist of a huge number of battery powered and low-cost devices, called sensor nodes. Each sensor node is equipped with sensing, data processing, and communicating components [2,6].

To provide secure communication within wireless sensor networks, it is essential that secret keys should be securely established between sensor nodes. The shared secret key may be later used to achieve some cryptographic goals such as confidentiality or data integrity. However, due to limited resources of sensor nodes, traditional schemes such as public key cryptography are impractical in sensor networks. Furthermore, the position of sensor nodes (hence, the neighbors of nodes) cannot be pre-determined since they are randomly deployed in unattended areas. Due to this restriction, most schemes are based on the pre-distribution of potential keys. However, this pre-distribution mechanism can be burdensome for non-critical commodity sensor networks.

---

* This work was supported by grant No. R01 − 2004 − 000 − 10704 − 0 from the Korea Science & Engineering Foundation.

V.R. Sirotiuk and E. Chávez (Eds.): ADHOC-NOW 2005, LNCS 3738, pp. 113–126, 2005.

Most key management schemes for sensor networks are designed on the assumption that attacks are launched even during the node deployment phase. This assumption is so strict that it is difficult to design a simple and efficient scheme. The attack models assumed by most schemes are as follows; nodes can be captured while being deployed, and all initialization traffics are monitored by attackers. However, not all sensor network applications are exposed to these attacks. Commodity sensor network applications have less security risk than mission critical applications because threats to commodity sensor networks are not pervasive or the value of the application itself may not be high enough to attract attackers. Assuming this loose attack model, an efficient key management scheme can be designed that achieves a reasonable level of security.

RELATED WORKS. When we consider the design of key distribution schemes, the simplest method is to embed a single network-wide key in the memory of all nodes before nodes are deployed. In this case, however, the entire network can be compromised if a single node is compromised. Another extreme method is that each node in a network of n nodes shares a unique pair-wise key with every other node in the network before deployment. This requires memory for $n - 1$ keys for each sensor node. These two methods are unsuitable for wireless sensor networks.

Perrig et al. presented SPINS [13], security protocols for sensor networks. In SPINS, each node shares a secret key with a base station and establishes its pair-wise keys through the base station. This architecture provides small memory requirement and perfect resilience against node capture. However, because the base station should participate in every pair-wise key establishment, SPINS requires significant communication overhead and does not support large scale networks.

Eschenauer and Gligor recently presented a random key pre-distribution scheme (hereafter cited as the EG scheme) for pair-wise key establishment [11] in which a key pool is randomly selected from the key space and a key ring, a randomly selected subset from the key pool, is stored in each sensor node before deployment. A common key in two key rings of a pair of neighbor nodes is used as their pair-wise key. The EG scheme has been subsequently improved by Chan et al. [7], Liu and Ning [12], and Du et al. [9,10].

Chan, Perrig, and Song presented three mechanisms for key establishment in [7]. First, the q-composite random key pre-distribution scheme has improved resilience against node capture under a small scale attack, while the scheme is still vulnerable to a large scale attack. Second, the multi-path reinforcement scheme improves security, whereas it requires high communication overhead. Third, the random pair-wise scheme provides perfect resilience against node capture. However, in a large network, the third scheme is impractical since each node requires large memory storage.

Du et al. presented a multi-space pair-wise key scheme [10]. This scheme, a combination of the EG scheme and the Blom's scheme [4], significantly enhances security. Similarly, Liu, Ning, and Li presented a key pre-distribution scheme [12] combining the EG scheme and the Blundo's scheme [5]. Additionally, Du et al. proposed a new key management scheme using deployment knowledge [9].

In this scheme, the sensor nodes are first pre-arranged in a sequence of small groups and these groups are deployed sequentially. Although the EG scheme is practical and its variants [7,9,10,12] have improved resilience against node capture, these require a significantly high cost for non-critical commodity sensor network applications.

Recently, Anderson, Chan, and Perrig presented a simple key scheme (hereafter cited as the basic ACP scheme) for non-critical commodity sensor network applications [1]. To make the application more flexible and usable, they defined a real attack model (hereafter cited as the ACP attack model). In their ACP attack model, during key setup, an attacker does not have any physical access and is unable to execute active attacks. Also, she is able to monitor only a small proportion of communications of sensor nodes during this phase. However, after the key setup, all attacks are possible. The basic ACP scheme does not require an infrastructure to load potential keys into the sensor node, unlike the pre-distribution schemes [7,9,10,11,12]. The authors also proposed Secrecy Amplification which gives an improvement of about 20 % from the basic ACP scheme. Unfortunately, this method has a vulnerability under the ACP attack model.

CONTRIBUTIONS. In this paper, we propose a new key scheme for non-critical commodity sensor networks to improve the security of the basic ACP scheme [1]. The main contributions of our approach can be summarized as follows.

- Improved security. The basic ACP scheme is highly secure under the ACP attack model without any additional communication overhead. However, Secrecy Amplification after the basic ACP scheme has a vulnerability under the ACP attack model and requires additional communication overhead. We propose a modified scheme which has no vulnerability and gives an improvement from the basic ACP scheme.
- Provable security. We prove the security of our scheme for non-critical commodity sensor networks. The proof is based on the one defined by Bellare et al. [3]. Assuming that a pseudorandom function is secure, our scheme has provable security.
- Low cost and Scalability. Our scheme is suitable to large scale networks because it does not require a base station or a large memory overhead. To be scalable, SPINS [13] requires significant communication overhead because of the participation of a base station in key establishment, and the random pair-wise scheme by Chan et al. [7] requires a large memory overhead for each node.

ORGANIZATION. The rest of the paper is as follows. Section 2 shows the notation used in the schemes. We give a definition of the ACP attack model in Section 3 and an overview of the ACP scheme in Section 4. We propose our scheme and analyze its performance and security in Section 5. Section 6 discusses non-critical commodity sensor network applications. Finally, we conclude our paper in Section 5.

2 Notation

We list the notations used in the paper below;

Notation	Description
n	the number of sensor nodes
d	the expected number of neighbor sensor nodes within communication radius of a given node
R	communication radius of each sensor node
W_i	i-th white sensor node which is not compromised
B_i	i-th black sensor node which is pre-deployed by an attacker
$\|$	concatenation operator
$\oplus$	exclusive or operator
K_{ij}	pair-wise key between W_i and W_j
$H(\cdot)$	one way hash function
$E(K, \cdot)$	symmetric encryption function using key K
$F(K, \cdot)$	pseudorandom function using key K

3 Attack Model

The following attack model was defined in [1] for commodity environments. First, the chance of physical node capture is acceptably low during initial key setup phase, but nodes are vulnerable to node capture after key setup. Second, during key setup, the attacker can monitor only a small proportion α of the communications in the sensor networks using some black nodes which are pre-deployed, but she can monitor all communications after key setup. Last, active attacks such as jamming or flooding are not launched during this phase. In other words, attacks are assumed to be rare during key setup, but the attack model is just the same as others after this phase is finished. The following table summarizes the ACP attack model.

Table 1. The ACP attack model

	During key setup	After key setup
Physical access (node capture)	no	yes
Monitor communications	α %	100%
Active attacks (jamming or flooding)	no	yes

4 ACP Scheme [1]

In the following, we describe the basic ACP scheme and Secrecy Amplification which enhances the security of the basic ACP scheme. After that, we will explain the vulnerability of Secrecy Amplification under their attack model.

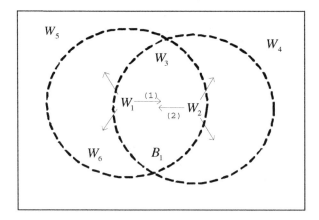

Fig. 1. The basic ACP scheme

In the Fig. 1, W_1, W_2, and W_3 are white nodes which are neighbor each other. The left circle depicts the communication boundary of W_1, and the right one is for W_2. B_1 is a pre-deployed black node. In this deployment, W_3 and B_1 can listen all communications of W_1 and W_2, while B_1 cannot listen to traffics from W_3.

4.1   The Basic ACP Scheme

The basic ACP scheme consists of two phases: (1) Once deployed, a white node W_1 generates and broadcasts a random key K_1. Its neighbor white node W_2 can obtain the message. (2) W_2 generates a random pair-wise key K_{12} and broadcasts an encrypted message $E(K_1, W_2 \parallel K_{12})$ using K_1. Finally, two nodes W_1 and W_2 have a common pair-wise key K_{12}.

In an area with no black nodes, the secure link between W_1 and W_2 is not compromised. However, if a black node B_1 sniffs all the messages of W_1 and W_2 during key setup (See Fig. 1), the pair-wise key between W_1 and W_2 will be compromised. In an example of [1], if there is one black node per 100 white nodes, and d is 4, the percentage of compromised links will only be 2.4 %.

4.2   Secrecy Amplification

Suppose that two neighbor nodes W_1 and W_2 have a pair-wise key K_{12} after key setup. Here, they can amplify the security of the key using their common neighbor node W_3. Secrecy Amplification after the basic ACP scheme consists of four phases: (1) W_1 sends an encrypted message $E(K_{13}, W_1 \parallel W_2 \parallel N_1)$ to W_3, where N_1 is an unpredictable nonce generated by W_1, and K_{13} is a pair-wise key between W_1 and W_3. (2) W_3 decrypts the encrypted message and sends another encrypted message $E(K_{23}, W_1 \parallel W_2 \parallel N_1)$ to W_2, where K_{23} is a pair-wise key between W_2 and W_3. W_2 decrypts the encrypted message and obtains N_1. Thus, W_1 and W_2 have the common nonce N_1 and can update their pair-wise key

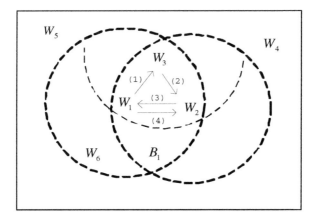

Fig. 2. Secrecy Amplification after the basic ACP scheme

as $K'_{12} = H(K_{12} \parallel N_1)$. (3) Next, to confirm the pair-wise key, W_2 sends an encrypted message $E(K'_{12}, N_1 \parallel N_2)$ to W_1, where N_2 is an unpredictable nonce generated by W_2. (4) Finally, W_1 sends an encrypted message $E(K'_{12}, N_2)$ to W_2 and two nodes confirm the updated pair-wise key with their nonces.

Suppose an adversary has pre-deployed a black node B_1, which is placed within the common communication range between W_1 and W_2, but is placed out of the communication range of W_3 (See Fig. 2). The black node can know K_{12}, but neither K_{13} nor K_{23} during the basic ACP scheme. After key setup, although the attacker can monitor all communication, she cannot know K_{13} or K_{23}. Thus, in Secrecy Amplification, it becomes hard for her to compute the updated key K'_{12}. Comparing with the basic ACP scheme, Secrecy Amplification can give an improvement of about 20 % [1]. Also, this can be extended by adding other neighbor nodes W_4 and W_5 to improve the security. However, the extension requires additional communication overhead.

4.3   A Vulnerability of Secrecy Amplification

Suppose that an attacker obtaining K_{12} using B_1 wants to know K'_{12}. In the ACP attack model, after key setup, the attacker is allowed to monitor all communications and to capture sensor nodes in order to obtain secret information stored within memory of the nodes. After Secrecy Amplification, if W_3 is captured by the attacker, K_{13} will be exposed to her. Also, since she has monitored the encrypted message $E(K_{13}, W_1 \parallel W_2 \parallel N_1)$, she can find the nonce N_1 and compute K'_{12} as $H(K_{12} \parallel N_1)$.

5   Our Scheme

The security model in our scheme is the ACP attack model. Our scheme eliminates the vulnerability of Secrecy Amplification after the basic ACP scheme. It is also possible to improve security without requiring additional cost for communication. Furthermore, the security of our scheme can be formally proved.

5.1 Modified Key Scheme

Our scheme consists of three phase: (1) A deployed white node, W_1, generates a random key K_1 and broadcasts it to its neighbor nodes. We suppose W_1, W_2 and W_3 are also neighbors each other. (2) The node W_2 generates a random key K_2 and broadcasts an encrypted message $E(K_1, W_3 \| K_2)$. Similarly, (2)' W_3 generates a random key K_3 and broadcasts an encrypted message $E(K_1, W_2 \| K_3)$. Here, two phases (2) and (2)' have no dependency each other, thus two messages of (2) and (2)' can be sent in any sequence. Now, nodes W_1, W_2, and W_3 can obtain K_2 and K_3 because they all know K_1 and catch all communications among them. Two nodes W_1 and W_2 set $K_{12} = F(K_3, F(K_2, K_1))$ as their pair-wise key, and two nodes W_1 and W_3 set $K_{13} = F(K_2, F(K_3, K_1))$ as their pair-wise key. After key setup, all the nodes securely delete K_1, K_2, and K_3 from their memory.

5.2 Security Analysis

The security of our scheme is compared with one of the ACP scheme in terms of the condition that pair-wise keys are exposed by a black node. To obtain the pair-wise key K_{12} in our scheme, a black node should be within the boundary where it can sniff all the communication among W_1, W_2, and W_3. This is also required in Secrecy Amplification after the basic ACP scheme to obtain the updated key K_{12}'. It means that the condition to be attacked in our scheme is the same as that of Secrecy Amplification. Thus, our scheme is also expected to improve security similar to Secrecy Amplification.

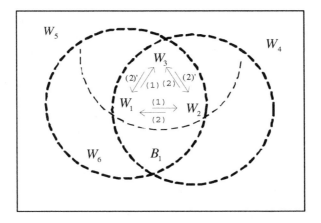

Fig. 3. Our scheme

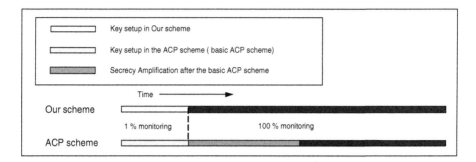

Fig. 4. Pair-wise key setup phase

Our scheme achieves the similar level of expected security improvement as Secrecy Amplification after the basic ACP scheme, while removing the vulnerability of Secrecy Amplification. Since our scheme requires a similar communication cost with the basic ACP scheme, it can finish within a similar time as the basic ACP scheme. In our scheme, to compute K_{12}, some black node must be pre-deployed to sniff communications among W_1, W_2, and W_3 during key setup. However, in the basic ACP scheme, she needs to sniff only the communications between W_1 and W_2. If Secrecy Amplification is included in key setup, the vulnerability of Secrecy Amplification will be removed because some black node, to monitor the encrypted message $E(K_{13}, W_1 \parallel W_2 \parallel N_1)$, must be pre-deployed to sniff all the communications among W_1, W_2, and W_3 during key setup. However, the time taken for initial key setup will be increased (See Fig. 4).

Now, we will apply the same method used for the vulnerability analysis of Secrecy Amplification. We suppose that K_1 and K_2 are exposed to a black node B_1 which is placed within the communication boundary between W_1 and W_2, but it is placed out of the communication boundary of W_3 (See Fig. 3). Also, we suppose that an attacker captures W_3 after key setup. Hence, she obtains K_{13} from the captured node. In this scenario, she attempts to acquire K_{12} by computing either directly or indirectly. First, if she computes K_3 from $K_{13} = F(K_2, F(K_3, F_1))$, she could acquire K_{12} by directly computing $F(K_3, F(K_2, F_1))$. However, this is impossible due to the property of one-way functions. Second, if there are some methods to indirectly compute K_{12} using only K_1, K_2, and K_{13} excluding K_3, our scheme will be insecure. For example, in case of the simple RSA signature scheme, it is possible to acquire a new valid signature by indirectly computing without using a signature secret key s as follows; After an attacker acquires two signatures $M_1{}^s$ and $M_2{}^s$, a new valid signature $(M_1 M_2)^s$ is computed.

To prove the security of our scheme, first of all, we review the security model of a pseudorandom function defined by Bellare et al. [3]

Definition 1. Suppose $F : \{0,1\}^k \times \{0,1\}^\ell \to \{0,1\}^L$ is some function family. Then for any distinguisher A we let

$$\mathbf{Adv}_F^{\mathrm{prf}}(A) \overset{\mathrm{def}}{=} \mid \mathsf{Pr}[\ f \overset{R}{\leftarrow} F : A^f = 1] - \mathsf{Pr}[\ f \overset{R}{\leftarrow} Rand^{\ell \to L} : A^f = 1]\ \mid$$

We associate to F an insecurity function $\mathbf{Adv}_{\mathsf{F}}^{\mathsf{prf}}(\cdot, \cdot)$ defined for any integers $q, t \geq 0$ via

$$\mathbf{Adv}_{\mathsf{F}}^{\mathsf{prf}}(q, t) \stackrel{\text{def}}{=} \max_A \{\mathbf{Adv}_{\mathsf{F}}^{\mathsf{prf}}(A)\}$$

The maximum is over all distinguishers A that make at most q oracle queries and have "running time" at most t. ∎

We define the security of our scheme (M) against the attack in Section 4.3. After key setup, an attacker is allowed to monitor all communications and to mount a chosen node capture attack (cna) in which it can obtain pair-wise keys of captured nodes. If it outputs a pair-wise key of an uncorrupted node, then it is considered successful.

Definition 2. Suppose $F : \{0,1\}^k \times \{0,1\}^\ell \to \{0,1\}^L$ is some function family. The M is secure if the following advantage is negligible in the security parameter k.

Experiment $\mathbf{EXP}_{\mathsf{M}}^{\mathsf{cna}}(B)$
$\quad K_1, K_2, K_3 \stackrel{R}{\leftarrow} \{0,1\}^k$
$\quad$ Compute $K_{13} = F(K_2, F(K_3, K_1))$
$\quad B \leftarrow K_1, K_2, K_{13}$
$\quad \hat{K}_{12} \leftarrow B$
$\quad$ If $F(K_3, F(K_2, K_1)) = \hat{K}_{12}$ then return 1 else return 0

Then an *advantage* of B is defined as

$$\mathbf{Adv}_{\mathsf{M}}^{\mathsf{cna}}(B) \stackrel{\text{def}}{=} \Pr[\mathbf{EXP}_{\mathsf{M}}^{\mathsf{cna}}(B) = 1]$$

We define an *advantage* of the M for any integers $t' \geq 0$. via

$$\mathbf{Adv}_{\mathsf{M}}^{\mathsf{cna}}(t') \stackrel{\text{def}}{=} \max_B \{\mathbf{Adv}_{\mathsf{M}}^{\mathsf{cna}}(B)\}$$

The maximum is over all forgers B such that the "running time" of B is at most t'. ∎

The following theorem means that she cannot compute K_{12} from K_1, K_2, and K_{13} if the pseudorandom function is secure.

Theorem 1. Let $F : \{0,1\}^k \times \{0,1\}^\ell \to \{0,1\}^L$ be a family of functions, and q, t, and $t' \geq 1$ be integers. Let M be our scheme. Then

$$\mathsf{Adv}_{\mathsf{M}}^{\mathsf{cna}}(t') \leq \mathsf{Adv}_{\mathsf{F}}^{\mathsf{prf}}(q, t) + \frac{1}{2^L}$$

where $q = 2$ and $t = t' + O(k)$.

Proof) Let B be an attacker attacking M. We construct a forger A_B attacking F. Consider the following experiment.

Distinguish A_B^f

 $K_1, K_2 \overset{R}{\leftarrow} \{0,1\}^k$
 $y \leftarrow f(K_1)$
 Compute $K_{13} = F(K_2, y)$
 Run attacker B
  $B \leftarrow K_1, K_2, K_{13}$
  $\hat{K}_{12} \leftarrow B$
 Compute $F(K_2, K_1)$
 $z \leftarrow f(F(K_2, K_1))$
 If $z = \hat{K}_{12}$ then return 1 else return 0

Here A_B is running B and we have

$$\Pr[\, f \overset{R}{\leftarrow} F : A_B^f = 1] \geq \mathbf{Adv}_M^{cna}(B)$$

$$\Pr[\, f \overset{R}{\leftarrow} Rand^{\ell \to L} : A_B^f = 1] = \frac{1}{2^L}$$

$$\mathbf{Adv}_F^{prf}(A_B) \geq \mathbf{Adv}_M^{cna}(B) - \frac{1}{2^L}$$

Inequality of the theorem is obtained as follows:

$$\mathbf{Adv}_M^{cna}(t') = \max_B \{\mathbf{Adv}_M^{cna}(B)\}$$

$$\leq \max_B \{\mathbf{Adv}_F^{prf}(A_B)\} + \frac{1}{2^L}$$

$$\leq \max_A \{\mathbf{Adv}_F^{prf}(A)\} + \frac{1}{2^L}$$

$$= \mathbf{Adv}_F^{prf}(q, t) + \frac{1}{2^L}.$$

Here, A_B makes two oracle queries $q = 2$ and the running time of A_B is $t = t' + O(k)$.   ■

5.3 Performance Analysis

The communication cost is still similar to that of the basic ACP scheme. Moreover, our scheme improves security without additional communication overhead, whereas Secrecy Amplification after the basic ACP scheme requires additional one.

  To compute the pair-wise key, our scheme requires addition computation cost of two pseudorandom functions for each node. One of the primary motivations for pseudorandom functions is to model block ciphers. Here, the computation cost of the pseudorandom function is the same as one of the symmetric encryption functions. Two pseudorandom function computations for each node are added in order to design our scheme as one having provable security. If we do not require the security proof, the pair-wise keys may be defined by $K_{12} = H(K_1, K_2, K_3)$, while still being intuitively secure.

5.4 Common Neighbor Nodes

In our scheme, if two neighbor nodes have two or more common neighbors, they choose one of them. On the other hand, if two neighbor nodes have no common neighbor node, they may compute $K_{12} = F(K_2, K_1)$ as their pair-wise key. However, this case has a low probability as computed by (1).

$$Pr[no\ common\ neighbor\ node] = \left(1 - \frac{2}{\pi}\ \tan^{-1}\sqrt{8} + \frac{\sqrt{8}}{9\pi}\right)^{d-1} \qquad (1)$$

Here, the probability will be 0.03 if d is 5.

6 Practical Sensor Network Applications and Benefits

So far, we suggested the modified key infection scheme and analyzed the security and performance of our scheme. Now, we will look at a possible application area of our scheme and summarize its economic aspects.

6.1 Two Major Categories of Sensor Network Applications

Sensor network applications in extremely hostile and mission critical environments. One application area where sensor networks would be pervasively adopted is for military or mission critical environments. Sensor nodes equipped with chemical material detectors might be placed into a chemically polluted area in order to sense and send dispersed chemical material information, and measure damage, thus they help the command force deploy proper anti-chemical aids. Sensor nodes aided with visual monitoring capacity can be dropped into combat areas where it is impossible or even too dangerous for soldiers to operate directly. Some sensor nodes would be deployed in order to monitor activities of the terrorists for a long period.

The sensed data from sensor nodes are invaluable to the deployer, but the counterparts do not want to be monitored or leak their information. This reasoning leads to the conclusion that the monitored parties will try to protect themselves proactively. They will monitor their areas in order to detect the existence of any unknown sensor nodes or any suspicious activities - for example, RF signals. Once detecting any suspicious sensor nodes, they will launch attacks against the sensor nodes, or capture some nodes physically. Depending on the type of attacks, the resulting impact on the sensor networks might be sabotaging, exposing secret keys, injecting data to mislead the overall operation, or even exposing location of the command center to the enemy. Failure to detect the sensor networks decreases chances of success and survival, hence, it is good to assume that very active attacks against sensor networks exist all through the life of sensor networks; from the physical deployment stage to the end of the nodes' life.

When highly active attacks are expected, a high degree security mechanism is imperative to ensure confidentiality, integrity, and availability. In order to build secure sensor networks, key management schemes, secure routing protocols, and other countermeasures should be designed and implemented considering the active attack model, where various attacks are possible from the beginning physical deployment phase of sensor networks. Currently, most sensor network security researchers design key management protocols based on this attack model.

Sensor network applications in less hostile, commodity environments.
Another application area of sensor networks is in commodities. Market and technology researchers anticipate that sensor networks will be used pervasively in our daily life and industrial areas innovating the way we collect data. Some applications will monitor factory instrumentation, pollution levels, freeway traffic, and the structural integrity of buildings [8]. Sensor networks are expected to detect and thus prevent forest fire. Endangered animals can be tracked by pervasively deployed sensor nodes giving useful data back to the zoologists. Other sensor nodes would be used as key components in home networks. They would measure and provide the current temperature and humidity. They could even be used to protect homes from burglars by detecting unexpected movement.

Threats or attempts to attack this type of sensor network are not serious compared to those networks used in mission critical environments. However, this does not mean that commodity sensor networks need no security at all. Sensor nodes are still exposed to physical node capture. Sensed data could be eavesdropped by adversaries and replayed at some later time. False data might be injected to make the sensor networks malfunction. Adversaries might be able to induce what he wants from traffic of sensor nodes. The risks from successful attacks range from short-time malfunctioning to endangering human lives. In other words, when the desired level of security cannot be achieved, the economic value of sensor network applications are also decreased.

So, how much security should sensor networks have? If commodity sensor network applications are hardened by the same security mechanisms used for critical applications, the variety of applications will be decreased. There always exist trade-offs between the level of security and usability. Thus, the proper level of security should be considered when designing key management for sensor networks.

Reasonable threats, or an attack model, in commodity sensor network applications should be first defined in order to determine the proper level of security mechanisms, especially key management. Threats during key setup phase rarely exist. This is because attacks during that phase are possible only if either the attack is aimed to a specific target or the attacker has been waiting for an anonymous victim. However, the probability of being a random victim of attackers is low enough to accept the risk during the deployment phase. As noted in [1], we expect that most commodity sensor network applications are categorized as this type. This assumption enables the applications to be more flexible and usable, and also gives more opportunities to keep key management simple and efficient.

6.2   Economic Benefits and Advantages

The assumption of our scheme, a loose attack model during key setup, makes our key management more efficient, practical, and even simpler. Our key management scheme can save resources such as CPU utilization, memory, and communications. Thus, more resources can be allocated to other components of the sensor nodes - routing, data processing, and application specific algorithms or protocols. Also, node lifetime will increase due to less resource consumption. We can focus more on the efficiency of operating sensor nodes after deployment which is significantly important and needs more attention. Economically, the price of sensor nodes could potentially decrease since they consume fewer resources. Low price is one of the key factors allowing sensor nodes to be used pervasively.

7   Conclusion

We have proposed a modified scheme which has no vulnerability under the ACP attack model [1]. Our scheme has the following properties. First, the security of our scheme has an improvement above the basic ACP scheme without additional communication overhead. Second, we prove the security of our scheme for non-critical commodity sensor networks. Finally, our scheme is suitable to a large network because it does not require a base station or large amount of memory for each node. Therefore, our scheme can be pervasively used for practical sensor network applications.

References

1. R. Anderson, H. Chan, and A. Perrig, "Key Infection : Smart Trust for Smart Dust", In 12th IEEE International Conference on Network Protocols, October. 2004.
2. I. F. Akyildiz, W. Su, Y. Sankarasubramaniam, and E. Cayirci, "A survey on sensor networks", IEEE Communications Magazine, Vol. 40, No. 8, pp. 102-114, August 2002.
3. M. Bellare, J. Kilian, and P. Rogaway, "The security of the cipher block chaining message authentication code", Journal of Computer and System Sciences, Vol. 61, No. 3, pp. 362-399, December 2000.
4. R. Blom, "An optimal class of symmetric key generation systems", In Proceedings of EUROCRYPT '84, LNCS Vol. 209, pp. 335-338, 1985.
5. C. Blundo, A. D. Santis, A. Herzberg, S. Kutten, U. Vaccaro, and M. Yung, "Perfectly-secure key distribution for dynamic conferences", In Proceedings of CRYPTO '93, LNCS Vol. 740, pp. 471-486, 1993.
6. D. W. Carman, P. S. Kruus, and B. J. Matt, "Constraints and approaches for distributed sensor network security", NAI Labs Technical Report 00-010, September 2000.
7. H. Chan, A. Perrig, and D. Song, "Random key predistribution schemes for sensor networks", In IEEE Symposium on Security and Privacy, pp. 197-213, May 2003.
8. H. Chan and A. Perrig, "Security and Privacy in Sensor Networks", IEEE Computer, Vol.36, No.10, pp. 103-105, October 2003.

9. W. Du, J. Deng, Y. S. Han, S. Chen, and P.K. Varshney, "A Key Management Scheme for Wireless Sensor Networks Using Deployment Knowledge", In Proceedings of the IEEE INFOCOM '04, pp. 586-597, March 2004.

10. W. Du, J. Deng, Y. S. Han, P.K. Varshney, J. Katz, and A. Khalili, "A Pairwise Key Pre-distribution Scheme for Wireless Sensor Networks", Accepted by ACM Transactions on Information and System Security, 2005.

11. L. Eschenauer and V. D. Gligor, "A key-management scheme for distributed sensor networks", In Proceedings of the 9th ACM conference on Computer and communications security, pp. 41-47, November 2002.

12. D. Liu, P. Ning, and R. Li, "Establishing Pairwise Keys in Distributed Sensor Networks", To appear in ACM Transactions on Information and System Security, April 2004.

13. A. Perrig, R. Szewczyk, V. Wen, D. Cullar, and J. D. Tygar, "SPINS: Security protocols for sensor networks", In Proceedings of the 7th Annual ACM/IEEE Internation Conference on Mobile Computing and Networking, pp. 189-199, July 2001.

Playing CSMA/CA Game to Deter Backoff Attacks in Ad Hoc Wireless LANs[*]

Jerzy Konorski

Gdansk University of Technology,
ul. Narutowicza 11/12, 80-952 Gdansk, Poland
jekon@eti.pg.gda.pl

Abstract. The IEEE 802.11 MAC protocol is vulnerable to selfish backoff attacks exploiting the constituent CSMA/CA mechanism. Administrative prevention of such attacks fails in wireless ad-hoc LANs which cannot mandate stations' behavior. We take a game-theoretic approach whereby stations are allowed to maximize their payoffs (success rates). Using a fairly accurate performance model we show that a noncooperative CSMA/CA game then arises with a payoff structure characteristic of a Prisoners' Dilemma. For a repeated CSMA/CA game, a novel SPELL strategy is proposed. If the stations are rational players and wish to maximize a long-term utility, SPELL deters a single attacker by providing a disincentive to deviate from SPELL.

1 Introduction

In the IEEE 802.11 MAC protocol [10], the CSMA/CA mechanism has a transmitting station back off for a random time upon a frame collision. A selfish *backoff attack* consisting in systematic selection of incorrectly short backoff times brings an attacker an unfairly large long-term bandwidth share and is easy to launch with user-accessible software [3]. Existing approaches to such attacks rely on culprit identification via anomaly detection and subsequent selective penalization using administrative leverage or reputation schemes [13], [17], random jamming [5], [16], or forced randomness of backoff times (entailing a major overhaul of IEEE 802.11) [6]. Behind all these approaches is the need for station authentication, which ad hoc networks typically lack. We address the problem in a game-theoretic framework and offer disincentives to potential attackers, while not affecting CSMA/CA operation and permitting station anonymity. The paper thus contributes in the still underexplored area of game-theoretic design of distributed MAC protocols resilient to selfish behavior (cf. [1], [2], [5], [12], [14], [16], [20], [21]).

---

[*] Effort sponsored by the Air Force Office of Scientific Research, Air Force Material Command, USAF, under grant nmber FA8655-04-1-3074. The U.S Government is authorized to reproduce and distribute reprints for Governmental purpose notwithstanding any copyright notation thereon. The views and conclusions contained herein are those of the author and should not be interpreted as necessarily representing the official policies or endorsements, either expressed or implied, of the AFOSR or the U.S. Government.

V.R. Sirotiuk and E. Chávez (Eds.): ADHOC-NOW 2005, LNCS 3738, pp. 127–140, 2005.
© Springer-Verlag Berlin Heidelberg 2005

In Sec. 2 we describe the model of a wireless LAN. Stochastic performance under a backoff attack is studied in Sec. 3. In Sec. 4 we define one-shot and repeated CSMA/CA games and in Sec. 5 propose a strategy for the latter that asymptotically punishes backoff attacks, while permitting to revert to honest behavior at any time. Sec. 6 concludes and outlines further research.

2 Network Operation

Consider a full-coverage ad hoc wireless LAN with N stations using the IEEE 802.11 DCF MAC-layer protocol [10] to access a single-channel. We assume negligible inter-station propagation delays and perfect channel and station operation. Furthermore we assume that: 1) the network operates under saturation i.e., each station is always ready to transmit user packets in the form of DATA frames, and 2) a station remains anonymous to non-recipients in that, based on a set of frames sensed on the medium, no non-recipient can either (a) deduce the identity of the sender or recipients, or even (b) reliably detect that any two frames have a common sender or recipient. Assumption 1 motivates selfish behavior: it is only under heavy load that the stations get interested in achieving larger-than-fair bandwidth shares. Part (a) of assumption 2 rules out selective punishment for a backoff attack, whereas part (b) renders backoff attacks undetectable by means of statistical traffic analysis. This assumption is justified by the fact that a station's physical identity (location) is untraceable due to mobility and the lack of tracking devices within the network, whereas its logical identity (e.g., IP address) can be obscured by fictitious MAC addresses (with any true identity information possibly encrypted end-to-end in the DATA frame payload). It seems reasonable to design countermeasures against backoff attacks under assumption 2, as they then make safe "upper bounds" for countermeasures relying on station identification.

Backoff attacks exploit CSMA/CA, the MAC-layer collision avoidance mechanism, which operates as follows. If *basic access* method is used then, upon a DATA frame transmission, a station waits until the medium has been sensed idle for a predefined DIFS interval. Then it sets a local *backoff counter* to a random value between 0 and $CW-1$, where CW is the current *contention window*. Initially, CW is set to a minimum w_{min}. The backoff counter is decremented each time the medium is sensed idle for a predefined *slot* interval, with the countdown frozen whenever the medium is sensed busy and resumed after it is sensed idle for another DIFS interval. When the backoff counter has reached zero the station understands it has captured a "virtual token" and transmits a DATA frame. If successful, the frame is responded to by a recipient's ACK frame and CW is reset to w_{min}. Otherwise i.e., if several stations simultaneously capture the "virtual token," their DATA frames collide, but each of them will have inferred a collision from the lack of ACK response. Subsequently it will double CW (unless it has reached a maximum w_{max}), set the backoff counter to a random value between 0 and $CW-1$, and start another countdown. Consecutive collisions beyond a certain limit cause a transmission abort; we neglect this feature in our model. A SIFS interval, shorter than DIFS, is specified to guarantee uninterrupted DATA+ACK exchange. If *RTS/CTS access* method is used, a station whose backoff counter has reached zero transmits a short RTS frame. A successful RTS frame is responded to by a recipient's CTS frame and a DATA frame transmission follows. If two or more stations simultaneously transmit RTS frames, they will infer a collision, double their *CW*s and set their backoff counters as above. A SIFS interval guarantees uninterrupted RTS+CTS+DATA+ACK exchange.

We complete the model by assuming that 3) at each station, the local w_{min} and w_{max} are configurable by user accessible software. Assumption 3 enables a backoff attack; see [17] and vendor data for details of manipulating these parameters in some existing network adapters. We rule out, however, selfish or malicious tampering with any other IEEE 802.11 parameters or functions e.g., DIFS or NAV; such tampering can be prevented or detected using simple means [3]. Define L so that $w_{max} = w_{min} \cdot 2^L$, thus the backoff scheme is fully characterized by a pair $w = <w_{min}, L>$ (e.g., $w = <16, 6>$ is recommended for 54 Mb/s OFDM-based or 1 Mb/s FHSS-based PHY layer and $w = <32, 5>$ is recommended for 11 Mb/s DSSS-based PHY layer).

3 Stochastic Performance Model

Below we develop two models to analyze saturated CSMA/CA under backoff attack using a station's success rate as a performance measure.

3.1 Performance of Saturated CSMA/CA

Bianchi [4] introduces a Markovian model of CSMA/CA under saturation, assuming a common configuration $<w_{min}, L>$ at all stations. His approach relies on an "independence hypothesis" whereby each station perceives the presence of the other stations through a constant probability c of a foreign transmission in any slot during backoff countdown. At each station, c can be interpreted as the rate of RTS or DATA frame transmission attempts by other stations, or as the inferred collision rate (percentage of own unsuccessful frame transmission attempts). Upon a slight modification of the model to reflect backoff freezing upon sensing the medium busy (cf. [23]), one arrives at the steady-state frame transmission rate at a station:

$$t(c) = \frac{1-c}{(w_{min}+1)/2 + (w_{min}/4) \cdot \sum_{l=1}^{L}(2c)^l - c}, \tag{1}$$

By the "independence hypothesis," c can be determined as the unique solution of

$$c = 1 - [1 - t(c)]^{N-1}. \tag{2}$$

The solution of (2) and the corresponding transmission rate, $\hat{c}$ and $\hat{t}$, determine each station's achieved bandwidth share [4], which is, however, PHY layer, access method, and DATA frame length specific. To remove dependence on these facets we measure a station's performance by its *success rate*:

$$b = \frac{\hat{t} \cdot (1-\hat{c})}{\hat{T}}, \tag{3}$$

where $\hat{T} = 1 - (1 - \hat{t})^N$ is the total frame transmission rate. Thus b represents the rate of successful frame transmissions per busy backoff slot. Clearly, b is indicative of the achieved bandwidth share.

3.2 Extension of Bianchi's Model

Suppose that each station n can configure its backoff scheme individually with $w_n = <w_{n,min}, L_n>$ possibly departing from the IEEE 802.11 standard. The smaller are $w_{n,min}$ and L_n, the "more selfish" is station n i.e., the more it deprives the other ones of transmission opportunity. In fact, unless $w_n = <1, L>$ is allowed, the most selfish backoff configuration is $w_n = <2, 0>$, whereby station n backs off for one slot only, or does not back off at all regardless of inferred collisions. Any "less selfish" configuration could be outperformed by stations adopting $<2, 0>$. Note that $w_n = <1, L>$ is impractical, as it would cut off all the other stations, including those with which station n might wish to communicate. We discuss $w_n = <1, 0>$ in Sec. 5.3.

We focus on a scenario where x stations out of N adopt a self-optimized backoff scheme configuration, while the other $N - x$ adopt the standard configuration prescribed for the present PHY layer. Thus x stations are *selfish* with $w_n = w_s = <2, 0>$, and $N - x$ stations are *honest* with $w_n = w_h$, where w_h is significantly "larger" than w_s (e.g., $w_h = <16, 6>$ or $<32, 5>$). As the inferred collision rates may differ at honest and selfish stations, we write c_h and c_s. An obvious extension of Bianchi's model leads to expressions for $t_h(c_h)$ and $t_s(c_s)$ similar to (1), and to a fixed-point relationship in c_h and c_s analogous to (2):

$$
\begin{aligned}
c_h &= 1 - [1 - t_h(c_h)]^{N-x-1}[1 - t_s(c_s)]^x \\
c_s &= 1 - [1 - t_s(c_s)]^{x-1}[1 - t_h(c_h)]^{N-x}
\end{aligned}, \quad x = 1, \ldots, N-1 \qquad (4)
$$

(the cases $x = 0$ and $x = N$ reduce to (2)). Numerical analysis shows that for $w_s = <2, 0>$ and any w_h prescribed for various PHY layers, (4) admits a unique solution $(\hat{c}_h, \hat{c}_s)$; let $(\hat{t}_h, \hat{t}_s)$ be the corresponding frame transmission rates. The success rates of a selfish and honest station are respectively $b_h = \hat{t}_h \cdot (1 - \hat{c}_h)/\hat{T}$ and $b_s = \hat{t}_s \cdot (1 - \hat{c}_s)/\hat{T}$, where $\hat{T} = 1 - (1 - \hat{t}_s)^x (1 - \hat{t}_h)^{N-x}$.

3.3 Mixed Model

Since it hinges upon the "independence hypothesis," the above extension of Bianchi's model becomes the less accurate, the more frequent and correlated are frame collisions. Given that selfish stations configure w_{min} and L much smaller than the standard recommends, such an extension yields inaccurate frame transmission and collision rates at selfish stations, particularly if they are few. Assuming $w_s = <2, 0>$, a *mixed model* described below better captures CSMA/CA under saturation, as it alleviates the "independence hypothesis" within the selfish stations.

The mixed model replaces all x selfish stations one "selfish aggregate" representing. The latter is modeled by a one-dimensional Markov chain with state

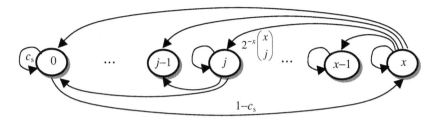

Fig. 1. State transitions for the "selfish aggregate"

space $\{0, ..., x\}$ depicted in Fig. 1. Each state encodes the number of selfish stations with expired backoff counters, hence attempting to transmit an RTS or DATA frame in the present slot. For legibility, transitions from generic and boundary states are only shown. From state $k > 0$, states $j = 0, ..., k$ can be directly reached; each such transition corresponds to exactly j out of k previously transmitting stations setting the backoff counter to zero for an immediate retransmission attempt. The self-loop at state 0 reflects all selfish stations freezing their backoff counters at one as they sense a frame transmission attempt by some honest stations; this happens with probability c_s, now interpreted as the rate of frame transmission attempts by all the honest stations. Finally, the transition from state 0 to state x occurs when all the selfish stations decrement their backoff counters upon sensing the medium idle. Instead of finding the steady-state probabilities we note that for the selfish stations,

$$t_s(c_s) = \frac{\Gamma_x}{\Gamma_x + c_s/(1-c_s)+1}, \tag{5}$$

$$b_s = \frac{2 \cdot (\Psi_x/x) \cdot (1-c_s)}{\Gamma_x + c_s/(1-c_s)}, \tag{6}$$

where Γ_x is the first-passage time from state x to state 0 and Ψ_x is the probability that the passage does not omit state 1. The $c_s/(1 - c_s) + 1$ term in the denominator of (5) is the average sojourn time in state 0 (due to honest stations' transmission attempts followed by an idle slot), and the 2 in the numerator of (6) is the average sojourn time in state 0 (consecutive successful attempts by a selfish station). The following recurrence relationships can be deduced from Fig. 1:

$$\Gamma_x = 1 + 2^{-x} \sum_{j=0}^{x} \binom{x}{j} \Gamma_j, \qquad x \geq 1 \tag{7}$$

$$\Psi_x = 2^{-x} \sum_{j=0}^{x} \binom{x}{j} \Psi_j, \qquad x \geq 2 \tag{8}$$

with $\Gamma_0 = 0$, $\Psi_0 = 0$, and $\Psi_1 = 1$. Recurrences of this type arise in the analysis of incomplete digital trees and can be solved using Poisson transforms [19]. Calculating along the lines of [18], [19] we get for $x \geq 1$ and the initial conditions stated:

$$\Gamma_x = \sum_{j=0}^{x} (-1)^{j+1} \binom{x}{j} \frac{1}{1-2^{-j}}, \tag{9}$$

$$\Psi_x = x\left[1+\sum_{j=1}^{x-1}(-1)^j\binom{x-1}{j}\frac{1}{2-2^{-j}}\right].\tag{10}$$

The success rates of honest stations obtain as in Sec. 3.2 using (1) and solving

$$
\begin{aligned}
c_h &= 1-[1-t_h(c_h)]^{N-x-1}[1-t_s(c_s)]\\
c_s &= 1-[1-t_h(c_h)]^{N-x}
\end{aligned},
\qquad x = 1, \ldots, N-1 \tag{11}
$$

(the cases $x = 0$ and $x = N$ reduce to (2)). Like (4), (11) admits a unique solution.

Further we indicate the dependence of success rates on N and x. Table 1 presents $b_h(N, x)$, $b_s(N, x)$, and the total success rate $b_\Sigma(N, x) = (N - x)\cdot b_h(N, x) + x\cdot b_s(N, x)$ obtained from the mixed model ($N = 10$ and 20, $w_h = $ <16, 6>). The results were validated via Monte Carlo simulation, with 95% confidence intervals within 1% of the sample averages. Analysis using extended Bianchi model was also carried out. Table 1 shows relative errors of $b_s(N, x)$ produced by both models with respect to simulation (shaded entries). Note that while the mixed model yields a good match with simulation, it generally underestimates $b_s(N, x)$; still, the approximation is far better than that of extended Bianchi model.

Table 1 shows that for any N and $x < N$, $b_s(N, x + 1)$ and $b_h(N, x)$ decrease in x and

$$b_s(N, x + 1) > b_h(N, x),\tag{12}$$

$$b_s(N, N) < b_h(N, 0).\tag{13}$$

That is, a station always benefits by playing selfish, yet too many selfish stations become worse off than they would be if all of them were playing honest. The critical number of selfish stations at which that happens is close to $N/2$. Before we consider the above results from a game-theoretic angle, let us point to three features invariant across a wide range of N. First, $b_h(N, 0)$ is distinctly nonzero, while $b_h(N, x) \approx 0$ for $x > 0$; second, $b_s(N, x)$ varies significantly with x if x is not too large; finally, $b_\Sigma(N, x)$ varies much less significantly with N than with x. Using these and by observing its own and the total success rate, each station n can infer x with a certain granularity, even though it has no means of knowing N or w_m for $m \neq n$, and the

Table 1. Success rates obtained from the mixed model

x	$b_h(N, x)$ and $b_s(N, x)$ (%) with relative errors (%)								$b_\Sigma(N, x)$ (%)	
	N=10				N=20				N=10	N=20
0	8.1		mix	ext.B	3.8		mix	ext.B		
1	0.2	94.7	-4.4	-5.8	0.2	89.6	-8.7	-11.7	96.5	93.1
2	0.1	24.4	-2.5	29.5	0.1	23.6	-5.5	24.8	49.4	48.8
3	0.1	14.9	-1.4	16.2	0.1	14.6	-3.8	12.7	45.1	44.7
4	0	10.2	-0.9	10.8	0	10.0	-2.8	7.0	41.1	40.8
5	0	7.6	-0.7	7.0	0	7.5	-2.7	2.9	38.0	37.7
10		3.1	0.1	-4.3	0	3.0	-1.2	-9.2	30.5	30.4
20						1.3	0.3	-20.7		25.4

success rates may be observed inaccurately. If $w_n = w_s$, observation of b_n permits to distinguish the cases $x = 1, ..., x = x^*$, and $x > x^*$, where $x^* \geq 5$. If $w_n = w_h$, a similar distinction follows from the observed b_Σ, while observation of b_n yields a distinction between $x = 0$ and $x > 0$. (Note that even in our anonymous setting, b_Σ is observable via monitoring of successful DATA+ACK exchanges.)

4 CSMA/CA Game

Given the choice between w_h and w_s, each station pursues a maximum success rate independently of (i.e., not seeking binding agreements with) the other stations; yet the result depends not only on its own choice, but also the other stations'. Thus a noncooperative N-player *CSMA/CA game* arises, in which success rates are the payoffs. Below we recall a few basic notions of game theory [9] in the context of our network model. First we consider a one-shot game, in which selection of w_h or w_s is a single act performed simultaneously at all the stations.

Definition 1. A CSMA/CA game is a triple $(\{1,...,N\}, W, b)$, where $\{1,...,N\}$ is the set of stations, $W = \{w_h, w_s\}$ is the set of feasible actions (backoff scheme configurations), and $b: W^N \to \mathbf{R}^N$ is a payoff function. Each station n selects $w_n \in W$ and subsequently receives a payoff (success rate) $b_n(w)$ dependent on the action profile $w = (w_n, w_{-n})$, where w_{-n} represents the opponent profile i.e., the actions selected by all the stations besides n.

Definition 2. An action w is a *best reply* to w_{-n} if $b_n(w, w_{-n}) \geq b_n(w', w_{-n})$ for all w' $\in W$. Let $BR(w_{-n})$ denote the set of best replies to w_{-n}. A *Nash equilibrium* (NE) is an action profile $w = (w_n, w_{-n})$ at which $w_n \in BR(w_{-n})$ for all $n = 1, ..., N$.

A NE is an action profile where each station has selected a best reply to the opponent profile, hence one from which no station has an incentive to deviate unilaterally. Such an outcome is expected when all the stations are rational (i.e., only interested in maximizing own payoffs) and their rationality is common knowledge [9].

Definition 3. An action profile $w \in W^N$ is *fair* if $b_1(w) = ... = b_N(w)$. A fair action profile w is *efficient* if $b_n(w) > b_n(w')$ for any other fair action profile w'.

In the CSMA/CA game, $b_n(w_n, w_{-n}) = b_h(N, x)$ if $w_n = w_h$ and $b_n(w_n, w_{-n}) = b_s(N, x)$ if $w_n = w_s$, where x is the cardinality of $\{n = 1, ..., N | w_n = w_s\}$. As seen from Table 1, the only fair action profiles are all-w_h and all-w_s (all stations selecting w_h and w_s, respectively), of which only the former is efficient, cf. (13). Thus all-w_h is a desirable outcome from the viewpoint of global design as it corresponds to a cooperative (live-and-let-live) scenario. Unfortunately, it is not a NE: as seen from (12), $BR(w_{-n}) = \{w_s\}$ for any n and any w_{-n}, implying that the unique (and inefficient) NE is all-w_s. It follows that the CSMA/CA game is a multiplayer Prisoners' Dilemma [22].

We now wish to model stations that learn from past experience and, in pursuit of a longer-term goal, may incline to honest play in search for efficient action profiles. Consider a *repeated CSMA/CA game*, each stage of which is an instance of one-shot

CSMA/CA game. In stage $k = 1, 2, \ldots$, station n selects $w_n^k \in \{w_s, w_h\}$ so that the resulting action profile is $w^k = (w_1^k, \ldots, w_N^k)$, whereas the received stage payoffs are

$$b_n^k = \begin{cases} b_s(N, x^k), \text{if } w_n^k = w_s \\ b_h(N, x^k), \text{if } w_n^k = w_h, \end{cases} \tag{14}$$

with $x^k = |\{m \mid w_m^k = w_s\}|$. Station n's *strategy* σ_n specifies for each k the probability of selecting $w_n^{k+1} = w_s$ given the current *play path* $(w^1, \ldots, w^k)$ i.e., $\sigma_n: \Pi \to [0, 1]$, where Π is the set of all finite-length play paths. Along with the strategy profile $\sigma = (\sigma_n, \sigma_{-n})$, any current play path $\pi \in \Pi$ induces a probability distribution $\mu(\sigma_n, \sigma_{-n}; \pi)$ of future b_n^k, and hence the average stage payoff $E_{\mu(\sigma_n, \sigma_{-n}; \pi)} b_n^k$. The long-term *utility* station n wishes to maximize is a liminf-type asymptotic [11]:

$$u_n(\sigma_n, \sigma_{-n}; \pi) = \liminf_{k \to \infty} E_{\mu(\sigma_n, \sigma_{-n}; \pi)} b_n^k, \tag{15}$$

where $\liminf_{k \to \infty} a_k = \lim_{k \to \infty} \inf\{a_k, a_{k+1}, \ldots\}$. An ideal strategy σ^* satisfies

$$b_h(N, 0) = u_n(\sigma^*, (\sigma^*, \ldots, \sigma^*); \pi) \geq u_n(\sigma, (\sigma^*, \ldots, \sigma^*); \pi) \tag{16}$$

for all $n = 1, \ldots, N$, and any strategy σ and play path $\pi \in \Pi$. The equality in (16) implies that ultimately the live-and-let-live (all-w_h) scenario prevails, whereas the inequality states that the strategy profile all-σ^* is a *subgame perfect* NE [9].

5 SPELL Strategy

In this section we present a strategy that satisfies (16), called *Selfish Play to Elicit Live-and-let-Live* (SPELL). A station deviating from SPELL cannot improve its utility if the others play SPELL. If, however, it does deviate and finds its payoffs worsening (perhaps after an initial improvement), it can revert to SPELL at any time, ending up in a live-and-let-live scenario.

5.1 Description

SPELL is configured with integer parameters Y and M, and a family P_r, $r = 1, 2, \ldots$, of complementary probability distribution functions over positive integers. The play proceeds in *spells*, each lasting a random number of stages drawn from P_r i.e., Pr[spell lasts $\geq i$ stages] = $P_r(i)$. If a spell lasts at most Y stages then w_h is selected throughout; otherwise w_s is selected in all but the last Y stages (Fig. 2). A station playing SPELL maintains an r-counter and a q-counter. Occasionally, q is disengaged, whereupon the station selects w_h in each stage until q is engaged again; this is controlled by the x^k inferred from b_n^k and b_Σ^k as explained in Sec. 3 (a stage should span enough CSMA/CA contentions to produce valid b_n^k and b_Σ^k estimates). When engaged, q is set to a random number drawn from P_r, decremented after each stage and upon reaching zero set to another random number drawn from P_r, so that w_s is selected when $q > Y$ and w_h otherwise. Until the station quits the game (e.g., has nothing to transmit), r can only be incremented. At the start of the game SPELL initializes r and engages q, and subsequently cycles through the following steps:

1) play out successive spells using q and P_r until $x^k \leq M$, whereupon disengage q;
2) play honest until $x^k > 0$, whereupon engage q;
3) increment r.

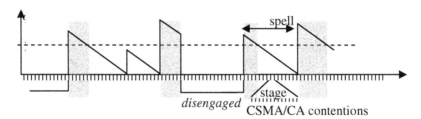

Fig. 2. SPELL operation (shaded areas indicate stages where w_s is selected)

If no station deviates from SPELL, step 2 is executed indefinitely, once entered, producing a persistent all-w_h. Otherwise, step 3 is executed at times and the idea is to favor longer spells as r increases, thereby punishing a deviator. As $0 < M \leq x^*$ should hold (cf. Sec. 3.3), only $x^k = 0$, 1..M, and $> M$ need to be distinguished.

5.2 Properties

Observe that at station n, a play path up to stage k only affects the current counter values, r_n^k and q_n^k. Thus checking (16) for all $\pi \in \Pi$ amounts to checking for all possible (r_n^k, q_n^k). Let supp(P_r) = $\{i \mid P_r(i) - P_r(i + 1) > 0\}$ be the support of P_r.

Proposition: $\sigma^* =$ SPELL satisfies the equality and inequality in (16) if, respectively, (i) for each $r = 1, 2, \ldots$, supp(P_r) is finite and contains relatively prime numbers, and (ii) for each $i = 1, 2, \ldots$, $P_r(i)$ increases and tends to one as r increases.

Proof: To prove (i) assume that from stage k on all the stations play SPELL, the current settings being (r_n^k, q_n^k), $n = 1,\ldots,N$. If some stations are executing step 2 in stage k then either $x^k = 0$, so that in stage $k + 1$ all the other stations join in step 2 and the assertion follows trivially, or $x^k > 0$, in which case all the stations execute step 1 in stage $k + 1$. Since no station deviates from SPELL, step 3 is never executed, so the r_n^k remain constant. It follows that $(q_1^l,\ldots,q_N^l)_{l=k}^\infty$ is an N-dimensional Markov chain with a finite state space. States where $|\{n \mid q_n^l > Y\}| \leq M$ are absorbing (the stations enter step 2). Moreover, they are accessible from all the other states, rendering the latter transient. To see that note that the value q_n^k at station n returns with positive probability after any number of stages of the form $\sum_{i \in \text{supp}(P_{r_n^k})} l_i \cdot i$, where l_i are nonnegative integers. By our assumption, the set of such numbers includes all consecutive integers above a certain minimum [7]. Hence, for a large enough K, the transition from $(q_1^k,...,q_N^k)$ to any $(q_1^{k+K},...,q_N^{k+K})$ occurs whenever q_n^k has returned after $K-(q_n^k - q_n^{k+K})$ stages i.e., with positive probability. Let the random variable $S(q_1^k,...,q_N^k)$ represent the number of stages to absorption, given the settings in stage k. Then $\Pr[S(q_1^k,...,q_N^k) \le s]$ tends to one as s increases [8]. The proof concludes by noting that there exists a bounded $\overline{b}$ such that $E_{\mu(\sigma_n,\sigma_{-n};\pi)} b_n^{k+s} = b_h(N,0) \cdot \Pr[S(q_1^k,...,q_N^k) \le s] + \overline{b} \cdot (1 - \Pr[S(q_1^k,...,q_N^k) \le s])$.

To prove (ii) let station N (say) deviate from SPELL, causing the other stations to enter step 3 of SPELL. If this happens only finitely many times, the assertion follows from part (i), so assume otherwise. For any i and $\varepsilon > 0$ there exists an $\overline{r}(i)$ such that $P_{\overline{r}(i)}(Y+i) > \sqrt[N-1]{1-\varepsilon}$. Let the random variable $\overline{k}$ represent the stage in which the SPELL stations enter step 3 for the $\overline{r}^{\text{th}}$ time, and pick any $k > \overline{k}$. Then with probability not less than $\vartheta(i) = \sum_{j=1}^{\infty} \Pr[k - \overline{k} = j][P_{\overline{r}(i)}(Y+j)]^{N-1}$ all the SPELL stations still select w_s in stage k. If station N also selects w_s to maximize its payoff (by (12)) then $E_{\mu(\sigma_N,\sigma_{-N};\pi)} b_N^k = \vartheta(i) \cdot b_s(N,N) + [1 - \vartheta(i)] \cdot \overline{b}$, where $\overline{b}$ is bounded. Taking a sufficiently large i one gets $\vartheta(i)$ arbitrarily close to $1 - \varepsilon$, so that as k increases, $E_{\mu(\sigma_N,\sigma_{-N};\pi)} b_N^k$ tends to $b_s(N, N)$. By (13), this is less than $b_h(N, 0)$. □

Since (i) and (ii) above are in terms of P_r only, neither M nor Y is critical to the correctness of SPELL. This leaves room for performance optimizations, cf. Sec. 5.4.

5.3 Modified SPELL

With $w_{n,\min} = 1$ excluded, all-w_h is a unique NE of the CSMA/CA game. Is it reasonable to launch a backoff attack by not invoking the backoff scheme at all i.e., selecting $w_n = w_g = <1, 0>$? Clearly, this would leave all stations besides n with a zero success rate, while station n would land 100%, provided that it were the only one to select $<1, 0>$ (hence "g" for greedy); otherwise it too would get zero. Suppose that the payoffs, now denoted b', reflect not only the success rates, but also transmission cost. Let this cost be negligible except when a station is not the only one to select w_g (i.e., spends all the power on frame collisions), in which case it perceives a "success rate" of $b_C < 0$. Let x, y and $N - x - y$ stations select w_s, w_g, and w_h, respectively. Then $b'_n = b_n$ if $y = 0$; otherwise $b'_n = 0$ if $w_n = w_h$ or $w_n = w_s$, $b'_n = 100\%$ if $w_n = w_g$ and $y = 1$, and $b'_n = b_C$ if $w_n = w_g$ and $y > 1$. Taking $W' = \{w_g, w_s, w_h\}$ we redefine the CSMA/CA game as $(\{1,...,N\}, W', b')$. We see that the game is no longer a Prisoners' Dilemma: any action profile with $y = 1$ (and no other) is a NE. Such degenerate action profiles, however, are not as compelling as is the unique all-w_s NE in the original game. Consider that the stations may seek a best reply to their beliefs as to the opponents' imminent play; the outcome of the game then depends on the stations' sophistication. For example, first-order sophistication might consist in selecting a best reply to the opponents' best replies to the current profile, rather than to the current profile itself (see [15] for a more systematic exposition and generalization). For a given action profile $(w_n, \boldsymbol{w}_{-n})$, denote by $O^{BR}(\boldsymbol{w}_{-n})$ the set of opponent profiles v in which $v_m \in BR(\boldsymbol{w}_{-m})$ for $m \neq n$ i.e., the opponents have selected best replies to the current action profile. A *first-order sophistication equilibrium* (1SE) is any action profile $\boldsymbol{w}$ in which $w_n \in \bigcup_{v \in O^{BR}(\boldsymbol{w}_{-n})} BR(v)$ for all $n = 1, \ldots, N$. While in the original game the only 1SE coincides with the NE, in the present game the set of equilibria is much larger: any action profile is a 1SE. Thus a strategy coping with possible selections of w_g must not be simply a replica of SPELL with $W = \{w_g, w_h\}$, since w_s is likely to be played too.

To cope with w_g we propose to modify SPELL as follows. First note that based on the observed b_n^k and b_Σ^k, a station n selecting $w_n^k = w_h$ can distinguish the cases $x^k + y^k = 0$ and $x^k + y^k > 0$ at the end of stage k. Likewise, if $w_n^k = w_s$ then $y^k > 0$, $(y^k = 0$ and $x^k \leq M)$, and $(y^k = 0$ and $x^k > M)$ can be distinguished, and if $w_n^k = w_g$ then $y^k = 1$ and $y^k > 1$ can be distinguished. Having initialized r and engaged q, modified SPELL performs steps 1 through 3 below. They are similar to those of SPELL except that in step 1, w_s is selected when $q > Y$ and w_h otherwise, whereas in step 3, w_g is selected when $q > Y$ and w_s otherwise.

1) play out successive spells using q and $\boldsymbol{P}_r$ until either $(y^k = 0$ and $x^k \leq M)$, whereupon disengage q and go to step 2, or $y^k > 0$, whereupon engage q, increment r, and go to step 3;
2) play honest until $x^k + y^k > 0$, whereupon engage q, increment r, and go to step 1;
3) play out successive spells using q and $\boldsymbol{P}_r$ until $y^k \leq 1$, whereupon disengage q and go to step 2.

Again, if no station deviates from modified SPELL then step 2 is ultimately executed. Note that in step 1, detection of a station selecting w_g brings about a painful punishment: all SPELL stations enter step 3 and toggle between w_g and w_s until $y^k \leq 1$, which may take quite long to happen. Like SPELL, modified SPELL satisfies (16); the proof is similar to that in Sec. 5.2.

5.4 Performance

Sample payoff trajectories ($E_{\mu(\sigma_n, \sigma_{-n}; \pi)} b_n^k$ vs. k) produced by modified SPELL have been obtained via Monte Carlo simulation to illustrate the validity of (16). $N = 10$ and $b_C = -5\%$ were kept fixed, and stage payoffs for $y = 0$ were taken from Table 1. The probability distributions $\boldsymbol{P}_r$ were uniform over $\{1, \ldots, D_r\}$, with $D_r = D_0 + r \cdot d$. For each station, D_0 was drawn at random. The depicted curves emerged after averaging over 1000 runs with the same initial spell counter settings.

Fig. 3a illustrates convergence to $b_h(N, 0) = 8.1\%$ when all stations play modified SPELL. The initial settings were chosen at random (with about a third of stations executing each of the three steps of SPELL); this imitates an arbitrary play path prior to the start of simulation. The combination of Y and d is far more critical to the speed of convergence than M. Taking too small Y and too large d causes a long punishment for what the SPELL stations perceive as deviations from SPELL (e.g., when $x^k + y^k > 0$ while own spell counter is disengaged), and what in fact is the impact of the initial settings. This may lead to extremely slow convergence to $b_h(N, 0)$.

In Fig. 3b, one station persistently deviates from modified SPELL, while the other initially execute step 2. To upper bound the utility of a conceivable deviator strategy, we have experimented a somewhat unrealistic "good reply" to SPELL that relies on ideal prediction, *prior* to stage k, of the opponent profile in stage k. It selects w_g when all the other stations are to select w_h; otherwise always selects w_s except when no other station is to select w_g, and M or less are to select w_s, in which case it selects w_h. That is, the deviator selects w_g sparingly for fear of punishment, and tries not to prevent entering step 2 with a view of landing a 100% success rate in the next stage.

That such a strategy works well against less smart strategies is visible in the two upper curves: against a "deficient" SPELL with $d = 0$, the deviator achieves a stable success rate of around 15% or 25% depending on Y i.e., twice or three times the fair success rate $b_h(N, 0)$. Yet when $d > 0$, the deviator fares much worse than it would if

6 Conclusion

We have studied a selfish *backoff attack* in a simplified form, whereby the configuration of the backoff scheme at each station is restricted to greedy, selfish and honest. Regarding stations' success rates as payoffs we have shown that a noncooperative CSMA/CA game then arises with a payoff structure characteristic of a Prisoners' Dilemma. The fact that the unique NE of such a game is inefficient and that the success rates decrease as the number of attackers increases permits to design a simple strategy for the repeated CSMA/CA game, called SPELL. Assuming that the stations are rational players and wish to maximize a long-term utility, SPELL deters a single attacker by inclining to greedy or selfish play, yet at any time retains the ability to end up in a live-and-let-live scenario with all the stations select the honest configuration. Note that SPELL can be implemented without affecting the IEEE 802.11 MAC standard. Among the issues not addressed in this paper is the ability of SPELL to deter multiple attackers. It seems obvious that a simultaneous backoff attack by multiple stations can only be beneficial if some cooperation between them exists. This is not unthinkable and deserves further research.

References

1. E. Altman, R. El Azouzi, T. Jimenez: Slotted Aloha as a Game with Partial Information. Computer Networks 45 (2004) 701–713
2. E. Altman, A. Kumar, D. Kumar, R. Venkatesh: Cooperative and Non-Cooperative Control in IEEE 802.11 WLANs. INRIA Tech. Rep.5541 (2005)
3. J. Bellardo, S. Savage: 802.11 Denial-of-Service Attacks: Real Vulnerabilities and Practical Solutions. In: Proc. USENIX Security Symposium (2003)
4. G. Bianchi: Performance Analysis of the IEEE 802.11 Distributed Coordination Function. IEEE J. Selected Areas Commun. SAC-18 (2000) 535-547
5. M. Cagalj, S. Ganeriwal, I. Aad, J.-P. Hubaux: On Selfish Behavior in CSMA/CA Networks. In: Proceedings of IEEE Infocom '05, Miami, FL (2005)
6. A. A. Cardenas, S. Radosavac, J. S. Baras: Detection and Prevention of MAC Layer Misbehavior in Ad Hoc Networks. Univ. of Maryland Tech. Rep. ISR TR 2004-30 (2004)
7. J. Chang: Stochastic Processes Notes. www.stat.yale.edu/~chang/251/mc.pdf
8. W. Feller: An Introduction to Probability Theory and its Applications. Wiley 1966
9. D. Fudenberg, J. Tirole: Game Theory. MIT Press 1991
10. IEEE Standard for Information Technology – Wireless LAN Medium Access Control (MAC) and Physical Layer (PHY) Specifications, ISO/IEC 8802-11 (1999)
11. V. Knoblauch: Computable Strategies for Repeated Prisoner's Dilemma. Games and Economic Behavior 7 (1994) 381-389
12. J. Konorski: Multiple Access in Ad-Hoc Wireless LANs with Noncooperative Stations. Springer-Verlag LNCS 2345 (2002) 1141–1146
13. P. Kyasanur, N. H. Vaidya: Detection and handling of MAC layer Misbehavior in Wireless Networks. In: Proc. Int. Conf. on Dependable Systems and Networks (2003)
14. A. B. MacKenzie, S. B. Wicker: Game Theory and the Design of Self-configuring, Adaptive Wireless Networks. IEEE Comm. Magazine 39 (2001) 126-131
15. P. Milgrom, J. Roberts: Adaptive and Sophisticated Learning in Normal Form Games. Games and Economic Behaviour 3 (1991) 82-100

16. O. Queseth: Cooperative and Selfish Behaviour in Unlicensed Spectrum Using the CSMA/CA Protocol.In: Proc. Nordic Radio Symposium (2004)
17. M. Raya, J.-P. Hubaux, I. Aad: DOMINO: A System to Detect Greedy Behavior in IEEE 802.11 Hotspots. In: Proc. MobiSys Conference (2004)
18. R. Rom, M. Sidi: Multiple Access Protocols Performance and Analysis. Springer 1991
19. W. Szpankowski: Average Case Analysis of Algorithms on Sequences. Wiley 2001
20. G. Tan, J. Guttag: The 802.11 MAC Protocol Leads to Inefficient Equilibria. In: Proceedings of IEEE Infocom '05, Miami, FL (2005)
21. S. H. Wong, I. J. Wassell: Application of Game Theory for Distributed Dynamic Channel Allocation. In: Proc. 55th IEEE Vehicular Technology Conference (2002) 404-408
22. X. Yao: Evolutionary Stability in the n-Person Iterated Prisoners' Dilemma. BioSystems 39 (1996) 189–197
23. E. Ziouva, T. Antonakopoulos: CSMA/CA Performance under High Traffic Conditions: Throughput and Delay Analysis. Computer Comm. 25 (2002) 313-321

Design of a Hard Real-Time Guarantee Scheme for Dual Ad Hoc Mode IEEE 802.11 WLANs

Junghoon Lee, Mikyung Kang, and Gyungleen Park

Dept. of Computer Science and Statistics, Cheju National University,
690-756, Ara-1 Dong, Jeju Do, Republic of Korea
{jhlee, mkkang, glpark}@cheju.ac.kr

Abstract. This paper proposes and analyzes a message scheduling scheme and corresponding capacity allocation method for the distributed hard real-time communication on dual Wireless LANs. By making the superframe of one network precede that of the other by half, the dual network architecture can minimize the effect of deferred beacon and reduce the worst case waiting time by half. The effect of deferred beacon is formalized and then directly considered to decide polling schedule and capacity vector. Simulation results executed via ns-2 show that the proposed scheme improves the schedulability by 36 % for real-time messages and allocates 9 % more bandwidth to non-real-time messages by enhancing achievable throughput for the given stream sets, compared with the network whose bandwidth is just doubled.

1   Introduction

Wireless communication technology is gaining a wide-spread acceptance for distributed systems and applications in recent years[1]. As both speed and capacity of wireless media such as WLAN (Wireless Local Area Network) increase, so does the demand for supporting time-sensitive high-bandwidth applications such as broadband VOD (Video On Demand) and interactive multimedia[2]. One of the promising application areas of wireless technology is a wireless sensor network, where the periodically sampled data are delivered to the appropriate node within a reasonable deadline to produce meaningful data[3]. That is, the message has a hard real-time constraint that it should be transmitted within a bounded delay (as long as there is no network error). Otherwise, the data is considered to be lost, and the loss of hard real-time message may jeopardize the correctness of the execution result or system itself. After all, a real-time message stream needs the guarantee from the underlying network that its time constraints are always met in advance of the system operation.

The IEEE 802.11 was developed as a MAC standard for WLAN[4]. The standard consists of a basic DCF (Distributed Coordination Function) and an optional PCF (Point Coordination Function). In fact, PCF, which essentially needs the BS (Base Station) installed, is not widely implemented in WLAN because it is not designed for the distributed environment and its centralized feature constrains the operation of WLAN. As contrast, the DCF is based on

V.R. Sirotiuk and E. Chávez (Eds.): ADHOC-NOW 2005, LNCS 3738, pp. 141–152, 2005.

CSMA/CA (Carrier Sense Multiple Access with Collision Avoidance) protocol for the transmission of asynchronous non-real-time messages, aiming at improving their average delivery time and network throughput. However, a real-time guarantee can be provided only if a deterministic contention-free schedule is developed[5], and the polling mechanism is the most common way to achieve this goal. Though there are so many options to elect the poller station, we will denote the poller node as PC (Point Coordinator), assuming that any station can play a role of PC. For example, the first real-time station in the polling list may have the responsibility to initiate CFP (Contention Free Period)[6]. As a result, WLAN runs both CFP and CP (Contention Period) phases alternatively.

Hard real-time guarantee depends strongly on the underlying polling policy, that is, how polling interval starts, how PC picks the node to poll next and how long a node transmits for each poll. As an example, weighted round-robin scheme makes the coordinator poll each node one by one, and the polled node transmits its message for a time duration, or weight, predefined according to its traffic characteristics. While this scheme makes the guarantee mechanism simple and efficient, it suffers from poor utilization due to polling overhead as well as *deferred beacon problem*[7]. Deferred beacon problem means a situation that a non-real-time message puts off the initiation of some CFPs behind the scheduled start time. Though the maximum amount of deferment is bounded, it seriously degrades the schedulability of real-time messages.

In order to overcome this poor schedulability, we are to propose a network architecture and corresponding network access scheme based on dual wireless LANs. As the wireless channel supports multiple frequency bands, it is not unusual for a group of components belonging to a common control loop to be linked by two or more networks. The dual link system can reduce the worst case waiting time for each node to be blocked until it can send its own message, improving the guarantee ratio for real-time messages as well as allocating more network bandwidth to non-real-time messages.

This paper is organized as follows: After issuing the problem in Section 1, Section 2 introduces the related works focusing on real-time communications on the wireless medium. Section 3 describes network and message models, respectively. Section 4 proposes a new bandwidth allocation scheme for IEEE 802.11 WLAN based on the round-robin polling policy, and this scheme is extended to the dual networks in Section 5. Section 6 shows and discusses performance measurement results and Section 7 finally concludes this paper with a brief summarization and the description of future works.

2 Related Works

Several MAC protocols have been proposed to provide bounded delays for real-time messages while providing a reasonable performance for a non-real-time data over a wireless channel. However, these protocols are typically based on a frame-structured access comprised of a contention part and reservation part, demanding a time synchronization among the nodes on a network, so they make it impossible or unrealistic to apply to the IEEE 802.11 WLAN. For example, Choi and Shin suggested a unified protocol for real-time and non-real-time communications in wireless networks[8]. In their scheme, a BS polls a real-time mobile station according to the non-preemptive EDF (Earliest Deadline First) policy. The BS also polls the non-real-time message according to the modified round-robin scheme rather than a standard CSMA/CA protocol to completely eliminate message collision.

M. Caccamo and et. al have proposed a MAC that supports deterministic real-time scheduling via the implementation of TDMA (Time Division Multiple Access), where the time axis is divided into fixed size slots[5]. Referred as *implicit contention*, their scheme makes every station concurrently run the common real-time scheduling algorithm to determine which message can access the medium. However, for this implicit contention, every node must schedule all messages in the network, making it difficult to scale to large networks. Such schemes cannot be ported to the IEEE 802.11 WLAN standard, as they ignored the mandatory CSMA/CA part of WLAN. In addition, the access right is given to a station not by the actual message arrival but just by the estimation. The discrepancy may invalidate the guaranteed schedule.

Most works that conform to the IEEE standard are aiming at just enhancing the ratio of timely delivery for soft multimedia applications, rather than providing a hard real-time guarantee. DBASE (Distributed Bandwidth Allocation/Sharing/Extension) is a protocol that supports both synchronous and multimedia traffics over IEEE 802.11 *ad hoc* WLAN[6]. The basic concept is that each time a real-time station transmits a packet it will also declare and reserve the bandwidth demanded at the next CFP. Every station collects this information and then calculates its actual bandwidth at the next cycle. This scheme can be ported to WLAN standard, but it does not provide a hard real-time guarantee as it does not directly consider the maximal bandwidth request. Moreover, unless a packet is correctly received by all member stations, they can have different schedules.

As another example, in BB (Black Burst) contention scheme, stations sort their access rights by jamming the channel with pulses of energy before sending their packet[9]. Since packets must be transmitted repeatedly in a constant interval, sending burst of energy for each packet will waste considerable network bandwidth. Moreover, the BB contention is not a regular scheme defined in IEEE 802.11 standard. In addition, a distributed fair scheduling is proposed for a wireless LAN by N. Vaidya, and et. al[10]. Its essential idea is to choose a backoff interval that is proportional to the finish tag of a packet to be transmitted. With fair scheduling, different flows share a bandwidth in proportion to their weights. However, this scheme cannot provide hard real-time guarantee due to unpredictable collisions between two or more packets that have same tag.

In the non-academic effort, IETF has produced new drafts, EDCF (Enhanced DCF) and HCF (Hybrid Coordination Function) to replace the CSMA/CA based and centralized polling based access mechanism, respectively[11]. No guarantees of service are provided, but EDCF establishes a probabilistic priority mechanism to allocate bandwidth based on traffic categories. According to HCF, a hybrid controller polls stations during a contention-free period. The polling grants a station a specific start time and a maximum transmit duration.

3   Network Model

3.1   Ad Hoc Mode IEEE 802.11 WLAN

In IEEE 802.11 WLAN, the independent basic service set (IBSS) is the most basic type, where IEEE 802.11 stations are able to communicate directly. This mode of operation is possible when 802.11 WLAN stations are close enough to form a direct connection without preplanning. This type of operation is referred to as an *ad hoc* network. For example, in smart-rooms and hot-spot networks, wireless access-enabled stations in a small area share the medium[12]. As the area is small, the wireless hosts pervade through the entire network and are all within each other's transmission range. Channeling all data through a single intermediate node, such as a base station, is inefficient in this network.

The PC periodically attempts to initiate CFP by broadcasting a *Beacon* at regular intervals calculated according to a network parameter of *CFPRate*. Round-robin is one of the commonly used polling policies in CFP and a polled station always responds to a poll immediately whether it has a pending message or not, as shown in Fig. 1. Every node is polled once per a polling round, and a polling round is either completed within one superframe, or spread over more than one superframe. If the CFP terminates before all nodes have been completely polled, the polling list is resumed at the next node in the ensuing CFP cycle.

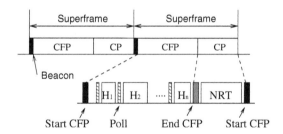

Fig. 1. Time axis of wireless LAN

As shown in Fig. 2, the transmission of the beacon by the coordinator depends on whether the medium is idle at the time of TBTT (Target Beacon Transmission Time). Only after the medium is idle the coordinator will get the priority due to the shorter IFS (InterFrame Space). Thus the delivery of a beacon frame can get delayed if another packet has already occupied the network. This situation can sometimes disable the timely delivery of hard real-time messages, as shown in Fig. 2. The last part of message, M, can be delivered after its deadline has already expired. This figure also demonstrates that the maximum amount of deferment coincides with the maximum length of a non-real-time packet. The probability of beacon deferment depends on the arrival time distribution of the non-real-time message. This probability gets higher as more non-real-time traffic flows in the network, since a non-real-time message longer than remaining CP time occupies the network more likely. Regardless of how often the beacon is delayed, any deferment can do harm to the hard real-time message that cannot tolerate any deadline miss unless it is caused by the network error.

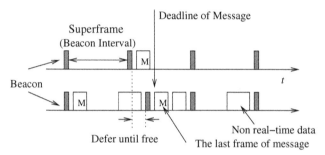

Fig. 2. Deferred beacon problem

3.2 Message Model

The traffic characteristics such as period and message size are usually known before the system operation. That is, the stream set does not change, or changes infrequently even if it changes. If the stream set is fixed, the network schedule is calculated offline as in the typical real-time system. In addition, each node can send PC an association request message to get associated with the PC via CP period. This message contains a PCF related real-time traffic information, such as period and transmission time. Every time the stream set changes, a new schedule, or bandwidth allocation procedure is performed and broadcasted to the participating nodes with the support of mechanisms such as mode change[13].

In real-time communication literature, the term real-time traffic typically means *isochronous* (or synchronous) traffic, consisting of message streams that are generated by their sources on a continuing basis and delivered to their respective destinations also on a continuing basis[13]. There are n streams of real-time messages, S_1, S_2, ..., S_n, while each stream can be modeled as follows: A message arrives at the beginning of its period and must be transmitted by the end of period. The period of stream, S_i, is denoted as P_i, while the maximum length of a message as C_i. The first message of each stream arrives at time 0. The destination of message can be either within a cell or outside a cell. The outbound messages are first sent to the router node such as AP and then forwarded to the final destination. A proper routing and reservation protocol can provide an end-to-end delay guarantee[2].

4   Bandwidth Allocation

4.1   Basic Assumptions

As is the case of other works, we begin with an assumption that each stream has only one stream, and this assumption can be generalized with the virtual station concept[14]. By allocation, we mean the procedure of determining capacity vector, $\{H_i\}$, for the given superframe time, F, as well as message stream set, $\{S_i(P_i, C_i)\}$. It is desirable that the superframe time is a hyperperiod of each stream's period and it is known that a message set can be made harmonic by reducing some periods by at most half[1]. At the other extreme, F may be a fixed network parameter that can harldy changable. So we assume that the superframe time is also given in priori, focusing on the determination of capacity vector. H_i limits the maximum time amount for which S_i can send its message. A polling round may spread over more than one superframe, however, for simplicity, we assume that a polling round completes within a single superframe and the assumption will be eliminated later. PC attempts to initiate CFP every F interval. When its timer expires, PC broadcasts a beacon frame after waiting short IFS from the instant it detects the medium free. Besides, the error control issues like packet retransmission, are not considered in this paper, because they are different problems and out of the scope of this paper. We just assume that some existing error control or QoS degrading schemes can be integrated into our framework[15]. Consequently, though there are many issues one needs to consider in wireless networks, we mainly focus on a significant performance issue, that is, timeliness[5].

4.2   Allocation Procedure

Allocation procedure determines capacity vector, $\{H_i\}$, for the given superframe time and message stream set. Fig. 1 in Section 3.1 also illustrates that the slot size is not fixed, namely, $H_i \neq H_j$ for different i and j. At this figure, a message of size C_i is generated and buffered at regular intervals of P_i, and then transmitted by H_i every time the node receives poll from PC. Let δ denote the total overhead within a superframe including polling latency, IFS, exchange of beacon frame, and the like, while D_{max} the maximum length of a non-real-time data packet. For a minimal requirement, F should be sufficiently large enough to make the polling overhead insignificant. In addition, if P_{min} is the smallest of set $\{P_i\}$, F should be less than P_{min} so that every stream can meet at least one superframe within its period. For each superframe, not only the start of CFP can be deferred by up to D_{max}, but also at least a time amount as large as D_{max}, should be reserved for a data packet so as to be compatible with WLAN standard. After all, the requirement for the superframe time, F, can be summarized as follows:

$$\sum H_i + \delta + 2 \cdot D_{max} \leq F \leq P_{min} \tag{1}$$

The number of polls a stream meets is different period by period. Meeting hard real-time constraints for a station means that even in the period which meets the smallest number of polls, the station can transmit message within its deadline. Fig. 3 analyzes the worst case available time for S_i. In this figure, a series of superframes are contained in P_i and each period can start at any spot of the superframe. Intuitively, the station meets the smallest number of polls in the period which starts just after the end of its slot.

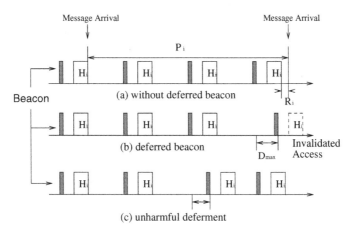

Fig. 3. Worst case analysis

In this figure, R_i is the residual obtained by dividing P_i by F, namely, $R_i = P_i - \lfloor \frac{P_i}{F} \rfloor \cdot F$. Without the deferred beacon, the CFP starts regularly at each time interval of F. In that case, for S_i, the least bound of network access within P_i is $\lfloor \frac{P_i}{F} \rfloor$, as illustrated in Fig. 3(a). This figure also implies that S_i requires at least 3 H_i's within P_i to meet its time constraint. On the contrary, if we take into account the deferred beacon, the deferred start of the last superframe may deprive S_i of one access, missing the message deadline, if R_i is less than D_{max}, as shown in Fig. 3(b). If R_i is greater than D_{max}, the number of available time slots is not affected by the delayed start of superframe. It doesn't matter whether an intermediate superframe is deferred or not as shown in Fig. 3(c). As a result, the minimum value of available transmission time, X_i is calculated as Eq. (2). Namely,

$$
\begin{aligned}
X_i &= (\lfloor \tfrac{P_i}{F} \rfloor - 1) \cdot H_i \qquad if (P_i - \lfloor \tfrac{P_i}{F} \rfloor \cdot F) \le D_{max} \\
X_i &= \lfloor \tfrac{P_i}{F} \rfloor \cdot H_i \qquad\qquad Otherwise
\end{aligned}
\tag{2}
$$

For each message stream, X_i should be greater than or equal to C_i ($X_i \ge C_i$). By substituting Eq. (2) for this inequality, we can obtain the least bound of H_i capable of meeting the time constraint of S_i.

$$
\begin{aligned}
H_i &= \frac{C_i}{(\lfloor \frac{P_i}{F} \rfloor - 1)} \qquad if (P_i - \lfloor \tfrac{P_i}{F} \rfloor \cdot F) \le D_{max} \\
H_i &= \frac{C_i}{\lfloor \frac{P_i}{F} \rfloor} \qquad\quad Otherwise
\end{aligned}
\tag{3}
$$

The allocation vector calculated by Eq. (3) is a feasible schedule if it satisfies Ineq. (1). By this, we can determine the length of CFP (T_{CFP}) and that of CP (T_{CP}) as follows:

$$T_{CFP} = \sum H_i + \delta, \quad T_{CP} = F - T_{CFP} \geq D_{max} \tag{4}$$

This calculation is easily fulfilled just with simple arithmetic operations. In addition, the size of time slot is different for each stream, resulting in a better network utilization compared to other schemes based on fixed size slots. Finally, as this allocation scheme generates a larger T_{CP} for the given F, the network can accommodate more non-real-time messages.

4.3   Miscellanies

In case a polling round spreads over more than one superframe, say k superframes, each one can be marked as F_1, F_2, ..., F_k. The size of each superframe is F, while each includes its own CP duration and performs only a part of polling round. S_i receives poll once a $k \cdot F$ and the allocation formula can be modified by replacing F with $k \cdot F$ in Eq. (2). But the condition remains intact which checks whether a stream will be affected by a deferred beacon. After all, Eq. (2) can be rewritten as follows:

$$\begin{aligned} X_i &= (\lfloor \tfrac{P_i}{k \cdot F} \rfloor - 1) \cdot H_i \quad & if(P_i - \lfloor \tfrac{P_i}{F} \rfloor \cdot F) \leq D_{max} \\ X_i &= \lfloor \tfrac{P_i}{k \cdot F} \rfloor \cdot H_i \quad & Otherwise \end{aligned} \tag{5}$$

If a node has no pending message when it receives a poll, it responds with a null frame containing no payload. The rest of the polling in the superframe can be shifted ahead without violating the time constraint of their real-time messages if and only if all the subsequent nodes have their messages to send in that superframe. The CP in such a superframe, say *reclaimable superframe* can start earlier than those of other normal superframes. In addition, we can improve the probability of bandwidth reclaim by rearranging the polling order.

5   Scheduling on Dual Networks

The poor utilization of the polling scheme stems from the fact that a message cannot be transmitted on demand, as there is no way for a node to request a immediate poll to the coordinator. The node should wait up to one superframe time in the worst case. However, dual network architecture can reduce this waiting time by half in round-round style network. It means the environment where each node has two radio and can operate on each channel. Hence, the dual network architecture is analogous to the dual processor system, as both network and processor can be considered as an active resource. The jobs or messages are scheduled on two equivalent processors or networks. Priority-driven scheduling on multiple resources induces scheduling anomaly that less tasks can meet their deadlines even with more resources. In addition, it is known that off-line scheduling for this environment is a NP-hard problem[13]. Contrary to priority-driven scheduling, the transmission control scheme proposed in Section 4 can easily calculate the feasible and efficient schedule for the dual or multiple resource systems.

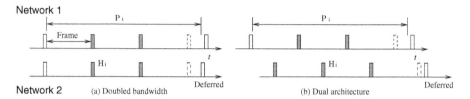

Fig. 4. The time axis of dual wireless LAN

It is desirable for the two networks to have a common coordinator that schedules and synchronizes them. A general non-PC node sends its message on any poll from either network. If we make a superframe progress simultaneously or randomly as shown in Fig. 4(a), S_i may simultaneously lose one access on each network, two in total. Beacon deferment on one channel is independent of that on the other, so the amount of delay is different on each channel. For a period, a node may lose 0, 1, or 2 accesses, and the real-time scheduling system should consider the worst case access scenario. This case is analogous to the case when network bandwidth is just doubled. So X_i is formalized as shown in Eq. (6).

$$
\begin{aligned}
X_i &= 2 \cdot \lfloor \tfrac{P_i}{F} \rfloor \cdot H_i & if(R_i \geq D_{max}) \\
X_i &= 2 \cdot (\lfloor \tfrac{P_i}{F} \rfloor - 1) \cdot H_i & otherwise
\end{aligned}
\tag{6}
$$

On the contrary, if we make one network precede the other by $\frac{F}{2}$ as illustrated in Fig. 4(b), one loss can be saved, improving the worst case behavior. In addition, the first poll on the second network is not lost, either. As F is reduced by half, so X_i can be calculated by replacing F in Eq. (2) with $\frac{F}{2}$. However, though D_{max} is smaller than F, it can be larger than $\frac{F}{2}$. In that case, the second network also loses the last access due to deferred beacon. So X_i is formalized as in Eq. (7).

$$
\begin{aligned}
X_i &= \lfloor \tfrac{2 \cdot P_i}{F} \rfloor \cdot H_i & if(R_i > D_{max}) \\
X_i &= (\lfloor \tfrac{2 \cdot P_i}{F} \rfloor - 1) \cdot H_i & if(R_i \leq D_{max} \leq R_i + \tfrac{F}{2}) \\
X_i &= (\lfloor \tfrac{2 \cdot P_i}{F} \rfloor - 2) \cdot H_i & if((R_i + \tfrac{F}{2}) \leq D_{max})
\end{aligned}
\tag{7}
$$

In addition, Fig. 5 illustrates how many polls may be lost according to the value of D_{max} when there are m networks.

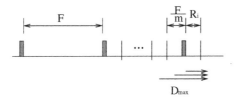

Fig. 5. Access times in case of m networks

The figure shows the last two superframes of the first network, with the access times of the other networks also marked in the time axis. The access times are evenly spaced by $\frac{F}{m}$. If R_i is greater than D_{max}, the least bound on the number of access times is not affected irrespective of deferred beacon. However, D_{max} is greater than R_i, one access can be lost if the start of CFP is delayed. One more access is lost when D_{max} exceeds $R_i + \frac{F}{m}$ on the $(m-1)$-th network. After all, each time D_{max} increases by $\frac{F}{m}$, one more access is lost. After all, X_i can be generalized into the case of more than 3 networks.

$$
\begin{aligned}
X_i &= \lfloor \tfrac{m \cdot P_i}{F} \rfloor \cdot H_i & if(D_{max} \geq R_i) \\
X_i &= (\lfloor \tfrac{m \cdot P_i}{F} \rfloor - 1 - \lfloor \tfrac{D_{max}}{\lfloor \frac{F}{m} \rfloor} \rfloor) \cdot H_i \; otherwise
\end{aligned}
\tag{8}
$$

6   Performance Measurement

TDMA-base schemes can be competitors to the proposed scheme in the sense that they can provide hard real-time guarantee[1][5][8], but, the comparison with them is nearly impossible because it requires so many assumptions on slot size, superframe operation, way to handle deferred beacon, and so on. Moreover, TDMA-based schemes generally did not consider CP interval, and their guarantee mechanism assumes the ideal condition of no deferred beacon. We measured the performance of the proposed scheme in the view of schedulability and achievable throughput via simulation using ns-2[16]. In the experiments, every time variable is aligned to the superframe time, and an initiation of superframe is deferred by from 0 to D_{max} exponentially.

For the first experiment on schedulability, we have generated 2000 stream sets whose utilization ranges from 0.68 to 0.70 for each network. The number of streams in a set is chosen randomly between 5 and 15. The period of each stream ranges from $5.0F$ to $10.0F$, while its message length from $0.3F$ to $5.0F$. We measured the schedulability, the ratio of schedulable stream sets to all generated sets, changing D_{max} from $0.0F$ to $0.14F$. Fig. 6 plots schedulability of proposed dual network architecture comparing with that of the network whose bandwidth is just doubled. The schedulability begins to drop abruptly at a certain point for each curve. The proposed scheme can schedule every stream set until D_{max} is $0.042F$ while that of doubled bandwidth network remains at 0.24. That is, the breakdown point shifts to the right, minimizing the protocol overhead of round-robin polling mechanism.

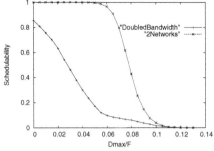

Fig. 6. Schedulability vs. D_{max}

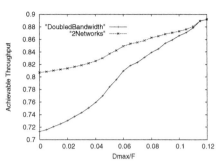

Fig. 7. Achievable throughput vs. D_{max}

We define *achievable throughput* as the virtual throughput for a given stream set assuming no collision even in CP. This can be calculated as the sum of utilization of real-time message stream set and ratio of average length of CP to F. Overallocation is the difference between the allocation, $\sum H_i$, and the actual requirement, U. After all, this parameter means how much overallocation is reduced to give more bandwidth to non-real-time messages. We address that our scheme provides tighter and smaller H_i's, as Eq. (7) is significantly weaker constraint than Eq. (6). According to the increase of D_{max}, more bandwidth is assigned to non-real-time traffic, because D_{max} is its minimum requirement of CP.

7 Conclusion

This paper has proposed and evaluated a bandwidth allocation scheme that enhances the schedulability of hard real-time streams via dual wireless network architecture, strictly observing IEEE 802.11 WLAN standard. By exploiting the round-robin style polling mechanism, network schedule and bandwidth allocation become much simpler, and they can check efficiently whether a stream is affected by deferred beacons. Additionally, by making the start times of both networks different by $\frac{F}{2}$ under the control of common coordinator, the proposed scheme improves the least bound of network access time for each message stream. Simulation results show that the capacity vector decided by the proposed scheme schedules more stream sets than the network whose bandwidth is just doubled, and that it can invite more non-real-time message to occupy the network, for the given experiment parameters.

As a future work, we consider an error control mechanism as well as a resource reclaiming schemes respectively. As for error control mechanism, we think that the architecture of dual networks can give a certain type of priority in coping with network errors, and the priority may be decided by the weight or urgency of messages, power management factors, and so on. In addition, we are now developing a bandwidth reclaiming scheme which allocates unused CFP time interval to non real-time traffic by extending the CP period. Grouping some nodes to be polled in a superframe and rearranging the polling order in that group can maximize the number of reclaimable superframes.

Acknowledgment

This research was supported by the MIC (Ministry of Information and Communication), Korea, under the ITRC (Information Technology Research Center) support program supervised by the IITA (Institute of Information Technology Assessment).

References

1. Carley, T., Ba, M., Barua, R., Stewart, D.: Contention-free periodic message scheduler medium access control in wireless sensor/actuator networks. Proc. IEEE Real-Time Systems Symposium (2003) 298–307
2. Mao, S., Lin, S., Wang, Y., Panwar, S. S., Li, Y.: Multipath video transport over wireless ad hoc networks. IEEE Wireless Communications (2005)
3. Li, H., Shenoy, P., Ramamritham, K., Scheduling communication in real-time sensor applications. Proc.10th IEEE Real-time Embedded Technology and Applications Symposium (2004) 2065–2080
4. IEEE 802.11-1999: Part 11-Wireless LAN Medium Access Control (MAC) and Physical Layer (PHY) Specifications. (1999)
5. Caccamo, M., Zhang, L., Sha, L., Buttazzo, G.: An implicit prioritized access protocol for wireless sensor networks. Proc. IEEE Real-Time Systems Symposium (2002)
6. Sheu, S., Sheu, T.: A bandwidth allocation/sharing/extension protocol for multimedia over IEEE 802.11 ad hoc wireless LANS. IEEE Journal on Selected Areas in Communications **19** (2001) 2065–2080
7. Lindgren, A., Almquist, A., Schenen, O.: Quality of service schemes for IEEE 802.11: A simulation study. 9-th Int'l Workshop on Quality of Service (2001)
8. Choi, S., Shin, K.: A unified wireless LAN architecture for real-time and non-real-time communication services. IEEE/ACM Trans. on Networking **8** (2000) 44–59
9. Sobrino, J., Krishakumar, A.: Quality-of-service in ad hoc carrier sense multiple access wireless networks. IEEE Journal on Selected Areas in Communications **17** (1999) 1353–1368
10. Vaidya, N., Bahl, P., Gupta, S.: Distributed fair scheduling in a wireless LAN. 6-th Annual Int'l Conference on Mobile Computing and Networking (2000)
11. Mangold, S., et. al.: IEEE 802.11e wireless LAN for quality of service. Proceedings of the European Wireless (2002)
12. Shah, S. H., Chen, K., Nahrstedt, K.: Dynamic bandwidth management for single-hop ad hoc wireless networks. ACM/Kluwer Mobile Networks and Applications (MONET) Journal **10** (2005) 199-217
13. Liu J.: Real-Time Systems. Prentice Hall (2000)
14. Mukherjee, S., Saha, D., Sakena, M. C., Triphati, S. K.: A bandwidth allocation scheme for time constrained message transmission on a slotted ring LAN. Proc. Real-Time Systems Symposium (1993) 44–53
15. Adamou, M., Khanna, S., Lee, I., Shin, I., Zhou, S.: Fair real-time traffic scheduling over a wireless LAN. Proc. IEEE Real-Time Systems Symposium (2001) 279–288
16. Fall, K., Varadhan, K.: Ns notes and documentation. Technical Report. VINT project. UC-Berkeley and LBNL (1997)

Selective Route-Request Scheme for On-demand Routing Protocols in Ad Hoc Networks[*]

Seung Jae Lee[1], Joo Sang Youn[1], Seok Hyun Jung[2], Jihoon Lee[3], and Chul Hee Kang[1]

[1] Department of Electronics and Computer Engineering Korea University,
5-Ka, Anam-Dong, Sungbuk-Ku, Seoul 136-701, Korea
{mountain3, ssrman, chkang}@widecomm.korea.ac.kr
[2] LG research center woomyen-dong seucho-Ku Seoul 137-724, Korea
hyoni@lge.com
[3] I-Networking Lab, Samsung Advanced Institute of Technology,
San 14-1, Nongseo-Ri, Kiheung-Eup, Yongin, Kyungki-do 449-712 Korea
Vincent.lee@samsung.com

Abstract. The on-demand routing protocols are appealing because of their lower routing overhead in bandwidth restricted mobile ad hoc networks, compared with Table-driven routing protocol. They introduce routing overhead only in the presence of data packets that need routes. However, the control overhead of on-demand routing protocols is increased by node mobility, node geographic density and traffic pattern density. In fact, this is undesirable feature for the scalable routing protocols whose control overhead should be under control to keep up with increasing offered load. The fundamental cause of these drawbacks is produced by flooding RouteRequest (RREQ) packet. As a solution for such a drawback of current on-demand routing schemes, we propose *Selective Request Scheme (SR scheme)*. In this protocol, the stability of local network can be improved by re-floods or discards RREQ. Therefore in dynamic environment and limited bandwidth, SR Scheme assists in discovery the robust route and reduces the number of flooded RREQ packet. We demonstrate the effectiveness of our enhancement by applying it to Ad hoc On-demand Distance Vector Routing (AODV). Simulation results show that proposed idea significantly reduces the control overhead and improves the performance and scalability of the routing protocols.

1 Introduction

Multi-hop ad hoc networks have been in the spotlight because they are self-creating, self-organizing, and self-administering without using any kind of infrastructure. With the absence of fixed infrastructure in ad hoc networks, nodes can communicate with one another in a peer-to-peer fashion. And each node in an ad hoc network is capable

[*] This paper was supported by the Samsung Advanced Institute of Technology, under the project "Research on Core Technologies ant its Implementation for Service Mobility under Ubiquitous Computing Environment".

V.R. Sirotiuk and E. Chávez (Eds.): ADHOC-NOW 2005, LNCS 3738, pp. 153–163, 2005.
© Springer-Verlag Berlin Heidelberg 2005

of moving independently and functioning as a router that discovers and maintains routes and forwards packets to other nodes. A key component of ad hoc networks is an efficient routing protocol, since all of the nodes in the network act as routers.

Current studies of ad hoc routing protocols can be classified into two approaches – Table-driven and on demand routing [3] [7]. Table driven protocols, which are commonly used in the wired networks, maintain routes by exchanging the route table periodically. In ad hoc networks, the routing table must be updated frequently enough to handle the dynamic topology changes. This constraint may involve a large amount of routing overhead caused by intensive exchanges of the route tables regardless of the actual needs for routes. On the other hand, the principal aim of on demand routing approaches (e.g., AODV, DSR) is to reduce the routing overhead with dynamic maintenance of routes. On- demand routing protocols can reduce control overhead at the expense of setup latency due to the route search. When node requires a route to a destination, a node initiates a route discovery process by flooding RouteRequest (RREQ) packet. It is continued until RREQ packet reaches the destination or the intermediate node with a route to destination. Each node that has received the RREQ packet is in charge of forwarding the packet to reach the destination no matter how much load this node currently has. Upon the reception of a route query packet, the destination sends the route reply packet to the source via the reverse path of the shortest route through which a route request packet passes.

If the demands of route queries are not high, on demand routing protocols work more successfully and effectively than table driven routing schemes. However on demand routing schemes have two drawbacks which degrade network performance. First, except for ABR (Associativity-Based Routing) [4], most on demand routing protocols have not considered the stability of the local network in route-discovery phase. Therefore route failure and reconstruction is frequently incurred. Second, production in common RREQ packet can aggravate congestion in the local network, because of generation of more routing packets. The fundamental cause of these drawbacks is that shortest path is only used as routing metric and the new RREQ is unconditionally broadcasted. The above observation motivate us to investigate SR schemes on demand routing protocols with following characteristics

First, in dynamic environment, SR scheme assists in discovering the robust route by limiting Route Request packet in unstable local network.

Second, SR scheme reduce the number of flooded RREQ packets which produce the congestion of the local network. Also it can reduce the waste of limited bandwidth.

Third, most of on-demand routing protocols can optionally make use of the SR scheme, because of using in the route-discovery phase.

Simulation results demonstrate that our approach improves the performance of AODV protocol. Also we feel sure that will improve the performance when our approach is applicable to DSR(Dynamic Source Routing protocol), ABR.

The organization of the rest part of the paper is as follows. We will present a brief review of the related works and motivations in Chapter 2, and described he detailed scheme in chapter 3, thereafter we demonstrate the contributions of our work through simulation results in Chapter 4. Finally, we draw out conclusions and discussion about the future work in Chapter 5.

2   Related Work and Motivation

Routing based on the stability of the wireless links has been proposed in [4], [13]. The Associativity-Based Routing (ABR) [4] is an on-demand routing protocols that consists of the following three phases-route discovery, route reconstruction, and route deletion phase. The novelty of this scheme is the route selection criteria based on degree of association stability. *Associativity* is measured by recording the number of control beacons that a node receives from its neighbors. During a route discovery phase, a query (= RREQ) arriving at the destination contains the associativity ticks along the route. By selecting nodes with high associativity ticks, the best route among multiple paths can be used for communication. The selected route is expected to be long-lived, thus it has less chance of route re-discovery in dynamic environment. However, ABR doesn't restrict to forward RREQ packet in intermediate node. It uses stability information only during the route selection process at the destination node. So it shares the same problem such as flooding RREQ packet. In addition to, the destination should wait for the arrival of all the possible query packets to select the best route.

The signal stability-based adaptive routing protocol Routing (SSA) [13] is an on-demand routing protocols that uses signal stability as the prime factor for finding stable routes. This protocol is beacon-based, in which the signal strength of the beacon is measured for determining link stability. The signal strength is used to classify a link as stable or unstable. However the main disadvantage of this protocol is that it puts a strong RREQ forwarding condition which results in RREQ failures. A failed RREQ attempt initiates a similar path-finding process for a new path without considering the stability criterion.

Such multiple flooding of RREQ packets consumes a significant amount of bandwidth in the already bandwidth-constrained network, and also increases the path setup time. Another disadvantage is that the strong links criterion increases the path length, as shorter paths may be ignored for more stable paths.

Each On-demand routing protocols has different features and operations in detail. Different major features are shown with reference to RREQ format, management of routing information and support of asymmetric link [7]. In case of RREQ packets, In DSR and ABR, RREQ or BQ includes the address of intermediate nodes in route record field or IN ID (intermediate identifier) field. In AODV, RREQ doesn't have the identification of intermediate nodes. In case of routing information, No routing information is maintained at the intermediate nodes in DSR. However, both ABR and AODV maintain routing table in intermediate nodes. Finally, in case of support of asymmetric link, DSR may support asymmetric link with piggyback the REPLY on a new RREQ. However, AODV and ABR can only support symmetric link. Surely, besides these differences, on-demand routing protocols have many differences such as use of beacon, format of data packet and etc... Nevertheless, most on-demand protocols have several common points. First, the route is established by flooding RREQ and Route Reply (RREP) packet. Second, when receiving the new RREQ, each node unconditionally re-floods RREQ. Finally, after destination node receives RREQs, it selects the shortest path as the best route. These common features make several drawbacks in on-demand routing protocols. First, it makes frequent route failure that node mobility isn't considered by route discovery process. Second, unconditional flooding RREQ incurs RREQ overhead.

These weak points of on-demand routing protocols motivate our research.

3 Selective Request Scheme

When each node receives the new RREQ packet in route discovery process, it operates the SR Scheme based on predefined associativity. In that, each node receiving the new RREQ selectively broadcasts or discards the RREQ according to a criterion for SR Scheme. If the node discards RREQ from a source, source doesn't discover any route along the node. Therefore, SR Scheme can protect unnecessary route disconnection in advance and considerably reduce RREQ overhead. To perform the SR Scheme, this scheme should make use of following information.

- *Who send RREQ or BQ?*
- *How stable is the connection with neighbor?*

First information is acquired by received RREQ. And, latter information is known from neighbor table made/updated by beacon signal. And, owing to simple modification of route discovery process, SR Scheme is applicable to various classes of on-demand routing protocols. We consider SR Scheme in DSR, ABR, and AODV in this paper.

3.1 Up-Node Detection with RREQ

To perform the SR Scheme, we designate the node receiving RREQ as down-node and the node sending RREQ as up-node. When each node receives RREQ, it has to know the node which sends RREQ packet. If down-node receives RREQ from the particular up-node, it selects whether it broadcasts RREQ based on associativity between itself and particular up-node.

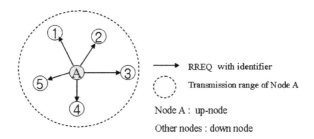

Fig. 1. Propagation of the RREQ - RREQ packet has the identification of Up-node

Table 1. Up-node identifier field (in RREQ)

DSR	ABR	AODV
Route record	IN (intermediate) ID	Marking method

The method to know the up-node differs from each on-demand routing protocol. In case of DSR, RREQ include *route record field* which has the identification for up-node [1]. Therefore, up-node is easily confirmed by route record in DSR RREQ packet. In case of ABR, because broadcast query (BQ) has the *IN ID (intermediate identifiers) field* [4], the detection of up-node is simple. Unfortunately, the RREQ packets of AODV don't have the identifier for up-nodes [4] [7]. However, this can be easily solved with *marking method*. i.e. before broadcasting RREQ, up-node marks its identification in RREQ packet.

3.2 Neighbor Table for SR Scheme

Neighbor information is a pre-requisite to the SR Scheme. Also it is used as the criterion for selection of flooding/discarding RREQ. The term neighbor indicates a node within the radio transmission range of another node. For link status sensing and neighborhood maintenance, each node periodically broadcasts a *hello beacon* to signify its existence. When neighbors receive a beacon, they update their neighbor table to reflect the presence of the sender. This mechanism was implemented as a MAC-level protocol [10] or a network layer mechanism [4]. This *neighbor table* is maintained by the soft state method. In that, if a node doesn't any beacon for a neighbor during predefined times, it deletes the neighbor in neighbor table. On the other hand, because ad hoc networks have limited bandwidth, the beaconing cost should be minimized by piggybacking technique. If a MAC-level protocol supports beaconing, the beaconing cost is more negligible since every RTS and CTS packet can also be used as beacons. A node's association with its neighbor can be measured by recording the number of received beacons from its neighbor [3]. Entries of the neighbor table are following:

- *Identification of neighbor node*
- *Assocociativity*: the connection relationship of a node with a neighbor. In that, this indicates the stability of the link between a node and a neighbor.
- *State of stability:* stable or unstable state. If $A>A_{TH}$, association between node A and neighbor node is stable. If $A<A_{TH}$, association is unstable state.

If the associativity of a neighbor exceeds a certain value, the association with the neighbor is the stable state. The certain value is the very *associtivity threshold (A_{TH})* which indicates periods of association stability. A_{TH} is a complex function of various physical parameters such as the transmission range, the maximum speed of mobile nodes, beacon interval, etc... That is further evaluated and possibly optimized. And associativity with each neighbor has a same value with each other. For example, node 2 has the value of 273 as associativity for node A.

Table 2. Neighbor table of node A ($A_{TH} = 250$)

Neighbor	Associativity (A)	State
1	75	Unstable
2	273	Stable
3	2	Unstable
4	405	Stable
5	358	Stable

3.3 Operation of SR Scheme

In on-demand routing protocols, when source node required a route to a destination, it initiates a route discovery process by broadcasting route request (RREQ) packet. And then, each node receiving new RREQ performs SR Scheme. The procedure of SR scheme is following:

- If a node (=down-node) receives the new RREQ, it performs SR Scheme.
- The down-node checks Up-node which has sent the new RREQ.
  - In case of DSR, using *route record* in RREQ
  - In case of ABR, using *IN ID* in BQ
  - In case of AODV, using *marked Up-node ID* in RREQ
- The down-node checks the associativity (A) with the up-node and compares A with A_{TH}.
- The down-node decides to re-flood RREQ *(criterion for selective request)*
  - In case of A < ATH, RREQ is discarded.
  - In case of A > ATH, RREQ is re-flooded.

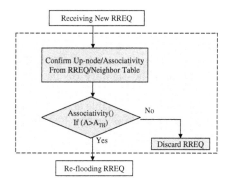

Fig. 2. The flow chart for SR Scheme operation

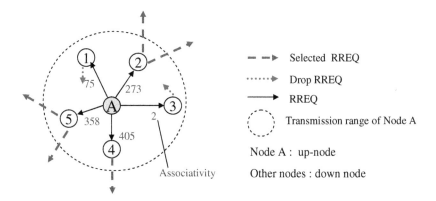

Fig. 3. Re-flooding RREQ after execution of SR Scheme

While each different advantage of on-demand routing protocols doesn't degrade, SR Scheme is applicable to on-demand routing protocols such as DSR, ABR, AODV, etc...

Figure 2 and 3 show the operation of SR Scheme. After receiving the new RREQ, each node executes SR Scheme and selectively rebroadcasts the RREQ into the network. As shown in Figure 2, node A has six neighbors in its transmission range. Thus, all neighbors (down-node) receive RREQ from node A (up-node) and perform SR Scheme based on their neighbor table. For example, when receiving the RREQ from node A, node 3 confirms node A as up-node, and it checks its neighbor table. As shown in Table 1, node 3 knows that the associativity with node A is smaller than its threshold (A_{TH} = 250). Therefore, node 3 don't re-flood RREQ packet.

4 Performance Evaluation

4.1 Simulation Model

A detailed simulation model based on *ns-2 is used* in our evaluation. In a recent paper the Monarch research group at Carnegie-Mellon University developed support for simulating multi-hop wireless networks complete with physical, data link, and medium access control (MAC) layer models on *ns-2*.

Data Link and Physical Layer Models. The distributed Coordination Function (DCF) of IEEE802.11 [11] for wireless LANs is used as the MAC layer protocol. The 802.11 DCF uses Request-To-Send (RTS) and Clear-To-Send (CTS) control packets [12] for unicast data transmission to a neighboring node. Data packet transmission is followed by an ACK. Broadcast data packet and the RTS control packets are sent using physical carrier sensing. An unslotted carrier sense multiple access (CSMA) technique with collision avoidance (CSMA/CA) is used to transmit these packets. The radio model uses characteristics similar to a commercial radio interface, Lucent's WaveLAN. WaveLAN is modeled as a shared media radio with a nominal bit rate of 2Mb/s and a nominal radio range of 250m. The protocols maintain a send buffer of 64packetes. It contains all data packets waiting for a route, such as packets for which route discovery has started, but no reply has arrived yet. To prevent buffering of packets indefinitely, packets are dropped if they wait in the send buffer for more than 30 sec. all packets (both data and routing) sent by the routing layer are queued at the interface queue until the MAC layer can transmit them. The interface queue has a maximum size of 50 packets and is maintained as a priority queue with two priorities each served in FIFO order. Routing packets get higher priority than data packets.

Traffic and Mobility Models. The mobility model is the random waypoint model in a rectangular field. The movement scenario is characterized by a pause time. Each node starts its movement from a random location to a random destination with a randomly chosen speed (uniformly distributed between 0-20 m/s). Once the destination is reached, another random destination is targeted after a pause. Every nodes repeats this behavior until the simulation ends. The *pause time* indicates the degree of mobility of the ad hoc network. Simulations are run for 1000 simulated seconds for 50 nodes in the 670m * 670m rectangular space.

Traffic sources are continuous bit rate (CBR). The source-destination pairs are spread randomly over the network. Only 512 byte data packets are use. The number of *source-destination pairs* is varied to change the offered load in the network.

The following parameters are varied in our simulation. The values are selected similarly to previous papers [7].

- The number of source-destination pairs: 10 and 40
- Pause time between the movements of each mobile node: 0, 30, 60, 120, 300, and 600 seconds

4.2 Simulation Results

We modify the routing mechanism of AODV to include function of our ideas, SR Scheme (Selective Request Scheme) and observe how much the performance of AODV is improved by our scheme. Also, we experiment the effect of associativity threshold, for 250 and 500. Surely, SR Scheme can be applied to other on-demand protocols with little restriction. However, because AODV routing protocol have a larger routing load than any other on-demand routing protocols, we apply SR Scheme to AODV. In particular, the major contribution to AODV's routing overhead is from route request (RREQ). Also, AODV don't consider the stability of network in route discovery.

Performance metrics. Three important performance metrics are evaluated:

- *Packet delivery fraction* – The ratio of the data packets delivered to the destinations to those generated by the CBR sources.
- *Normalized routing load* – The number of routing packets transmitted per data packet delivered at the destination.
- *Average end-to-end delay* – Difference between the time the agent in upper layer generates a data packet and the time the corresponding agent receives the data successfully. This includes all possible delays caused by route discovery latency, queuing delay at interface queue, propagation and retransmission delays in data link and physical layer.

Packet delivery fraction. The result of the packet delivery fraction is represented in Figure 4. For 10 sources, AODV and AODV-SR show similar results. However, AODV-SR certainly outperforms than AODV in 40 CBR sources. This is because a stable route is selected by SR Scheme and SR Scheme reduces overall routing load, in particular RREQ packets. On the other hand, when associativity threshold (A_{TH}) is 500 rather than 250, packet delivery fraction is slightly improved. This is because the larger A_{TH} don't make more frequently **route failure**.

Normalized routing load. Performance results on routing overhead are described in Figure 5 and 7. In spite of using additional hello beacons to make neighbor table, AODV with SR Scheme always has a lower routing load than AODV. As shown Figure 7(a), the major contribution to AODV's routing overhead is from RREQ packets. However, SR Scheme can considerably reduce the number of RREQs in route discovery process. As the mobility of nodes is higher and traffic sources are more, source nodes generate more RREQs for route reconstruction and route discovery. Therefore, in high mobility under high traffic load, AODV without SR Scheme incurs serious RREQ overhead. However, because AODV with SR Scheme selectively discards RREQ packets based on the associativity, it considerably reduces routing load. On the other hand, the large A_{TH} reduces more routing overhead than the small $A_{TH.}$

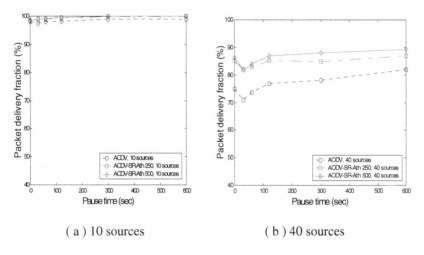

(a) 10 sources   (b) 40 sources

Fig. 4. Packet delivery fraction of AODV and AODV-SR

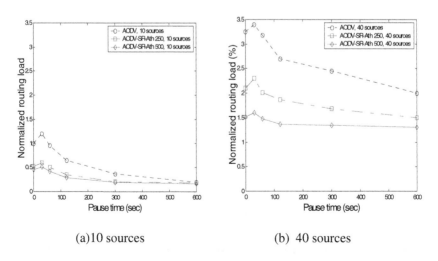

(a) 10 sources   (b) 40 sources

Fig. 5. Normalized routing load of AODV and AODV-SR

Average end-to-end delay. Average end-to-end delay includes all possible delays caused by route discovery latency, queuing delay at interface queue, propagation and retransmission delays in data link and physical layer. For 10 and 40 sources, the average delay of AODV and AODV-SR are very similar as shown in Figure 6. However, for a large pause time, AODV has smaller average delay than AODV-SR. the cause of this result is that queuing delay is a dominant factor in average end-to-end delay. Using SR Scheme keeps on the same route for a long time. Therefore, a large number of data packets are dropped in the interface queue. Nevertheless, the difference of average delay decrease at 40 sources and is negligible. Also, the processing delay for SR Scheme almost never affects average end-to-end delay.

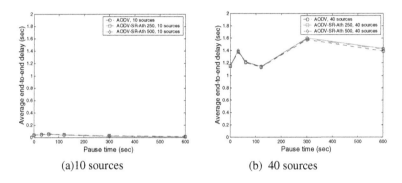

(a) 10 sources   (b) 40 sources

Fig. 6. Average end-to-end delay of AODV and AODV-SR

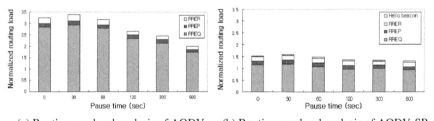

(a) Routing overhead analysis of AODV   (b) Routing overhead analysis of AODV-SR

Fig. 7. Routing overhead analysis of AODV and AODV-SR-A_{TH}=500 in 40 sources

5   Conclusion

With flooding-based on-demand route discovery in mobile ad hoc networks, many routing packets (i.e. RREQ) are propagated unnecessarily. To overcome the perform-ance limitations of on-demand routing protocols, we propose the novel routing scheme, Selective Request Scheme (SR Scheme). As each node receiving the new RREQ selectively re-floods RREQ based on associativity, source can find the robust route which doesn't incur a frequent route failure. As a result of execution of SR Scheme, RREQ packets are not transmitted along the low quality route which consists of unstable wireless links or mobile nodes. Therefore, SR Scheme can protect unnecessary route disconnection in advance and considerably reduce RREQ overhead. On the other hand, as proposed scheme is only used during route discovery process, it is applicable to most on demand routing protocols. We modify the routing mechanisms of AODV to include functions of our ideas, and observe how much the performance of AODV is enhanced by our scheme. In simulation results, SR Scheme considerably reduces the RREQ flooding overhead and improves packet delivery in dynamic environments as well. In the near future, we also plan to evaluate the effectiveness of our scheme in other on demand routing protocols such as DSR, ABR, etc.

References

1. D.B. Johnson and D.A. Maltz.: Dynamic Source Routing in Ad Hoc Network, Tornasz Imielinski and Hank Korth, Eds., Mobile Computing, Kluwer, (1996)
2. K. Chandran, S. Raghunathan, S. Venkatesan, and R. Prakash,: A Feedback Based Scheme For Improving TCP Performance in Ad Hoc Wireless Networks, ICDCS, (1995)
3. Elizabeth Royer and C.-K. Toh.: A Review of Current Routing Protocols for Ad Hoc Mobile Wireless Networks, in IEEE Personal Communications Magazine, First Quarter (1999)
4. C.-K. Toh.: Ad Hoc Mobile Wireless Networks: Protocols and Systems, Prentice Hall, (2000)
5. Charles E. Perkins,: Ad Hoc Networking, Addison-Wesley, (2001) 139-220
6. Seung-Joon Seok, Sung-Bum Joo and Chul-Hee Kang.: A Mechanism for Improving TCP Performance in wireless Environments, ICT 2001 at Bucharest Romania. June (2001).
7. Charles E. Perkins, Elizabeth M. Royer, Samir R. Das and Mahesh K. Marina,: Performance Comparison of Two On-Demand Routing Protocols for Ad Hoc Networks, IEEE Personal Communications, February (2001)
8. Yongjun Choi and Daeyeon Park, : Associativity Based Clustering and Query Stride for On-demand Routing Protocols in Ad Hoc Networks, Journal of Communications and Networks, vol.4, NO.1, March (2002)
9. S. Singh, M.Woo, and C. S. Raghavendra.: Power-Aware Routing in Mobile Ad Hoc Networks, in proceedings of ACM/IEEE MobiCom'98 Conference, October (1998)
10. Z. Haas and M. Pearlman.: The performance of query control schemes for the zone routing protocol, in proc. SIGCOMM'98, pp. 167-177, Sept. (1998)
11. IEEE.: Wireless LAN Medium Access Control (MAC) and Physical Layer (PHY) Specifications, IEEE Std. 802.11-1997, (1997)
12. V. Bharghavan et al.: MACAW: A Media Access Protocol for Wireless LANs, *Proc. ACM SIGCOMM '94*, Aug. (1994)
13. R.Dube, C. D. Rais, K. Y. Wang, and S. K. Tripathi.: Signal Stability-Based Adaptive Routing for Ad Hoc Mobile Networks, IEEE personal Communications Magazine, pp. 36-45, February (1997)

Enhancing the Security of On-demand Routing in Ad Hoc Networks⋆

Zhenjiang Li[1] and J.J. Garcia-Luna-Aceves[1,2]

[1] Computer Engineering, University of California, Santa Cruz,
1156 high street, Santa Cruz, CA 95064, USA
Phone:1-831-4595436, Fax: 1-831-4594829
{zhjli, jj}@soe.ucsc.edu
[2] Palo Alto Research Center (PARC), 3333 Coyote Hill Road,
Palo Alto, CA 94304, USA

Abstract. We present the Ad-hoc On-demand Secure Routing (AOSR) protocol, which uses pairwise shared keys between pairs of mobile nodes and hash values keyed with them to verify the validity of the path discovered. The verification processes of route requests and route replies are independently executed while symmetrically implemented at the source and destination nodes, which makes AOSR easy to implement and computationally efficient, compared with prior approaches based on digital signing mechanisms. By binding the MAC address (physical address) with the ID of every node, we propose a reliable neighbor-node authentication scheme to defend against complex attacks, such as wormhole attacks. An interesting property of AOSR is the "zero" communication overhead caused by the key establishment process, which is due to the exploitation of a Self-Certified Key (SCK) cryptosystem. Analysis and simulation results show that AOSR effectively detects or thwarts a wide range of attacks to ad hoc routing, and is able to maintain high packet-delivery ratios, even when a considerable percentage nodes are compromised.

1 Introduction

Ad hoc networks can be rapidly deployed in critical scenarios such as battle fields and rescue missions because a prior infrastructure is not needed. However, to be truly effective, communication over such networks should be secure.

The attacks to an ad hoc network can be classified into *external attacks* and *internal attacks* based on the information acquired by the attackers. External attacks are launched by malicious users who do not have the cryptographic credentials (e.g., the keys used by cryptographic primitives) that are necessary to participate in the routing process. On the other hand, internal attacks are originated by attackers who have compromised legitimate nodes, and therefore

---

⋆ This work was supported in part by the National Science Foundation under Grant CNS-0435522, by the UCOP CLC under grant SC-05-33 and by the Baskin Chair of Computer Engineering at University of California, Santa Cruz.

V.R. Sirotiuk and E. Chávez (Eds.): ADHOC-NOW 2005, LNCS 3738, pp. 164–177, 2005.

can have access to the cryptographic keys owned by those nodes. As a result, internal attacks are far more difficult to detect and not as defensible as external attacks. For a good description of potential attacks to ad hoc routing, please refer to [6, 3].

While security problems can be tackled at different layers of the network architecture, we focus on the routing process at the network layer in this paper. The main contribution of this paper is to present a secure on-demand routing protocol for ad hoc network, which is lightweight for the process of path discovery, and powerful enough to thwart a wide range of attacks to the ad hoc routing.

This paper is organized as follows. Section 2 reviews the Self-Certified Key (SCK) cryptosystem, which facilitates the establishment of shared keys between any two nodes in the network. Section 3 presents the ad hoc on-demand secure routing (AOSR) protocol we developed to enhance the security of ad hoc routing. A security analysis is presented in Sect. 4, and Sect. 5 provides the simulation evaluation of AOSR. Lastly, we conclude our work in Sect. 6.
1

2   Key Distribution

There are three cryptographic techniques that can be used to secure the ad hoc routing: hash functions, symmetric cryptosystems and asymmetric (or public key) cryptosystems. An asymmetric cryptosystem is more efficient in key utilization in that the public key of a node can be used by all the other nodes. However, an asymmetric cryptosystem is computationally expensive and the binding between a public key and its owner also need to be authenticated. On the other hand, a symmetric cryptosystem is computationally cheap, but a shared key must exist between any pair of nodes. Hash functions can be implemented quickly, and usually work together with symmetric or asymmetric algorithms to create more useful credentials such as a digital certificate or keyed hash value.

In AOSR, we use an asymmetric cryptosystem only to establish the pairwise shared keys between each pair of nodes, and we use hash values keyed with these shared keys to verify the routing messages exchanged amongst nodes, and as such achieve computational efficiency. Our approach is different from other protocols in that, the public key system we choose has the unique property that a centralized authority is required only at the initial network formation. After that, any pair of nodes can derive, and also update the correct shared keys between them in a non-interactive manner without the aid of an on-line centralized authority. Hence, there is "zero" communication overhead caused by key distribution and progression. In the following, we summarize the Self-Certified Key (SCK) cryptosystem [5], which is used for key establishment in AOSR.

---

[1] In our discussion, **k-MAC** refers to keyed-message authentication code (a keyed hash value), while **MAC** refers to media access control unless specified otherwise.

Initialization. A centralized authority (CA) Z is assumed to exist before the network formation. Z chooses large primes p, q with $q|(p-1)$ (i.e., q is a prime factor of $(p-1)$), a random number $k_A \in Z_q^*$, where Z_q^* is a multiplicative subgroup with order q and generator α; then Z generates its public/private key pair (x_Z, y_Z). We assume that the public key y_Z is known to every node in the network hereafter. To issue the private key for node A with identifier ID_A, Z computes the signature parameter $r_A = \alpha^{k_A} \pmod p$ and $s_A = x_Z \cdot h(ID_A, r_A) + k_A \pmod q$, where $h(\cdot)$ is a collision-free one-way hash function and $\pmod p$ means modulo p. Node A publishes the parameter r_A (also called guarantee) together with its identifier ID_A, and keeps $x_A = s_A$ as its private key. The public key of A can be computed by any node that has y_Z, ID_A and r_A according to

$$y_A = y_Z^{h(ID_A, r_A)} \cdot r_A \pmod p \tag{1}$$

We denote this initial key pair as $(x_{A,0}, y_{A,0})$

User-controlled key pair progression. Node A can update its public/private key pair *either synchronously or asynchronously*. In the synchronous setting, where A uses the key pair $(x_{A,t}, y_{A,t})$ in time interval $[t \cdot \Delta T, (t+1) \cdot \Delta T)$, node A can choose n random pairs $\{k_{A,t} \in Z_q^*, \ r_{A,t} = \alpha^{k_{A,t}} \pmod p\}$, where $1 \le t \le n$, and publishes guarantees $r_{A,t}$. Then the private keys of node A can progress as the following

$$x_{A,t} = x_{A,0} \cdot h(ID_A, r_{A,t}) + k_{A,t} \pmod q \tag{2}$$

The corresponding public keys are computed according to

$$y_{A,t} = y_{A,0}^{h(ID_A, r_{A,t})} \cdot r_{A,t} \pmod p \tag{3}$$

In the asynchronous setting, Node A can simply inform other nodes that it has updated to guarantee $r_{A,t}$, such that A's new public key $y_{A,t}$ can be computed based on (3)

Non-interactive pairwise key agreement and progression. Pairwise shared keys between any two nodes A and B can be computed and updated synchronously or asynchronously. This can be achieved as follows:

Node A:
$x_{A,t} = x_{A,0} \cdot h(ID_A, r_{A,t}) + k_{A,t}$
$y_{B,t} = y_{B,0}^{h(ID_B, r_{B,t})} \cdot r_{B,t} \pmod p$
$K_{A,t} = y_{B,t}^{x_{A,t}} \pmod p$
$K_t = h(K_{A,t})$

Node B:
$x_{B,t} = x_{B,0} \cdot h(ID_B, r_{B,t}) + k_{B,t}$
$y_{A,t} = y_{A,0}^{h(ID_A, r_{A,t})} \cdot r_{A,t} \pmod p$
$K_{B,t} = y_{A,t}^{x_{B,t}} \pmod p$
$K_t = h(K_{B,t})$

The pairwise shared keys obtained by Node A and Node B are equal because

$$h(K_{A,t}) = h\left(y_{B,t}^{x_{A,t}} \pmod p\right) = h\left(\alpha^{x_{A,t} x_{B,t}} \pmod p\right)$$
$$= h\left(y_{A,t}^{x_{B,t}} \pmod p\right) = h(K_{B,t}) \tag{4}$$

The SCK system has two features that make them convenient and efficient to establish a pairwise shared key between any pair of nodes.

Firstly, given N nodes in the network, assume that their IDs are globally known, in order to distribute their public keys, N guarantees are distributed, instead of N traditional certificates. The advantage is that, unlike a certificate based approach, these N guarantees can be published and need not to be certified (signed) by any centralized authority. This means that we can derive the public key of any node, and update the public/private key pair between any two nodes without the aid of an on-line CA (access to CA is only required at the initial network formation, as described above).

Secondly, given that N guarantees are already distributed to all nodes in the network, and the public key of the CA is known to everyone, then any two nodes can establish a pairwise shared key, also the updated keys thereafter, in a non-interactive manner. This means that no further negotiation message is needed for key agreement and progression, such that *zero* communication overhead for key negotiation is achieved. This is especially valuable to resource-constrained scenario such as MANETs.

3   Operations of AOSR

3.1   Assumptions

We assume that each pair of nodes (Node N_i and node N_j) in the network shares a pairwise secret key $K_{i,j}$, which can be computed by the key distribution scheme described in Sect. 2.

We also assume that the MAC (media access control) address of a node cannot be changed once it joins the network. Even though some vendors of modern wireless cards do allow a user to change the card's MAC address, we will see that this simple assumption can be very helpful in detecting some complicated attacks such as wormhole. Moreover, every node must obtain a certificate signed by the CA, which binds its MAC and ID (can be the IP address of this node), before this node joins the network. A node presents this to nodes that the node meets for the first time, and a node can communicate with its neighbor nodes only if its certificate has been verified successfully. The approach used to maintain neighbor-node information is presented Sec. 3.2. The notations used in this paper are summarized in Table 1.

3.2   Neighbor-Node List Maintenance

It is easy to maintain a neighbor-node list by performing a neighbor discovery protocol similar to that used in AODV [4], where periodical beacons are broadcast by a node to signal its existence to its neighbor nodes. However, there is no assurance in the fidelity of the information acquired based on such a naive scheme, largely because of the potential node identification spoofing. To authenticate the existence of a neighbor node, we require that the periodically broadcast

Table 1. Notations used in this paper

Name	Meaning
S, D, N_i	the IDs of nodes, by default, $S = source$ and $D = Destination$
$RREQ$	the type identifier for a route request $RREQ$
$RREP$	the type identifier for a route reply $RREP$
$RERR$	the type identifier for a route error report $RERR$
$QNum$	the route request ID, a randomly generated number
$RNum$	the route reply ID, and $RNum = QNum + 1$ for the same round of route discovery
HC	the value of hop-count, more specifically, $HC_{i \rightarrow j}$ represents the hop-count from node N_i to N_j
$QMAC$	the k-MAC (keyed message authentication code) used in $RREQ$
$RMAC$	the k-MAC used in $RREP$
$EMAC$	the k-MAC used in $RERR$
$K_{i,j}$	the pairwise key shared between nodes N_i and N_j, thus $K_{i,j} = K_{j,i}$
$\{NodeList\}$	Records the accumulated intermediate nodes traversed by messages ($RREQ$, $RREP$ or $RERR$), and represents a path from source S to destination D. For clarity, they are numbered increasingly from S to D, e.g., $\{S, N_1, N_2, N_3, ...N_{HC}\}$
$rT_{i \rightarrow j}$	the route from node N_i to node N_j

beacon message also carry a CA signed certificate, which bind the ID and MAC address of the sending node.

When a node receives a beacon from a node for the first time, the MAC address contained in the certificate is compared against the source MAC address in the header of the packet frame. This can be done because the MAC address field of a physical frame is always in clear text. If these two MAC addresses match, then the certificate is further verified given the public key of the CA. The ID and MAC address pair then are added into the local neighbor-node list if the above two verifications are successful. Because nodes can move, an active bit should be used to indicate whether a node is still within the radio range. The active bit is turned off if no beacon received from the peer node after a predefined timeout. A verified $< MAC, ID >$ pair can be kept on a neighbor list to save unnecessary repetitive computation on certificate verification. The authenticity of a node's ID can be ensured because the node's MAC address cannot be changed once the node joins the network, based on the assumption we made.

Every time a node receives a packet from a neighbor node, it first checks if this node is already on its neighbor-node list and also active. If so, it further checks if the source MAC address of this packet matches the MAC address paired with the node's ID stored by the local neighbor list. Only those packets passing these two checks are delivered to upper layers for further processing. For the purpose of brevity, in the following discussion, we assume that all the nodes participating in routing process have passed the identification verification at the MAC layer, and all packets are passed to upper layers only if they are authenticated successfully.

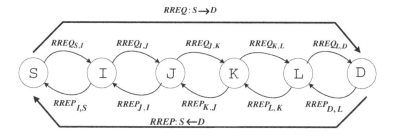

Fig. 1. Route discovery between source S and destination D

3.3   Route Discovery

An illustration of the message flow of the route discovery is shown in Fig. 1, which consists of *route request initialization, route request forwarding, route request checking at the destination D, and the symmetric route reply initialization, route reply forwarding and route reply checking at the source S.* The details are presented as follows.

Route Request Initialization. Node S (source) generates a route request $RREQ$ and broadcasts it when S wants to communicate with node D (destination) and has no route for D.

$$RREQ = \{RREQ, S, D, QNum, HC, \{NodeList\}, QMAC_{s,d}\} \qquad (5)$$

where $HC = 0$ and $\{NodeList\} = \{Null\}$, because no intermediate node has been traversed by $RREQ$ at the source S.

$QMAC_{s,d} = Hash(CORE, HC, \{NodeList\}, K_{s,d})$ is the k-MAC which will be further processed by intermediate nodes, and used by the destination D to verify the integrity of $RREQ$ and the validity of the path reported in $\{NodeList\}$; $CORE = Hash(RREQ, S, D, QNum, K_{s,d})$ serves as a credential of S to assure D that a $RREQ$ is really originated from S and its immutable fields are integral during the propagation.

Route Request Forwarding. A $RREQ$ received by an intermediate node N_i will be processed and further broadcast only if it has never been received (the IDs of nodes S, D and randomly generated $QNum$ uniquely identify the current round of route discovery). More specifically, N_i increases HC by one, appends the ID of the upstream node N_{i-1} into $\{NodeList\}$, and updates QMAC as

$$QMAC_{i,d} = Hash(QMAC_{i-1,d}, HC, \{NodeList\}, K_{i,d}) \qquad (6)$$

Checking $RREQ$ At the Destination D. Figure 2 shows the procedure conducted by the destination D in order to authenticate the validity of the path reported by $RREQ$.

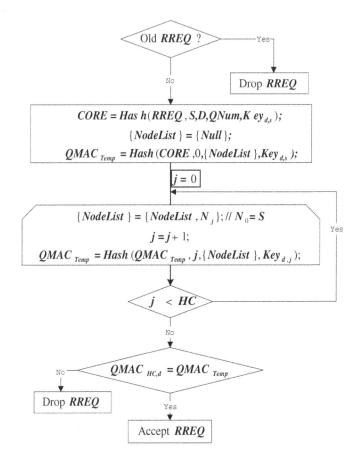

Fig. 2. Checking $RREQ$ at destination D

Basically, D repeats the computation done by all the intermediate nodes traversed by $RREQ$ from the source S to itself, which are recorded in $\{NodeList\}$, *using the shared keys owned by D itself*. Obviously, the number of hash operations need to perform by D equals to the HC as reported by $RREQ$.

If the above verification is successful, D can be assured that this $RREQ$ was really originated from S, and every node listed in $\{NodeList\}$ actually participated in the forwarding of $RREQ$. It is also easy for D to learn the hops traversed by $RREQ$ (i.e., $HC_{s \to d}$).

Note that the *route reply initialization, reverse forwarding of route reply and checking RREP at the source S are basically symmetric to that of $RREQ$*, and as such are omitted due to the space limitation.

3.4   Route Maintenance

A route error message ($RERR$) is generated and unicast back to the source S if an intermediate node N_i finds the downstream link is broken. Before accepting

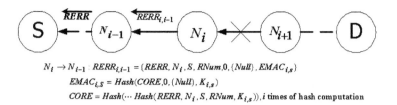

$$N_i \to N_{i-1} : RERR_{i,i-1} = (RERR, N_i, S, RNum, 0, (Null), EMAC_{i,s})$$
$$EMAC_{i,s} = Hash(CORE, 0, (Null), K_{i,s})$$
$$CORE = Hash(\cdots Hash(RERR, N_i, S, RNum, K_{i,s})), i \text{ times of hash computation}$$

Fig. 3. N_i generates $RERR$ when the downstream link fails

a $RERR$, we must make sure that (a)the node generating $RERR$ is really in the path to the destination; (b)the node reporting link failure should actually be there when it was reporting the link failure.

The process of sending back a $RERR$ from node N_i is similar to that of originating a route reply from N_i to the source S. Therefore, we only describe the main differences, and provide more details in Fig. 3.

A $RERR$ has a format similar to that of a $RREP$, except the type identifier and the computation of $CORE$, which is calculated as

$$CORE = Hash(...Hash(RERR, N_i, S, RNum, K_{i,s})) \qquad (7)$$

where i hash computations are performed and i equals the number of hops from node N_i to the destination D.

Intermediate nodes in the path back to the source only process and back-forward a $RERR$ received from the downstream nodes towards destination D according to their forwarding table, which ensures that node N_i is actually in the path to D when it is reporting the failure.

When the source S receives a $RERR$, it performs a verification procedure similar to that of $RREP$. The only difference is the computation of $CORE$, which is conducted as the following

$$CORE = Hash(...Hash(RERR, N_i, S, RNum, K_{s,i})) \qquad (8)$$

where $(HC_{s \to d} - HC_{reported\ by\ RERR})$ hash computations are performed.

4   Security Analysis

The network topology and notation used in the analysis are presented in Fig. 4. In the following, we consider $RREP$ only because the processing of $RREQ$ and the processing of $RREP$ are symmetric.

In AOSR, a route reply $RREP$ consists of immutable fields such as $RREP$, $RNum$, D, S; and mutable fields such as $RMAC$, HC and $\{NodeList\}$. As for immutable parts, they are secured by the one-way hash value $CORE$, which has $RREP$, D, S, $RNum$ and $K_{d,s}$ as the input. Any node cannot impersonate the initiator D to fabricate $RREP$ due to the lack of key $K_{d,s}$ known only to S and D; and any modification on these fields can also be detected by the source S because the k-MAC carried in the $RREP$ will not match that S can compute.

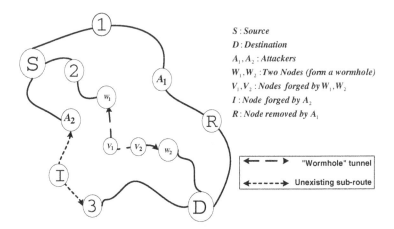

Fig. 4. Example network topology for security analysis

Unlike those immutable fields, $\{HC, \{NodeList\}, RMAC\}$ will be modified by intermediate nodes when a $RREP$ propagates back to S. In AOSR, the authenticity of $HC, \{NodeList\}$ and $RMAC$ is guaranteed by integrating HC and $\{NodeList\}$ into the computation of $RMAC$, in a way that no node can be added into $\{NodeList\}$ by the upstream neighbor, unless it actually back-forwards a $RREP$; and no node can be maliciously removed from $\{NodeList\}$ either. For instance, let us assume that the attacker A_1 attempts to remove node R from $\{NodeList\}$ and decrease HC by one. When receiving this $RREP$, S will recompute $RMAC$ according to the nodes listed in $\{NodeList\}$. Because the computation conducted by R, $RMAC_{r,s}$ more accurately, has been omitted, S can not have a match with the received $RMAC$. The reason is that the hash operation is based on one-way functions, and hence A_1 has no way to reverse the hash value $RMAC_{r,s}$ computed by node R. Another possible attack is for attacker A_2 to insert a non-existent node I into $\{NodeList\}$ and increase HC by one. To achieve this, A_2 needs to perform one more hash computation which requires $K_{i,s}$ as the input, which is impossible simply because $K_{i,s}$ is known only to S and I. For the same reason, A_2 cannot impersonate another node (Spoofing) and show in $\{NodeList\}$ either.

An attacker can also just *get and back-forward* a $RREP$ without doing any processing on $RREP$, hoping that the hops learn by the source is one hop less than the actual length. In AOSR, the ID of a downstream neighbor can only be added into $\{NodeList\}$ by its upstream neighbor. Therefore, to *get and back-forward* a $RREP$ is equivalent to insert another node (the one does get-and-forward) into $\{NodeList\}$, which will fail as we described before.

Because of the randomly generated $RNum$, AOSR is free of the *Wrap-around* problem, which can happen to other approaches using monotonically increased sequence numbers. It is also difficult for an adversary to predict the next $RNum$ to be used, which protects AOSR against simple replay, or other attacks to the

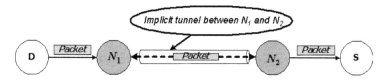

Dark colored nodes are attackers and white nodes are legal nodes

Fig. 5. Illustration of wormhole attack

sequence number, in which for example a sequence number can be maliciously raised to the maximum value and has to be reset.

Wormhole is a special attack that is notoriously difficult to detect and defend against. Wormhole usually consists of two or more nodes working collusively, picking up packets at one point of the network, tunnelling them through a special channel, then releasing them at another point far away. The goal is to mislead the nodes near the releasing point to believe that these packets are transmitted by a neighbor close by. A demonstrative scenario of wormhole is shown in Fig. 5. Wormhole is a big threat to ad hoc routing, largely because wrong topology information will be learnt by the nodes near the releasing point. As a result, data packets are more likely to be diverted into the tunnel, in which attackers can execute varied malicious operations, such as dropping data packets (black-hole attack), modifying packet contents or performing traffic analysis, and others.

Wormhole can be classified into two categories: at least one of the end nodes is outside attacker, or both end nodes are inside attackers. For outside attackers, they need to make themselves invisible to other nodes due to the lack of valid keys to participate in the routing process. Therefore, what they actually perform is sending packets from one end of the tunnel to the other end without any modification. On the other hand, inside attackers can "legally" participate in the routing process, and therefore are able to manipulate the tunnelled packets with much more possibilities.

The chained k-MAC values computed by all intermediate nodes during the route discovery, together with the authenticated neighbor information provided by the neighbor maintenance scheme, enable AOSR to detect wormhole and varied attacks derived from it. As an example, let us assume that node W_1 and W_2 in Fig. 4 are two adversaries who have formed a tunnel $Tul_{w_1 \leftrightarrow w_2}$. First, they can refuse to forward $RREP$, but this is not attractive because this will exclude them from the communication. Second, they can attempt to modify HC or $\{NodeList\}$, but this will be detected when the source S checks the $RMAC$ carried by $RREP$. They can also insert some non-existent nodes, like V_1, V_2, into $\{NodeList\}$, but this attempt will fail due to the lack of shared keys $K_{v1,s}$ and $K_{v2,s}$.

When two colluding nodes try to tunnel packets through a hidden channel, if only one of them is outside attacker, then any packet relayed by this "outsider" will be detected and dropped because the MAC address of the outside node does not match any ID maintained by the neighbor list of the receiving node near the releasing point (or does not exist at all). This can be done because a node's MAC address cannot be changed, any binding of a MAC address and an ID on the neighbor list has been authenticated, and the MAC address of a packet is always in clear text. For an instance, assume again that the node W_1 and W_2 in Fig. 4 are two outside attackers and form a tunnel $Tul_{w_1 \leftrightarrow w_2}$, and w_1 or w_2 is tunnelling a packet from node 2 to node D. This packet will not be accepted as a valid one by D because the MAC address shown in packet (the MAC address of W_2) does not match the MAC address of node 2 stored by node D's neighbor list, or node 2 is not on the list at all. This is also true when only one of the two end nodes is outside attacker.

The only kind variation of wormhole attacks, for which our protocol cannot detect, happens when both colluding nodes are inside attackers to the network, in which case they own all necessary valid cryptographic keys. To date, there is still no effective way to detect this kind of wormhole attacks. There exist other proposals for detecting and defending against wormhole [2], in which time synchronization is needed for the proposed *packet leashes* to work. Compared with packet leash and other proposals, our $< MAC, ID >$ binding scheme is simple to implement and provides almost the same defensive results.

5   Performance Evaluation

In order to evaluate the validity and efficiency of AOSR, we made an extension to the network simulator NS2 [1] and carried out a set of simulations. The simulation field is 1000 meters long and 250 meters wide, in which 30 nodes move around according to the random way-point model, and the velocity for each of them is uniformly distributed between 0 and 15m/sec. The traffic pattern being used is 15 random sessions (the source and destination are randomly chosen for each session), and each of them is a constant bit rate (CBR) flow at a rate of two packets per second, and 512 bytes per packet in size. The hash function (used for the computation of k-MAC) and digital signing function (used by the neighbor maintenance scheme) in our simulation are MD5 (128 bits) and RSA (1024 bits) from the RSAREF library. In this way, we take into account the cost and delay incurred by the cryptographic operations executed by AOSR, in addition to the normal routing overhead.

The metrics we use to measure the performance of AOSR are the following: *packet delivery ratio* is the total number of CBR packets received, over the total number of CBR packets originated, over all the mobile nodes in the network; *end-to-end packet delay* is the average elapsed time between a CBR packet is passed to the routing layer and that packet is received at the destination node; *route discovery delay* is the average time it takes for the source node to find a route to the destination node; *normalized routing overhead* is the total routing messages transmitted or forwarded over the total number of CBR packets received, over all the mobile nodes; *average route length* is the average length (hops) of the paths discovered, over all the mobile nodes in the network.

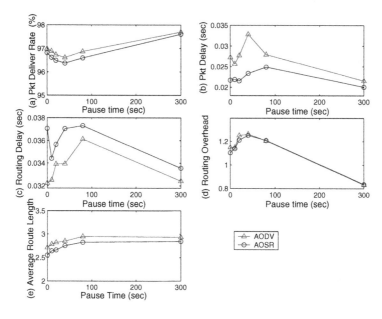

Fig. 6. Performance without attackers

Firstly, to test the overhead caused by cryptographic operations, we simulate and compare the performance difference between AOSR and AODV when there is no malicious users in the network. The results are shown in Fig. 6 as functions of pause time. For all points presented in Fig. 6, each of them represents the average over 16 random movement patterns at each pause time, and the same patterns are used by both AOSR and AODV. The simulation time is 300 seconds long, which means that nodes are stationary when the pause time is set to 300 seconds.

As shown in Figs. 6 (a) (b) and (c), the packet delivery ratio and packet delivery delay of AOSR are very close to that of AODV, and the average time for AOSR to find a route is just milliseconds longer than that of AODV. This indicates that AOSR is almost as efficient as AODV in route discovery and data delivery, and the computation of chained k-MAC and the maintaining of neighbor list in AOSR do not incur significant routing delay.

AOSR also does not cause higher routing overhead than AODV, as shown in Fig. 6(d), if no attack happens to the network. The reason is that, in this case, only normal signaling messages, such as route request, reply and error report messages, are exchanged amongst nodes. The average route length found by AOSR is a little shorter than that of AODV, as shown in Fig. 6(d). This is because AOSR requires all route requests to reach destination nodes, while AODV allows intermediate nodes to reply to a RREQ if they have a valid path cached for the specified destination, which may not be the shortest one at that moment. This also explains why the packet delivery delay of AOSR is smaller than that of AODV, as shown in Fig. 6(b).

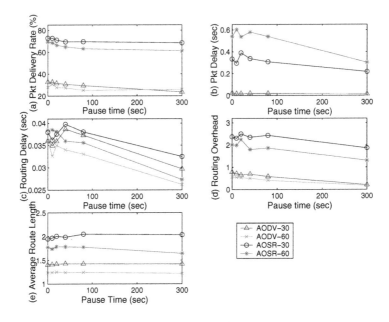

Fig. 7. Performance with 30% and 60% nodes compromised

Figure 7 shows the simulation results when 30% and 60% of the nodes in the network are compromised, and fabricate fake route replies to any route query by claiming that they are "zero" hop away from the specified destination node, in hope that the querying source node will send its succeeding data packets to them. After that, a compromised node will simply drop all the data packets received (black-hole attack).

The packet delivery ratio of AODV decreases drastically, as shown in Fig. 7(a), because most of the packets are sent to, so dropped by the compromised nodes. The average time to find a route and the average route length of AODV are much less than when there is no malicious node exists, as shown in Figs. 7 (c) and (e). The reason is obvious because a compromised node is very likely to see and therefore reply to the route request for a specified destination earlier than the destination itself or other nodes having valid routes. This also means that most of the successful packet deliveries happen only between neighbor nodes (i.e., only one-hop away).

On the other hand, AOSR is still be able to achieve over 65% packet delivery ratios for all pause time configurations even when 60% of the nodes are compromised, as shown in Fig. 7(a). Of course, this is achieved at the cost of more routing time to find a route, longer end-2-end packet delay and higher routing overhead, as shown in Figs. 7(b) (c) and (d), respectively. Lastly, because AOSR can not be misled by intermediate compromised nodes, and can find a route to the destination if there exists one, the average length of routes found by AOSR is longer than that found by AODV, as shown in Fig. 7(e).

6 Conclusion

In AOSR, the authentication processes of route requests and route replies are independent and symmetric, which is easy to implement and also allows both source and destination to verify a route reported. By using self-certified key (SCK), our protocol achieves *zero* communication overhead caused by key distribution, and needs the aid of a centralized administration (such as a certificate authority) only at the initial network formation. Pairwise shared keys between pairs of mobile nodes and hash values keyed with them are employed to authenticate routing messages and the validity of the paths discovered, which makes our protocol computationally efficient compared with prior proposals based on digital signing mechanisms. Lastly, by maintaining a neighbor-node list, on which the $< MAC, ID >$ pair of every neighbor is authenticated and updated in a timely manner, our protocol is also able to detect some complicated attacks, such as wormhole and its variations.

References

[1] The Network Simulator - NS2. http://www.isi.edu/nsnam/ns/.
[2] Y. Hu, A. Perrig, and D. Johnson. Packet Leashes: A Defense against Wormhole Attacks in Wireless Networks. In *Proceedings of IEEE INFOCOM, San Francisco, USA*, March 30 - April 3, 2003.
[3] Y. Hu, A. Perrig, and D. Johnson. Ariadne: A Secure On-demand Routing Protocol for Ad Hoc Networks. In *Proceedings of the 8th ACM International Conference on Mobile Computing and Networking (MobiCom)*, September 2002.
[4] C. E. Perkins and E. M. Royer. Ad Hoc On Demand Distance Vector Routing. In *Proceedings of the 2nd IEEE Workshop on Mobile Computing Systems and Applications, New Orleans, LA*, pages 90–100, February 1999.
[5] H. Petersen and P. Horster. Self-Certified Keys - Concepts and Applications. In *Proceedings of the 3rd Conference of Communications and Multimedia Security, Athens*, September 22-23, 1997.
[6] K. Sanzgiri, B. Dahill, B. N. Levine, E. Royer, and C. Shields. A Secure Routing Protocol for Ad Hoc Networks. In *Proceedings of the 10th Conference on Network Protocols (ICNP)*, 2002.

DPUMA: A Highly Efficient Multicast Routing Protocol for Mobile Ad Hoc Networks[*]

Rolando Menchaca-Mendez[1,**], Ravindra Vaishampayan[1],
J. J. Garcia-Luna-Aceves[1,2], and Katia Obraczka[1]

[1] Department of Computer Engineering, University of California, Santa Cruz,
1156 High Street, Santa Cruz, CA 95064, U.S.A.
{menchaca, ravindra, jj, katia}@soe.ucsc.edu
[2] Palo Alto Research Center (PARC), 3333 Coyote Hill Road,
Palo Alto, CA 94304, U.S.A.

Abstract. In this paper we present DPUMA, a mesh-based multicast routing protocol specifically designed to reduce the overhead needed to deliver multicast packets, saving bandwidth and energy, two of the scarcest resources in MANETS. The two main features of DPUMA are: (1) for each multicast group, it periodically floods a single control packet to build the mesh, elect the core of the mesh and get two-hop neighborhood information; and (2), it computes the mesh's k-dominating set to further reduce overhead induced by flooding the mesh when forwarding data packets. These two characteristics contrast with other protocols that blindly flood the net in different stages to construct their routing structure (mesh or tree), to elect the leader of the structure, and that exchange hello messages to get neighborhood information. Using detailed simulations, we show over different scenarios that our protocol achieves similar or better reliability while inducing less packet transmission overhead than ODMRP, MAODV and PUMA which is DPUMA's predecessor.

1 Introduction

Mobile Ad Hoc Networks (or MANETS), also called "networks without network", do not rely on a fixed infrastructure. In other words, any MANET node can act as traffic originator, destination, or forwarder. Hence, MANETs are well suited to applications where rapid deployment and dynamic reconfiguration are necessary. Examples of such scenarios are: military battlefield, emergency search and rescue, conference and conventions. The objective of a multicast protocol for MANETs is to enable communication between a sender and a group of receivers in a network where nodes are mobile and may not be within direct wireless transmission range of each other. These

---

[*] This work was supported in part by the National Science Foundation under Grant CNS-0435522, by the UCOP CLC under grant SC-05-33 and by the Baskin Chair of Computer Engineering at University of California, Santa Cruz.
[**] This author is supported by the Mexican National Council for Science and Technology (CONACyT) and by the Mexican National Polytechnic Institute (IPN).

V.R. Sirotiuk and E. Chávez (Eds.): ADHOC-NOW 2005, LNCS 3738, pp. 178–191, 2005.

multicast protocols have to be designed to efficiently use the available bandwidth as well as nodes' energy, two of the most precious and scarce resources in MANETs.

A multicast group is defined by a unique identifier that is known as the group identifier. Nodes may join or leave the multicast group (or multiple multicast groups) anytime while their membership remains anonymous. In the last five years, several MANET multicast protocols have been proposed (e.g. [1], [2], [3], [4], [5], [6], [7], [8], [9], [10], [11], [12], [13], [14]). In general, the approaches taken up to date can be classified by the way they construct the underlying routing structure; namely tree-based and mesh based protocols. The main objective of the routing structure is to efficiently deliver information to the members of the multicast group while avoiding non members.

A tree-based multicast routing protocol constructs and maintains either a shared multicast routing tree or multiple multicast trees (one per each sender) to deliver packets from sources to receivers of a multicast group. Recent examples of tree-based multicast routing protocols are the Multicast Ad hoc On-demand Distance Vector Protocol (MAODV) [5], and the Adaptive Demand-driven Multicast Routing Protocol (ADMR) [6]. This approach has proven to deliver adequate performance in wired networks [18], however, in the context of MANETs, where due to mobility topology changes are common, establishing and maintaining a tree or a set of trees requires continuous interchange of control messages which has a negative impact in the overall performance of the protocol.

On the other hand, a mesh-based multicast routing protocol maintains a mesh consisting of a connected sub-graph of the network which contains all receivers of a particular group and some relays needed to maintain connectivity. Maintaining a connected component is much less complicated than maintaining a tree and hence mesh-based protocols tend to be simpler and more robust. Two well-know representatives of this kind of protocols are the Core Assisted Mesh Protocol (CAMP) [2] and the On-Demand Multicast Routing Protocol (ODMRP) [3].

In this paper we present DPUMA, a protocol that further improves the reliability and efficiency of its direct predecessor PUMA [1]. DPUMA is mesh-based and was specifically designed to reduce the overhead needed to deliver multicast packets and thus save bandwidth and energy. The two main characteristics of DPUMA are: (1) for each multicast group, it periodically floods a single control packet (a multicast announcement) to build the mesh, elect the core of the mesh and to get two-hop neighborhood information; and (2) uses mesh's k-dominating set to further reduce overhead induced by flooding the mesh when transmitting a data packet. These two characteristics contrast with other protocols that, in different phases, blindly flood the network to deliver data packets, construct their routing structure (mesh or tree), elect the leader of the structure, and that exchange hello messages to get neighborhood information.

The remaining of this paper is organized as follows. Section 2 describes related work, both on multicast routing protocols for MANETs and on the distributed computation of connected dominating sets. Section 3 describes DPUMA, while section 4 presents a series of performance comparisons among PUMA, DPUMA, MADOV and ODMRP over different scenarios. Finally, in Section 4 we present concluding remarks and directions for future work.

2 Related Work

This section is divided in two subsections. The first one reviews some of the most representative multicast routing protocols, emphasizing on MAODV and ODMRP which are the protocols that we compare against DPUMA in Section 4. The second subsection reviews current approaches to compute connected dominating sets.

2.1 Multicast Routing Protocols for Ad Hoc Networks

As we mentioned in the previous section, MAODV maintains a shared tree for each multicast group consisting of receivers and relays. Sources acquire routes to the group on demand in a way similar to the Ad hoc on Demand Distance Vector (AODV) [13]. Each multicast group has a group leader who is the first node joining the group. The group leader is responsible for maintaining the group's sequence number which is used to ensure freshness of routing information. The group leader in each connected component periodically transmits a group hello packet to become aware of reconnections. Receivers join the shared tree by means of a special route request (RREQ) packet. Any node belonging to the multicast tree can answer to the RREQ with a route reply (RREP). The RREP specify the number of hops to the tree member. Upon receiving one or multiple RREP, the sender joins the group through the node reporting the freshest route with the minimum hop count to the tree. Data is delivered along the tree edges maintained by MAODV. If a node that does not belong to the multicast group wishes to multicast a packet it has to send a non-join RREQ which is treated similar in many ways to RREQ for joining the group. As a result, the sender finds a route to a multicast group member. Once data is delivered to a group member, the data is delivered to remaining members along multicast tree edges.

ADMR [6] maintains multicast source based trees for each sender of a multicast group. A new receiver performs a network-wide flood of a multicast solicitation packet when it wants to joint the multicast group. Each group source replies to the solicitation and the receiver sends a receiver join packet to each source who answered its solicitation. An individual source-based tree is maintained by periodic keep-alive packets from the source, which allow intermediate nodes to detect link breaks in the tree by the absence of data or keep-alive packets. A new sender also sends a network-wide flood to allow existing group receivers to send receiver joins to the source. MZR [11] like ADMR, maintains source-base trees. MZR performs zonal routing; and hence the flooding of control packets is less expensive.

In order to establish the mesh, ODMRP requires cooperation of nodes wishing to send data to the multicast group. Senders periodically flood a *Join Query* packet throughout the network. These periodic transmissions are used to update the routes. Each multicast group member on receiving a *Join Query*, broadcast a *Join Table* to all its neighbors in order to establish a *forwarding group*. Sender broadcast data packets to all its neighbors. Members of the forwarding group forward the packet. Using ODMRP, multiple routes from a sender to a multicast receiver may exits due to the mesh structure created by the forwarding group members. The limitations of ODMRP are the need for network-wide packet floods and the sender initiated construction of the mesh. This method of mesh construction results in a much larger mesh as well as numerous unnecessary transmissions of data packets compared to a receiver initiated approach. DCMP [10] is an extension to ODMRP that designates certain senders as cores and reduces the number of senders performing flooding. NSMP [12] is another extension to ODMRP aiming to restrict the flood of control packets to a subset of the entire network. However, DCMP and NSMP fail to eliminate entirely ODMRP's drawback of multiple control packet floods per group.

CAMP avoids the need for network-wide floods from each source to maintain multicast meshes by using one or more cores per multicast group. A receiver-initiated approach is used for receivers to join a multicast group by sending unicast join requests towards a core of the desired group. The drawbacks of CAMP are that it needs the pre-assignment of cores to groups and a unicast routing protocol to maintain routing information about the cores. This later characteristic may induce considerable overhead in a large ad hoc network.

2.2 Distributed Computation of Connected Dominating Sets

2.2.1 Basic Concepts
We use a simple graph $G = (V,E)$ to represent an ad hoc wireless network, where V represents a set of wireless mobile nodes and E a communication link between two nodes. An edge (u,v) indicates that in a particular time, both nodes u and v are within their transmission range, hence, connections of nodes are based on the geographic distances among them. Such graph is also called *unit disk graph* [20]. It is easy to see that the topology of this type of graphs vary over time due to node mobility.

For a given undirected graph $G = (V, E)$, a connected dominating set (CDS) in the graph is any set of connected vertices $V' \subseteq V$ such that each $v \in V - V'$ is adjacent to some vertex in V'. The problem of determining the minimum connected dominating set (MCDS) of a given graph is known to be NP-complete. Therefore, only distributed approximated algorithms running in polynomial-time are practical for wireless ad hoc networks.

In this way, if we compute a dominating set V' of a given network, only those nodes belonging to V' have to broadcast a packet in order to reach every node in the network, with the corresponding savings of V- V' messages. It is important to remark that distributed approximations that run in polynomial-time don't compute the minimum dominating set, however, in the context of ad hoc networks computing a bigger dominating set could be desirable in order to augment the reliability with which a packet is delivered. Here we have a tradeoff among reducing communication overhead, the information that every node must keep, the complexity induced by computing the CDS and reliability.

2.2.2 Distributed Computation of Dominating Sets
Lim and Kim [22] showed that the *minimum connected dominating set* (MCDS) problem can be reduced to the problem of building a *minimum cost flooding tree* (MCFT) and they proposed a set of heuristics for flooding trees that lead to two algorithms: *self-pruning* and *dominant pruning* (DP). They also showed that both algorithms perform better than *blind flooding*, in which each node broadcast a packet to its neighbors whenever it receives the packet along the shortest path from the source node, and that DP outperforms self-pruning.

Since then, many other approaches have been purposed to compute CDS and to improve communication protocols applying CDS (mainly for unicast communication in ad hoc networks). For example, enhancements to dominant pruning have been reported by Lou and Wu [20] who describe the *total dominant pruning* (TDP) algorithm and the *partial dominant pruning* (PDP) algorithm, and by Spohn and Garcia-Luna-Aceves [19] who presented the *enhanced dominant pruning* (EDP) algorithm which improves DP's performance. All these algorithms utilize information about nodes' two-hop neighborhood.

In this work we use an approach similar to the one used by Lim and Kim in their dominating pruning algorithm [22]. This approach uses a greedy set cover (GSC) strategy in order to compute the dominating set of each 2-hop neighborhood of the nodes involved in packet diffusion.

2.2.3 Greedy Set Cover

GSC recursively chooses one-hop neighbors that cover the most two hop neighbors and repeat the process until all two-hop neighbors are covered. As in [20] we use $N(u)$ to represent the neighbor set of u (including u) and $N(N(u))$ to represent the neighbor set of $N(u)$ (i.e., the set of nodes that are within two hops from u). A schematic representation of these sets is shown in figure 1. When node v receives a packet from node u, it selects a minimum number of forwarding nodes that can cover all the nodes in $N(N(v))$. u is the previous relaying node, hence nodes in $N(u)$ have already received the packet, and nodes in $N(v)$ will receive the packet after v rebroadcast it. Therefore, v just needs to determine its forwarding list k-$F(u,v)$ from $B(u,v) = N(v) - N(u)$ to cover all nodes in $U(u,v) = N(N(v)) - N(u) - N(v)$. The greedy *set cover* algorithm is as follows.

Compute: $F(u,v) = [f_1, f_2, \ldots, f_m]$ such that:

$$f_1 \in B(u,v) \text{ satisfying } \bigcup_{f_i \in F} (N(f_i) \cap U(u,v)) = U(u,v) \tag{1}$$

This forwarding list is derived by repeatedly selecting f_i that has the maximum number of uncovered neighbors in $U(u,v)$. These steps are presented as pseudocode in the following lines where Z is a subset of $U(u,v)$ of nodes covered so far. S_i is the neighbor set of v_i in $U(u,v)$ and K is the set of S_i. Selection process:

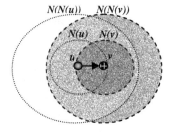

Notation:

- $N(u)$: One-hop neighborhood of u
- $N(N(u))$: Two-hop neighborhood of u
- $B(u,v)$: $N(v)$-$N(u)$
- $F(u,v) \subseteq B(u,v)$: Resulting set of the set cover algorithm. A set of nodes that cover $N(N(v))$-$N(u)$-$N(v)$

Fig. 1. Set of nodes involved in the computation of the CDS

1. Let $F(u,v)=[]$ (empty list), $Z = \phi$ (empty set), and $K = \bigcup S_i$, where $S_i = N(v_i) \cap U(u,v)$ for $v_i \in B(u,v)$

2. Find the set S_i whose size is maximum in K

3. $F(u,v) = F(u,v)\|v_k$, $Z = Z \cup S_i$, $K = K - S_i$, and $S_j = S_j - S_i$ for all $S_j \in K$

4. If no new node is added to $Z = U(u,v)$, exit; otherwise, goto step 2

3 Dominant PUMA

3.1 DPUMA Overview

DPUMA, as well as PUMA, supports the IP multicast service model of allowing any source to send multicast packets addressed to a given multicast group, without having to know the constituency of the group. Furthermore, sources need not join a multicast group in order to send data packets to the group.

Like CAMP, MAODV and PUMA, DPUMA uses a receiver initiated approach in which receivers join a multicast group using the address of a special node (core in CAMP or group leader in MAODV), without the need for network-wide flooding of control or data packets from all the sources of a group. Like MAODV, DPUMA eliminates the need for a unicast routing protocol and the pre-assignment of cores to multicast groups. DPUMA implements a distributed algorithm to elect one of the receivers of a group as the core of the group, and to inform each router in the network of at least one next-hop to the elected core of each group (mesh establishment). The election algorithm used in PUMA is essentially the same as the spanning tree algorithm introduced by Perlman for internetworks of transparent bridges [15]. For more information about the election algorithm and its performance, please refer to [1]. Within a finite time proportional to the time needed to reach the router farthest away from the eventual core of a group, each router has one or multiple paths to the elected core.

Every receiver connects to the elected core along all shortest paths between the receiver and the core. All nodes on shortest paths between any receiver and the core collectively form the mesh. A sender sends a data packet to the group along any of the shortest paths between the sender and the core. In the basic PUMA protocol, once a multicast message reaches a mesh member, it is flooded along the whole mesh. This can lead to unnecessary overhead because a given node can be covered by more than one neighbor and hence receive a multicast message more than once. In order to reduce this overhead, DPUMA incorporates the concept of dominating sets to dynamically determine a subset of one-hop nodes such that if these nodes broadcast the packet, it will be received by all mesh members in a two-hop neighborhood and eventually by all members in the mesh.

DPUMA uses a single control message for all its functions, the multicast announcement. Each multicast announcement specifies a sequence number, the address of the group (group ID), the address of the core (core ID), the distance to the core, a mesh member flag that is set when the sending node belongs to the mesh, a parent that states the preferred neighbor to reach the core and a list of neighbors who are mesh members. Successive multicast announcements have a higher sequence number than previous multicast announcements sent by the same core. With the information contained in such announcements, nodes elect cores, determine the routes for sources outside a multicast group to unicast multicast data packets towards the group, notify others about joining or leaving a group's mesh, maintain the mesh and get 2-hop information of nodes belonging to each multicast group.

3.2 Connectivity List, Two-Hop Neighborhood and Propagation of Multicast Announcements

A node that considers itself to be the core of a group transmits multicast announcements periodically for that group. As the multicast announcement travels through the network, it establishes a connectivity list at every node in the network. Using connectivity lists, nodes are able to establish a mesh, and route data packets from senders to receivers. Nodes also take advantage of this periodic dissemination of multicast announcements to include a list with the 1-hop neighbors who belong to a particular multicast group. Using this 1-hop neighbor information, nodes are able to get 2-hop information regarding each group.

A node stores the data from all the multicast announcements it receives from its neighbors in the connectivity list. Fresher multicast announcements from a neighbor (i.e., one with a higher sequence number) overwrite entries with lower sequence numbers for the same group. Hence, for a given group, a node has only one entry in its connectivity list from a particular neighbor and it keeps only that information with the latest sequence number for a given core.

Each entry in the connectivity list, in addition to storing the multicast announcement, also stores the time when it was received, and the neighbor from which it was received. The node then generates its own multicast announcement based on the best entry in the connectivity list. For the same core ID, only multicast announcements with the highest sequence number are considered valid. For the same core ID and sequence number, multicast announcements with smaller distances to the core are considered better. When all those fields are the same, the multicast announcement that arrived earlier is considered better. After selecting the best multicast announcement, the node generates the fields of its own multicast announcement in the following way:

- ☐ Core ID: core ID in the best multicast announcement
- ☐ Group ID: group ID in the best multicast announcement
- ☐ Sequence number: sequence number in the best multicast announcement
- ☐ Distance to core: One plus the distance to core in the best multicast announcement
- ☐ Parent: The neighbor from which it received the best multicast announcement
- ☐ One-hop neighborhood: All known one-hop neighbors that are part of the mesh
- ☐ Mesh membership status: The procedure to decide whether a particular node is a mesh member is described in the following section.

The connectivity list stores information about one or more routes that exist to the core. When a core change occurs for a particular group, then nodes clear the entries of their old connectivity list and builds a new one, specific to the new core. Figure 1 illustrates the propagation of multicast announcements and the building of connectivity lists. Solid arrows indicate the neighbor from which a node receives its best multicast announcement. In this example, Node 6 has three entries in its connectivity list for neighbors 5, 1, and 7 respectively. However it chooses the entry it receives from 5 as the best entry, because it has the shortest distance to the core and has been received earlier that the one from node 1. Node 6 uses this entry to generate its own multicast announcement, which specifies Core ID = 11, Group ID = 224.0.0.1, Sequence Number = 79, Distance to Core = 2 and Parent = 5.

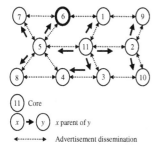

	Connectivity List at node 6		
	Core Id = 11, Group Id = 224.0.0.1, Seq. No 79		
Neighbor	Multicast Announcement		Time
	Distance to core	Parent	(ms)
5	1	11	12152
1	1	11	12180
7	2	5	12260

(11) Core

(x) ➔ (y)   x parent of y

◂·········▸   Advertisement dissemination

Fig. 2. Dissemination of multicast announcements and establishment of the connectivity lists

When a node wants to send data packets to the group it forwards them to the node from which it received its best multicast announcement. If that link is broken then it tries its next best and so on. Hence each node in the network has one or more routes to the core. The multicast announcement sent by the core has distance to core set to zero and parent field set to *invalid address*.

Multicast announcements are generated by the core every T seconds (in our simulations we set this value to three). After receiving a multicast announcement with a fresh sequence number, nodes wait for a short period (e.g. 100 ms) to collect multicast announcements from multiple neighbors before generating their own multicast announcement.

When multiple groups exist, nodes aggregate all the fresh multicast announcements they receive, and broadcast them periodically every *multicast announcement interval*. However, multicast announcements representing groups being heard for the first time, resulting in a new core, or resulting in changes in mesh member status are forwarded immediately, without aggregation. This is to avoid delays in critical operations, like core elections and mesh establishment.

3.3 Mesh Establishment and Maintenance

Initially only receivers consider themselves as mesh members and set the mesh membership flag to TRUE in the multicast announcements they send. Non-receivers consider themselves mesh-members if they have at least one mesh child in their connectivity list. A neighbor in the connectivity list is a mesh child if : (1) its mesh member flag is set, (2) the distance to core of the neighbor is larger than the node's own distance to core, and (3) the multicast announcement corresponding to this entry was received in within a time period equal to two multicast announcement intervals. Condition 3 is used to ensure that a neighbor is still in the neighborhood. If a node has a mesh child (therefore is a mesh member), then it means that it lies on a shortest path from a receiver to the core. The above scheme results in the inclusion of *all* shortest paths from the receiver to the core in the mesh.

Whenever a node generates a multicast announcement, it sets the mesh member flag depending on whether or not it is a mesh member at that point of time. In addition to generating a multicast announcement when it detects a core change, or when it receives a fresh multicast announcement, a node also generates a multicast announcement when it detects a change in its mesh member status. This could occur when a node detects a mesh child for the first time, or when a node that previously had a mesh child detects that it has no mesh children.

3.4 Forwarding Data Packets

The parent field of the connectivity list entry for a particular neighbor corresponds to the node from which the neighbor received its best multicast announcement. This field allows nodes that are non-members to forward multicast packets towards the mesh of a group. A node forwards a multicast data packet it receives from its neighbor if the parent for the neighbor is the node itself. Hence, multicast data packets move hop by hop, until they reach mesh members.

In the basic PUMA protocol, when a multicast message reaches a mesh member it is flooded along the whole mesh. Duplicated messages are discarded at receivers and relay nodes based on the message sequence number. This scheme can lead to unnecessary overhead because a given node can be covered by more than one neighbor and hence receive a message more than once. This mechanism is improved in DPUMA using the greedy set cover algorithm described in Section 2.2.3. In this way, nodes dynamically determine the set of nodes that should retransmit a packet in order to cover the entire two-hop neighborhood and eventually the whole mesh. In this paper we present two different variants of the algorithm to compute the dominating set. DPUMA-1k computes 1-dominating sets, meaning that each node in the 2-hop neighborhood must be covered by at least one 1-hop neighbor. DPUMA-2k computes 2-dominating sets meaning that whenever it is possible, nodes in the 2-hop neighborhood are covered by two 1-hop neighbors. Here we are experimenting with the trade-off between augmenting redundancy in order to achieve high reliability; and saving bandwidth and energy reducing the number of retransmissions. As our experimental results show, under many scenarios DPUMA-1k outperform PUMA, DPUMA-2k, ODMRP and MADOV. This is more evident under high load conditions where DPUMA-1k achieves higher reliability by reducing the number of collisions, while saving bandwidth. The algorithm followed by mesh members to disseminate a data packet is as follows:

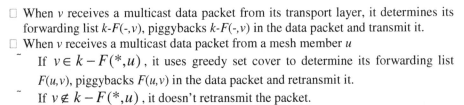

The routing of data packets from senders to receivers is also used to update the connectivity list. When a nonmember transmits a packet, it expects its parent to forward the packet. Because all communication is broadcast, the node also receives the data packet when it is forwarded by its parent. This serves as an implicit acknowledgment of the packet transmission. If the node does not receive an implicit acknowledgment within ACK TIMEOUT, then it removes the parent from its connectivity list.

4   Experimental Results

We compared the performance of DPUMA against the performance of PUMA, ODMRP and MAODV which are representatives of the state of the art in multicast routing protocols for MANETs. We used the discrete event simulator Qualnet [17] version 3.5. The distribution of Qualnet itself had the ODMRP code, which was used for ODMRP simulations. The MAODV code for Qualnet was obtained from a third party[1] who wrote the code independently of our effort following the MAODV IETF Specification [19]. We employed RTS/CTS when packets were directed to specific neighbors. All other transmissions used CSMA/CA. Each simulation was run for five different seed values. To have meaningful comparisons, all timer values (i.e., interval for sending JOIN requests and JOIN tables in ODMRP and the interval for sending multicast announcements in DPUMA) were set to 3 seconds. Table 1 lists the details about the simulation environment and the parameters of the MAODV code.

The metrics used for the evaluations are the average number of data packets delivered at receivers and the average number of data packets relayed. We didn't include the control overhead metric because in [1] it has been demonstrated that PUMA incurs far less overhead than ODMRP and MAODV. Since DPUMA uses the same signaling mechanism as PUMA, the same performance trend observed for PUMA still holds.

Table 1. Simulation environment and parameters used in the MAODV code

Simulation Environment		MAODV parameters	
Total Nodes	50	Allowed Hello Loss	2
Simulation time	100s	Grp. Hello Interval	5 s
Simulation area	1300×1300m	Hello Interval	1 s
Node Placement	Random	Hello life	3 s
Mobility Model	Random Waypoint	Pkt. Id save	3 s
Pause Time	10s	Prune Timeout	750 ms
Min-Max Vel.	0 – 10 m/s	Rev Route life	3 s
Transmission Power	15 dbm	RREQ retries	2
Channel Capacity	2000000 bps	Route Disc. Timeout	1 s
MAC protocol	IEEE 802.11	Retransmit timer	750 ms
Data Source	MCBR		
Num. of pkts. sent per src.	1000		

---

[1] We thank Venkatesh Rajendran for providing the simulation code for MAODV.

4.1 Packet Size and Traffic Load

In our first experiment we varied the packet size from 64 to 1024 bytes. This experiment shows the ability of DPUMA-1k to reduce the channel contention by reducing the number of nodes that broadcast a packet in a given time. There are two senders and one group composed of 30 nodes. Only one of the two senders belongs to the group. As it is shown in Figure 3, for packets of 256 bytes and larger DPUMA-1k achieves higher delivery ratios than the other protocols while incurring less retransmission overhead. From the same figure, we can notice that MAODV performs very well for small packet sizes but as the size increases the delivery ratio drops dramatically. Under this particular scenario, for packets of 256 bytes and larger the simulations for MAODV were unable to finish within the 100s.

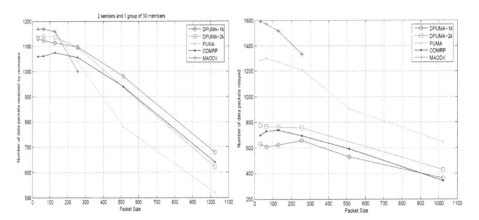

Fig. 3. Average number of data packets received at receivers and average number of data packets relayed when varying the packet size

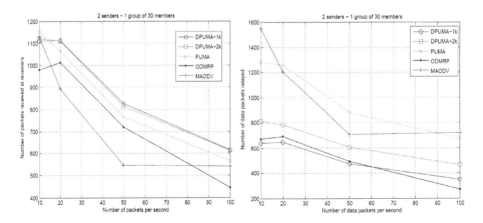

Fig. 4. Average number of data packets received at receivers and average number of data packets relayed when varying the number of packets per second sent by the senders

In our second experiment we varied from 10 to 100 the number of packets per second that the senders multicast to a group of 30 receivers. As in the previous case, only one of the senders belongs to multicast group. As shown in Figure 4, for light loads PUMA achieves higher delivery ratios than the other protocols but at the cost of a considerable retransmission overhead. However, when traffic load increases DPUMA-1k delivers a larger number of packets. As in the previous case, this is due to the fact that DPUMA is able to reduce collisions.

4.2 Number of Groups and Senders

In these two experiments we tested the ability of the protocols to handle different numbers of senders and groups. Figure 5 shows the results of varying the number of senders from 1 to 20 with a group composed of 20 members. Figure 6 shows the results of varying the number of groups from 1 to 5 with two nodes generating packets

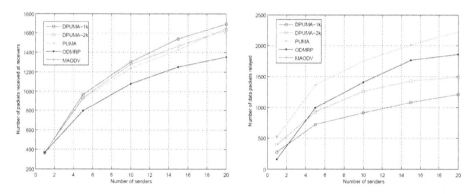

Fig. 5. Average number of data packets received at receivers and average number of data packets relayed when varying the number of senders

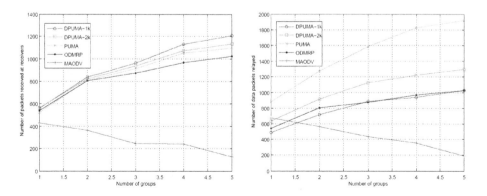

Fig. 6. Average number of data packets received at receivers and average number of data packets relayed when varying the number of groups

for each group. For both experiments, each sender generates 20 packets per second with a packet size of 256 bytes. In both cases DPUMA-1k achieves the highest delivery ratio while generating much less retransmissions. Our results confirm that mesh-based protocols (DPUMA, PUMA, and ODMRP) consistently tend to perform better than the tree-based protocol (MAODV). The main reason is that tree-based protocols expend too many packets trying to maintain the tree structure of the multicast group. In the experiments where we vary the number of senders, MAODV was unable to complete the simulation within 100s when having five or more senders. DPUMA and PUMA outperform ODMRP because the latter perform flooding of control packets per source per group which leads to congestion and to the reduction of reliability shown in the results.

5 Conclusions

In this paper we presented DPUMA, a mesh-based multicast protocol that carries out its three basic tasks (electing a core, establishment of the mesh and getting 2-hop neighborhood information) by flooding a single control packet per each multicast group. When diffusing a data packet over the mesh, nodes in DPUMA compute the k-dominating set to reduce the number of nodes that need to broadcast the packet in order to cover the whole mesh. We carried out a series of simulations were we show that under moderate and high loads, DPUMA achieves higher delivery ratios than ODMRP, MAODV and PUMA while incurring in less retransmission overhead.

Our current research focuses on the energy savings obtained by the reduced overhead of DPUMA and how this can be used to extend the lifetime of the whole MANET. We are also working on defining adaptive approaches to unify PUMA and DPUMA in order to get the reliability achieved by PUMA under light loads and the one achieved by DPUMA under high loads. We are also working on reducing the impact of mobility over the accuracy of the algorithms for computing DCDS.

References

1. R. Vaishampayan and J.J. Garcia-Luna-Aceves, Efficient and Robust Multicast Routing in Mobile Ad Hoc Networks , Proc. IEEE MASS 2004: The 1st IEEE International Conference on Mobile Ad-hoc and Sensor Systems, Fort Lauderdale, Florida, October 25-27, 2004
2. J. J. Garcia-Luna-Aceves and E.L. Madruga, "The core assisted mesh protocol," IEEE Journal on Selected Areas in Communications, Special Issue on Ad-Hoc Networks, vol. 17, no. 8, pp. 1380–1394, August 1999
3. S.J. Lee, M. Gerla, and Chian, "On-demand multicast routing protocol," in Proceedings of WCNC, September 1999
4. S.J. Lee, et-al, "A performance comparison study of ad hoc wireless multicast protocols," in Proceedings of IEEE INFOCOM, Tel Aviv, Israel, March 2000
5. E. Royer and C. Perkins, "Multicast operation of the ad hoc on-demand distance vector routing protocol," in Proceedings of Mobicom, August 1999
6. L. Ji and M. S. Corson, "A lightweight adaptive multicast algorithm", in Proceedings of IEEE GLOBECOM 1998, December 1998, pp. 1036–1042

7. L. Ji and M.S. Corson, "Differential destination multicast - a manet multicast routing protocol for small groups," in Proceedings of IEEE INFOCOM, April 2001
8. P. Sinha, R. Sivakumar, and V. Bharghavan, "Mcedar: Multicast core extraction distributed ad-hoc routing," in Proceedings of the Wireless Communications and Networking Conference, WCNC, September 1999, pp. 1313–1317
9. C.K. Toh, G. Guichala, and S. Bunchua, "Abam: On-demand associativity-based multicast routing for ad hoc mobile networks," in Proceedings of IEEE Vehicular Technology Conference, VTC 2000, September 2000, pp. 987–993
10. S.K. Das, B.S. Manoj, and C.S. Ram Murthy, "A dynamic core based multicast routing protocol for ad hoc wireless networks," in Proceedings of the ACM MobiHoc, June 2002
11. V. Devarapalli and D. Sidhu, "MZR: A multicast protocol for mobile ad hoc networks," in ICC 2001 Proceedings, 2001
12. S. Lee and C. Kim, "Neighbor supporting ad hoc multicast routing protocol," in Proceedings of the ACM MobiHoc, August 2000
13. C. Perkins and E. Royer, "Ad hoc on demand distance vector (AODV) routing," in Proceedings of the 2nd IEEE Workshop on Mob. Comp. Sys. and Applications, February 1999
14. Scalable Network Technologies, "Qualnet 3.5," http://www.scalablenetworks.com/
15. R. Perlman, "An algorithm for distributed computation of a spanning tree in an extended lan," in ACM Special Interest Group on Data Com. (SIGCOMM), 1985, pp. 44–53
16. E.M. Royer and C.E. Perkins., "Multicast ad hoc on demand distance vector (MAODV) routing," Internet-Draft, draft-ietf-draftmaodv-00.txt
17. Park and Corson, "Highly adaptive distributed routing algorithm for mobile wireless network," in Proceedings of IEEE INFOCOM, March 1997
18. Deering S. E., et-al, "The PIM Architecture for Wide-Area Multicast Routing", IEEE/ACM Transactions on Networking, Vol.4, No.2, April 1996
19. M.A. Spohn and J.J. Garcia-Luna-Aceves, "Enhanced Dominant Pruning Applied to The Route Discovery Process of On-demand Routing Protocols," Proc. IEEE IC3N 03: Twelfth Int. Conf. on Computer Com. and Networks, Dallas, Texas, October 20 - 22, 2003
20. Wei Lou, Jie Wu, "On Reducing Broadcast Redundancy in Ad Hoc Wireless Networks," IEEE Transactions on Mobile Computing, Vol. 1, Issue 2 (April 2002), Pages: 111 – 123
21. Wei Lou, Hailan Li, "On calculating connected dominating set for efficient routing in ad hoc wireless networks," Proc. of the 3rd int. workshop on Discrete algorithms and methods for mobile comp. and communications, Seattle, Washington, United States, Pages: 7 – 14
22. H. Lim and C. Kim, "Flooding in wireless ad hoc networks," Computer Communications, vol. 24, February 2001

Efficient Broadcasting in Self-organizing Multi-hop Wireless Networks

Nathalie Mitton and Eric Fleury

INRIA/ARES - INSA de Lyon - 69621 Villeurbanne Cedex
Tel: 33-(0)472-436-415
firstname.lastname@insa-lyon.fr

Abstract. Multi-hop wireless networks (such as ad-hoc or sensor networks) consist of sets of mobile nodes without the support of a pre-existing fixed infrastructure. For the purpose of scalability, ad-hoc and sensor networks may both need to
be organized into clusters and require some protocols to perform common global
communication patterns and particularly for broadcasting. In a broadcasting task,
a source node needs to send the same message to all the nodes in the network.
Some desirable properties of a scalable broadcasting are energy and bandwidth
efficiency, *i.e.*, message retransmissions should be minimized. In this paper, we
propose to take advantage of the characteristics of a previous clustered structure
to extend it to an efficient and scalable broadcasting structure. In this way, we
build only one structure for both operations (organizing and broadcasting) by applying a distributed clustering algorithm. Our broadcasting improve the number
of retransmissions as compared to existing solutions.

Keywords: multi-hop wireless networks, self-organization, broadcasting.

1 Introduction

Multi-hop wireless networks (such as ad-hoc or sensor networks) consist of sets of
mobile wireless nodes without the support of a pre-existing fixed infrastructure. They
offer unique benefits for certain environments and applications as they can be quickly
deployed. Each node acts as a router and may arbitrary appear or vanish. Protocols
must adapt to frequent changes of the network topology. Ad-hoc networks and sensor
networks are instances of multi-hop wireless networks. In ad-hoc networks, nodes are
independent and may move at any time at different speeds. In sensor networks, nodes
are more static and collect data they have to forward to specific nodes. For the purpose
of scalability, ad-hoc and sensor networks may both need to be organized into clusters and require some protocols to perform common global communication patterns as
for broadcasting. An organization is needed to allow the scalability in terms of number of nodes or/and node density without generating too much traffic (for routing, for
instance) neither too much information to store. A common solution is to adapt a hierarchical organization by grouping nodes into clusters and bind them to a leader. Such
an organization may allow the application of different routing schemes in and between
clusters. In a broadcasting task, a source node needs to send the same message to all the
nodes in the network. Such a functionality is needed, for example, when some queries

V.R. Sirotiuk and E. Chávez (Eds.): ADHOC-NOW 2005, LNCS 3738, pp. 192–206, 2005.

about the measures (in sensor networks) or a node location (in ad hoc networks) need to be disseminated over the whole network or within a cluster. Broadcasting in a cluster may also be useful for synchronizing nodes. The desirable properties of a scalable broadcasting are reachability, energy and bandwidth efficiency.

In this paper, we propose to take advantage of the characteristics of our previous wireless network clustered structure [9] to extend it to an efficient and scalable broadcasting structure. In this way, we build only one structure for both operations (organizing and broadcasting) by applying a distributed clustering algorithm. The resulting broadcasting, analyzed and compared to some other existing protocols, saves more retransmissions. The remainder of this paper is organized as follows. Section 2 presents some previous broadcasting solutions. Section 3 summarizes our previous work and highlights some characteristics of our cluster organization which might be useful for a broadcasting task. Section 4 presents the way we extend our cluster structure into a broadcasting structure and details our broadcasting scheme. Section 5 compares several broadcasting schemes by simulation and presents the results. Finally, we conclude in Section 6 by discussing possible future areas of investigation.

2   Broadcasting in Multi-hop Wireless Networks

The desirable properties of a scalable broadcasting are reachability, energy and bandwidth efficiency. Indeed, as in a wireless environment, a node wastes energy when transmitting as well as receiving a packet, the number of retransmissions and receptions should be minimized. In this paper, we only consider reliability at the network layer, *i.e.*, a broadcasting scheme is said reliable if every node connected to the source receives a broadcast packet in a collision free environment. In this section, we focus on the solutions proposed in the literature for network layer broadcasting schemes which are based on dominating set and use omni-directional antennas.

The easiest way to broadcast a message over a network is the blind flooding, *i.e.*, each node re-emits the message upon first reception of it. Obviously, this causes many collisions and wastes bandwidth and energy. Therefore, this broadcasting technique can not be envisaged over large scale or very dense networks. This gave birth to more intelligent broadcasting protocols which try to minimize the number of retransmissions by selecting a subset of nodes allowed to forward a message. This subset is called a dominating set. To obtain a reliable broadcasting scheme, each node in the network should be either in the dominating set (and is called an internal node) or neighboring at least one node in the dominating set. The main challenge is to find a connected dominating set which minimizes the number of these transmitters as well as the number of copies of a same message received by a node. I. Stojemovic and J. Wu [15] classify the broadcasting schemes according to the kind of dominating set they use: cluster-based, source-dependent dominating set and source-independent dominating set schemes. All of them provide a reliable broadcasting task with a relevant number of retransmissions saved, compared to the blind flooding.

In oldest solutions, cluster-based, [3,5,8], the idea is that every node which has the lowest Id or the highest degree (Linked Cluster Architecture-LCA protocol) in its 1-neighborhood becomes a cluster-head. If a non-cluster-head node can hear more than one cluster-head among its neighbors, it becomes a gateway. The dominating set is thus composed of both the cluster-heads and the gateways. From it, some optimizations have been proposed to localize the maintenance process and avoid the chain reaction which can occur in case of node mobility [4] or to limit the number of gateways [17].

In solutions based on source-dependent dominating set [7,13], the sending nodes select adjacent nodes that should relay the message. The set of relays of a node u is chosen to be minimal and such that each 2-hop neighbor of node u has at least one neighbor among the relays of u. Methods differ in details on how a node determines its forwarding list. The most popular of them is the one based on the Multi-Point Relay (MPR) of OLSR [13]. In OLSR, the MPR are also used for propagating the routing information. This kind of structure has thus a double use too.

In solutions based on source-independent dominating set, the set of internal nodes is independent of the source node. This is the case of our proposal. Many solutions have been proposed. A simple and efficient algorithm, the NES (*Neighbors Elimination Based Scheme*) [14,16], introduces the notion of *intermediate* nodes. Node A is *intermediate* if at least two of its neighbors are not direct neighbors. Two selection rules are then introduced to reduce the number of transmitter nodes. From it, several solutions have thus been derived [2,14].

3   Previous Work and Main Objectives

In this section, we summarize our previous clustering work on which our broadcasting scheme proposition relies. Only basis and features which are relevant for broadcasting are mentioned here. For more details or other characteristics of our clustering heuristic, please refer to [9,10,11]. For the sake of simplicity, let's first introduce some notations. We classically model a multi-hop wireless network, by a graph $G = (V, E)$ where V is the set of mobile nodes ($|V| = n$) and $e = (u, v) \in E$ represents a bidirectional wireless link between a pair of nodes u and v if and only if they are within communication range of each other. We note $\mathcal{C}(u)$ the cluster owning the node u and $\mathcal{H}(u)$ the cluster-head of this cluster. We note $\Gamma_1(u)$ the set of nodes with which u shares a bidirectional link. $\delta(u) = |\Gamma_1(u)|$ is the degree of u.

Our objectives for introducing our clustering algorithm were motivated by the fact that in a wireless environment, the less information exchanged or stored, the better. First, we want a cluster organization suitable for large scale multi-hop networks, *i.e.*, non-overlapping clusters not restricted to a given fixed radius/diameter but with a flexible radius (*The clustering schemes mentioned in Section 2 have a radius of* 1, *in [1,6] the radius is set a priori*) and able to adapt to the different topologies. Second, we want the nodes to be able to compute the heuristic from local information, only using their 2-neighborhood. *In [1], if the cluster radius is set to d, the nodes need to gather information up to d hops away before taking any decision.* Finally, we desire an organization robust and stable over node mobility, *i.e.*, that we do not need to re-compute for each single change in the topology. Therefore, we introduce a new metric called *density*. The notion of density characterizes the "relative" importance of a node in the network and within its neighborhood. The underlying idea is that this link density (noted $\rho(u)$) should smooth local changes down in $\Gamma_1(u)$ by considering the ratio between the number of links and the number of nodes in $\Gamma_1(u)$.

Definition 1 (density).

The density of a node $u \in V$ is $\rho(u) = \frac{|\{e=(v,w)\in E \mid w\in\{u\}\cup\Gamma_1(u) \ and \ v\in\Gamma_1(u)\}|}{\delta(u)}$.

On a regular basis, each node locally computes its density value and regularly locally broadcasts it to its 1-neighbors (*e.g.*, using `Hello` packets). Each node is thus able to compare its density value to its 1-neighbors' and decides by itself whether it joins one of them (the one with the highest density value) or it wins and elects itself as cluster-head. In case of ties, the node with the lowest Id wins. In this way, two neighbors can not be both cluster-heads. If node u has joined node w, we say that w is node u parent in the clustering tree (noted $\mathcal{P}(u) = w$) and that node u is a child of node w (noted $u \in Ch(w)$). A node's parent can also have joined another node and so on. A cluster then extends itself until it reaches another cluster. The cluster-head is the node which has elected itself. If none of the nodes has joined a node u ($Ch(u) = \emptyset$), u becomes a leaf. Thus, in this way, as every node chooses itself a parent among its 1-neighbors, a cluster can also be seen as a directed tree which root is the cluster-head. When building clusters, we then also build a spanning forest composed of as many directed acyclic graphs (DAG) as clusters.

3.1 Some Characteristics of This Clustering Algorithm

This algorithm stabilizes when every node knows its *correct* cluster-head value. It has been proved by theory and simulations to self-stabilize within an expected low, constant and bounded time [11]. It has also been proved that a cluster-head is aware of an information sent by any node of its cluster in a low, constant and bounded time. The number of clusters built by this heuristic has been showed analytically and by simulations to tend toward a low and constant asymptote when the number of nodes in the network increases. Moreover, compared to other clustering schemes as DDR [12] or Max-min d cluster [1], our cluster organization has revealed to be more stable over node mobility and arrivals and to offer better behaviors over non-uniform topologies (see [9]). Other interesting features for broadcasting obtained by simulations are gathered in Table 1. They are commented in Section 3.2.

Table 1. Some clusters and clustering trees characteristics

	500 nodes	600 nodes	700 nodes	800 nodes	900 nodes	1000 nodes
# clusters/trees	11.76	11.51	11.45	11.32	11.02	10.80
Diameter $D(\mathcal{C})$	4.99	5.52	5.5	5.65	6.34	6.1
Cluster-head eccentricity	3.01	3.09	3.37	3.17	3.19	3.23
Tree depth	3.27	3.34	3.33	3.34	3.43	3.51
% leaves	73,48%	74,96%	76,14%	76,81%	77,71%	78,23%
Non-leaves'degree(in trees)	3.82	3.99	4.19	4.36	4.51	4.62
Voronoi: Euclidean distance	84.17%	84.52%	84.00%	83.97%	83.82%	83.70%
Voronoi: # of hops	85.43%	84.55%	84.15%	83.80%	83.75%	83.34%

3.2   Objectives

As explained earlier, our clustering heuristic leads at the same time to the formation of a spanning forest. We thus propose to use these *"clustering trees"* as a basis for the broadcasting task. This broadcasting scheme is dominating set-based where the non-leaf nodes (internal nodes) belong to the dominating set. As mentioned in Table 1, a great proportion of nodes (about 75%) are actually leaves, therefore a broadcasting scheme based on this dominating set is expected to save many retransmissions. As the *clustering trees* form a spanning forest, the set of trees actually is a dominating set of the network but is not a connected dominating set as the trees are independent. So, to perform a reliable broadcasting task in the whole network, we need to connect these trees by electing some gateways. Our gateway selection is described in Section 4.

As already mentioned, each node only needs to know its 2-neighborhood to choose its parents, and to know whether it has been chosen as parent by one of its neighbors or it is a leaf. Thus, the forwarding decision of a non-gateway node is based on local state information. Only the gateway selection can be qualified of quasi-local (according to the classification of [17]) as only few nodes need information up to 4 hops away (tree depth). Thus, our broadcasting scheme does not induce a high costly maintenance. We propose to use this structure not only to perform a traditional broadcasting in the whole network but also for broadcasting in a cluster only. This kind of task might be interesting for clustered architectures when, for instance, a cluster-head needs to spread information only in its cluster like in sensor networks, for instance, where the base station may need to update devices or spread a query over them. The eccentricity of a node is the greater distance in number of hops between itself and any other node in its cluster. We can see in Table 1 that the tree depth is pretty low and close to the optimal we could expect which is the cluster-head eccentricity. This presents a good property for performing a broadcasting within our clusters. Indeed, none node is really far away from its cluster-head and can expect to receive quickly an information it would spread. Moreover, we computed the proportion of points closer to their cluster-head than any other one in Euclidean distance (Voronoi: Euclidean distance in Table 1) and in number of hops (Voronoi: # of hops in Table 1). Results show that a large part of nodes (more than 83%) lays in the Voronoi cell of their cluster-head whatever the process intensity. This characteristic is useful in terms of broadcasting efficiency as if the cluster-heads need to spread information over their own cluster, if most of the nodes are closer to the one which sends the information, we save bandwidth, energy and latency.

4   Our Contribution

In this section, we first propose an algorithm for the gateway selection, then we detail the two kinds of broadcasting: within a cluster and in the whole network.

4.1   Gateway Selection

A gateway between two neighboring clusters $\mathcal{C}(u)$ and $\mathcal{C}(v)$ actually is a pair of nodes $\langle x, y \rangle$ noted $\mathcal{G}ateway(\mathcal{C}(u), \mathcal{C}(v)) = \langle x, y \rangle$ such that $x \in \mathcal{C}(u)$, $y \in \mathcal{C}(v)$ and

$x \in \Gamma_1(y)$. In such a pair, we will say that node x is the gateway node and that node y is the mirror-gateway node of the gateway. If $\mathcal{C}(u)$ and $\mathcal{C}(v)$ are two neighboring clusters, we note the gateway $\mathcal{G}ateway(\mathcal{C}(u), \mathcal{C}(v)) = \langle x = GW(\mathcal{C}(u), \mathcal{C}(v)), y = GWm(\mathcal{C}(u), \mathcal{C}(v)) \rangle$, where $GW(\mathcal{C}(u), \mathcal{C}(v))$ is the gateway node and $GWm(\mathcal{C}(u), \mathcal{C}(v))$ is the mirror-gateway node. Note that $GW(\mathcal{C}(u), \mathcal{C}(v)) \in \mathcal{C}(u)$ and $GWm(\mathcal{C}(u), \mathcal{C}(v)) \in \mathcal{C}(v)$.

To select a gateway between two clusters, we thus need to define a pair of nodes. Our selection algorithm runs in two steps. The first step allows each frontier node to locally choose its "mirror(s)" in the neighboring cluster(s). We call a node u a frontier node if at least one of its neighbors does not belong to the same cluster than u. A frontier node and its mirror then form an eligible pair. The second step selects the most appropriate pair as the gateway. The algorithm tries to promote the selection of the nodes which already are internal nodes in order to minimize the size of the dominating set.

Mirror node selection. As seen in Section 3, as the density-based clustering algorithm uses the node Id as the last decision criterion, every node u might be aware in an expected bounded and low time, whether it exists among its neighbors a node v which does not belong to the same cluster than u. If so, node u is a frontier node and so is a possible gateway node for $\mathcal{G}ateway(\mathcal{C}(u), \mathcal{C}(v))$. Each frontier node u then selects its *mirror node* among its neighbors which do not belong to $\mathcal{C}(u)$. To do so, u first selects the non-leaf nodes, *i.e.*, the internal nodes/transmitters in every case and chooses among them the node with the highest density value. If every node $v \in \Gamma_1(u)$ such that $\mathcal{C}(u) \neq \mathcal{C}(v)$ is a leaf, u chooses the node with the lowest degree in order to limit the receptions induced by an emission of the mirror. In case of ties, the lowest Id decides. If u is a frontier node of the cluster $\mathcal{C}(v)$ ($\mathcal{C}(v) \neq \mathcal{C}(u)$), we note $m(u, \mathcal{C}(v))$ the mirror chosen by u in $\mathcal{C}(v)$. Note that if a node u is a frontier node for several clusters, it has to select a mirror for each cluster.

Algorithm 1 Mirror selection

For each frontier node u, *i.e.*, $\exists v \in \Gamma_1(u)$ s.t. $\mathcal{C}(v) \neq \mathcal{C}(u)$
  For each cluster $\mathcal{C}$ **for which** u **is a frontier node:** $\mathcal{C} \neq \mathcal{C}(u)$ **and** $\exists v \in \Gamma_1(u) \cap \mathcal{C}$.
    Select S the set of nodes such that $S = \mathcal{C} \cap \Gamma_1(u) \cap \{v \mid Ch(v) \neq \emptyset\}$.
    ▷ *u first selects the set of the non-leaf nodes as they are transmitters in every case.*
    if $(S \neq \emptyset)$ **then** Select S' the set of nodes such that $S' = \{v \mid v = max_{w \in S}\rho(w)\}$.
    ▷ *u collects internal nodes with the highest density in order to promote stability.*
    else ▷ *All the possibly mirrors of u are leaves.*
      $S = \{\mathcal{C} \cap \Gamma_1(u)\}$.
      Select S' the set of nodes such that $S' = \{v \mid v = min_{w \in S}\delta(w)\}$.
      ▷ *u collects the leaves with the lowest degree in order to minimize the receptions induced by the addition of this node in the dominating set.*
    end
    if $(S' = \{v\})$ **then** $m(u, \mathcal{C}) = v$.
    ▷ *There are no ties. S' contains only one node: the mirror of u.*
    else $m(u, \mathcal{C}) = v$ such that $Id(v) = min_{w \in S'}Id(w)$.
    ▷ *There are ties. u elects the node with the lowest Id.*
    end

Gateway selection. Once each frontier node has chosen its *mirror*, we have to choose the most appropriate pair as gateway. Once a gateway node u is elected as $GW(C(u),$ $C(x))$, we have $\mathcal{G}ateway(C(u), C(v)) = \langle GW(C(u), C(x)), mirror(GW(C(u),$ $C(x)))\rangle$. According to the taxonomy of [17], this step is quasi-local unlike the first one which is local. The gateways are *directed* gateways in the meaning that two clusters are linked by two gateways $\mathcal{G}ateway(C(u), C(v))$ and $\mathcal{G}ateway(C(v), C(u))$ which may be different.

The gateway selection we propose is distributed *i.e.*, a selection is performed at every level in the tree and tries to limit useless receptions by favoring internal nodes. Frontier nodes send to their parent the following information: their Id, whether they are leaves and whether they have a leaf as mirror. Each parent selects the best candidate among its children and sends the same information up to its own parent and so on, up to reach the cluster-head. Thus, the selection is semi-distributed as every internal node eliminates some candidates. In this way, only small size packets are forwarded from the frontier nodes to the cluster-head. As mentioned in Table 1, the mean degree of all the internal nodes (cluster-head included) is small and constant whatever the number of nodes, which induce a small and bounded number of messages at each level which is also bounded by a low constant [11].

Table 2. Number of gateways selected and used per cluster for a global broadcasting initiated at a randomly-chosen source in function of the intensity process

	500nodes	600nodes	700nodes	800nodes	900nodes	1000nodes
#clusters	11.93	11.64	11.36	11.30	11.14	10.72
#gw selected per cluster	5.86	6.02	6.16	6.20	6.22	6.26
#gw used per cluster	1.76	1.74	1.73	1.76	1.68	1.66

Let's express that v is in the subtree rooted in u (noted $v \in sT(u)$) if u is the parent of node v or if the parent of node v is in the subtree rooted in u: $\{v \in sT(u) \cap \Gamma_1(u)\} \Leftrightarrow$ $\{v \in Ch(u)\}$ or $\{v \in sT(u) \cap \bar{\Gamma}_1(u)\} \Leftrightarrow \{\mathcal{P}(v) \in sT(u)\}$.

The best candidate choice is performed as follows. For each of the neighboring clusters $C(x)$ of its subtree, an internal node u considers the set G of the candidate nodes (frontier nodes) ($G = \{v \in sT(u) \mid \exists w \in \Gamma_1(v) \mid C(w) \neq C(u)\}$). Then, it selects among them the subset $G' \subset G$ of the internal nodes, still in order to limit the

---

Algorithm 2 Gateway selection

For each internal node u:
    For each cluster $C \neq C(u)$ for which $\exists v \in sT(u)$ which is a frontier node:
        Gather the set G of candidate nodes: $G = \{v \in sT(u) \mid \exists w \in \Gamma_1(v) \mid C(w) = C\}$.
        Select $G' \subset G$ the set of nodes v s.t. $G' = G \cap \{v \mid Ch(v) \neq \emptyset\}$.
        ▷ *u first selects the set of the non-leaf nodes as they are transmitters in every case.*
        if $(G' \neq \emptyset)$ **then**
        ▷ *There are non-leaf candidates. u will favor the ones with a non-leaf mirror and/or the highest density.*

Select $G" \subset G'$ the set of nodes s.t. $G" = G' \cap \{v | Ch(m(v, \mathcal{C}) \neq \emptyset\}$.
if $(G" \neq \emptyset)$ **then** Select $Finalist \subset G"$ the set of nodes s.t. $Finalist = \{v | \rho(v) = max_{w \in G"} \rho(w)\}$.
▷ *Internal Node↔Internal Node Gateway.*
else Select $Finalist \subset G"$ the set of nodes s.t. $Finalist = \{v | \rho(v) = max_{w \in G'} \rho(w)\}$.
▷ *Internal Node↔Leaf Gateway.*
end
else
▷ *All candidates are leaves. u will favor the ones with a non-leaf mirror and/or the smallest degree.*
Select $G" \subset G$ the set of nodes s.t. $G" = G \cap \{v | Ch(m(v, \mathcal{C}) \neq \emptyset\}$.
if $(G" \neq \emptyset)$ **then** Select $Finalist \subset G"$ the set of nodes s.t. $Finalist = \{v | \delta(v) = min_{w \in G"} \delta(w)\}$.
▷ *Leaf↔Internal Node Gateway.*
else Select $Finalist \subset G"$ the set of nodes s.t. $Finalist = \{v | \delta(v) = min_{w \in G'} \delta(w)\}$.
▷ *Leaf↔Leaf Gateway.*
end
end
if $(Finalist = \{v\})$ **then** $Winner = v$.
else $Winner = v | Id(v) = min_{w \in Finalist} Id(w)$.
▷ *There are ties. u elects the node with the lowest Id.*
end
if $u = \mathcal{H}(u)$ **then** $Winner$ becomes the final gateway node: $\mathcal{G}ateway(\mathcal{C}(u), \mathcal{C}) = \langle Winner, m(Winner, \mathcal{C}) \rangle$ **else** Send the $Winner$ identity to $nF(u)$.
end

---

number of transmitter nodes. If G' is only composed of leaves, the selection is processed among all candidates of G. From there, u favors the nodes which mirror is a non-leaf node, then it uses either the density value (if remaining candidates are non-leaves) or degree (otherwise) to decide. At the end, in case of ties, the node with the lowest Id is elected as $GW(\mathcal{C}(u), \mathcal{C}(x))$. Note that, if $\mathcal{C}(u)$ and $\mathcal{C}(v)$ are two neighboring clusters, $\mathcal{G}ateway(\mathcal{C}(u), \mathcal{C}(v)) \neq \mathcal{G}ateway(\mathcal{C}(v), \mathcal{C}(u))$ in most cases.

4.2 Broadcasting Heuristic

We need to be able to perform three kinds of broadcasting tasks: local (broadcast to 1-neighbors), clusters and global (broadcast to all the nodes in the network). Therefore, we need the use of a broadcasting address in the packet[1].

When the broadcasting task is performed within a cluster $\mathcal{C}(u)$, the message is forwarded upon first reception by all the non-leaf nodes of $\mathcal{C}(u)$. When the broadcasting task is performed in the whole network, all the non-leaf nodes in the network forward the message as well as the gateway and mirror-gateway nodes under some conditions. As our gateways are directed, a gateway node $GW(\mathcal{C}(u), \mathcal{C}(w))$ forwards

---

[1] Such a broadcasting address is also used in IPv6.

Algorithm 3 Broadcasting algorithm

For all node u, **upon reception of a broadcast packet** P **coming from node** $v \in \Gamma_1(u)$
▷ *Node v is the previous hop and not necessary the broadcast source.*
    if (u receives P for the first time)
      if *global broadcasting*
        if ($Ch(u) \neq \emptyset$) **then** Forward ▷u *is an internal node.*
        else
            if ($\mathcal{C}(u) = \mathcal{C}(v)$ and $u = GW(\mathcal{C}(u), \mathcal{C}(w)) \forall w \in V$) **then** Forward
            ▷ *u is a gateway node and P is coming from its cluster.*
            end
            if ($\mathcal{C}(u) \neq \mathcal{C}(v)$ and $u = GWm(\mathcal{C}(v), \mathcal{C}(u))$) **then** Forward
            ▷ *P is coming from the cluster for which u is a mirror-gateway node.*
            end
        end
      else ▷ *It is question of a broadcasting within a cluster.*
      ▷ *P is forwarded only by internal nodes of the considered cluster.*
        if (($\mathcal{C}(v) = \mathcal{C}(u)$) and ($Ch(u) \neq \emptyset$)) **then** Forward **end**
    end

---

the message only if it is coming from its own cluster $\mathcal{C}(u)$ and a mirror-gateway node $GWm(\mathcal{C}(u), \mathcal{C}(w))$ forwards it only if it is coming from the cluster $\mathcal{C}(u)$ for which it is a mirror. But a mirror-gateway node $GWm(\mathcal{C}(u), \mathcal{C}(w))$ forwards a message coming from $\mathcal{C}(u)$ whatever the transmitter node in $\mathcal{C}(w)$ which is not necessarily the gateway node $GW(\mathcal{C}(u), \mathcal{C}(w))$ for which it is a mirror.

5   Simulation Results

We performed simulations in order to qualify the gateways elected and to compare the broadcasting tasks performed with our clustering and gateway selection schemes. This section details the results. All the simulations follow the same model. We use a simulator we developed which assume an ideal MAC layer. Nodes are randomly deployed using a Poisson point process in a 1×1 square with various levels of intensity λ. In such processes, λ represents the mean number of nodes per surface unit. The communication range R is set to 0.1 in all tests. In each case, each statistic is the average over 1000 simulations. When several algorithms are compared, they are compared for each iteration over the same node distribution.

5.1   Gateways: Election and Utilization

If we consider two neighboring clusters $\mathcal{C}(u)$ and $\mathcal{C}(v)$, we may have four different types of gateways between them:

- Leaf↔Leaf gateways: $GW(\mathcal{C}(u), \mathcal{C}(v))$ and $GWm(\mathcal{C}(u), \mathcal{C}(v))$ are both leaves. This kind of gateway is the more costly as it adds two transmitter nodes and thus induces more receptions.

- Leaf↔Internal Node gateways: $GW(\mathcal{C}(u), \mathcal{C}(v))$ is a leaf and $GWm(\mathcal{C}(u), \mathcal{C}(v))$ an internal node. This kind of gateway adds only one transmitter node. It's the less popular, as shown later by simulations.
- Internal Node↔Leaf gateways: $GW(\mathcal{C}(u), \mathcal{C}(v))$ is an internal node and $GWm(\mathcal{C}(u), \mathcal{C}(v))$ a leaf. This kind of gateway adds only one transmitter node.
- Internal Node↔Internal Node gateways: $GW(\mathcal{C}(u), \mathcal{C}(v))$ and $GWm(\mathcal{C}(u), \mathcal{C}(v))$ are both internal nodes. This kind of gateway is the one we try to favor since it does not add extra-cost at all as it does not add any transmitter neither induces any additional receptions. But, as we will see, they unfortunately are the less popular ones.

Table 2 shows the mean number of gateways a cluster has to elect and maintain and the mean number of gateways used when a global broadcasting task is performed. As we can note, the number of gateways to elect is reasonable and remains almost constant while process intensity increases. This shows a scalability feature of this heuristic. Nevertheless, this was predictable since, as we saw in Section 3, the number of clusters is constant from a certain amount of nodes, so is the mean number of neighboring clusters for a cluster and so the number of gateways to elect. Moreover, we also saw that the cluster topology was close to a Voronoi tessellation. Yet in a Voronoi tessellation, a cell has 6 neighbors in average. This is actually the mean number of gateways a cluster has to elect. Figure 1(a) gives the proportion of each kind of gateways selected. We can note that the two less elected kinds of gateways are the Leaf↔Internal Node and Internal Node↔Internal Node gateways. This is due to the fact that, by construction, most of frontier nodes are leaves. This also explains the great proportion of other kinds since, as soon as there is an internal node as a frontier node, it is elected (and thus Internal Node↔Leaf gateways are preferred to Leaf↔Internal Node ones). The more sparse the network, the less chance to find internal nodes on borders. So, the proportion of Internal node↔Leaf gateways increases with the intensity process while the proportion of Leaf↔Leaf gateways decreases.

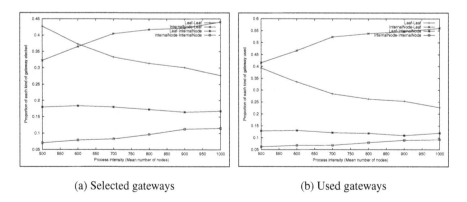

(a) Selected gateways   (b) Used gateways

Fig. 1. Proportion of each kind of selected gateways and used gateways per cluster as a function of the process intensity(+: Leaf↔Leaf; ×: Internal Node↔Leaf; *: Leaf↔Internal Node; □: Internal Node↔Internal Node;)

When a global broadcasting task is performed, all gateways are not necessary used since 2 neighboring clusters $\mathcal{C}(u)$ and $\mathcal{C}(v)$ are connected via 2 gateways $GW(\mathcal{C}(u), \mathcal{C}(v))$ and $GW(\mathcal{C}(v), \mathcal{C}(u))$ and in most cases, only one of them will be used. As Table 2 shows, the number of gateways used per cluster is quite constant and remains pretty low, always comprised between 1 and 2. This means that generally, either the broadcast enters a cluster and dies in it (in this case, it uses only one gateway), either it crosses it and thus uses two gateways (one to enter the cluster, one to leave it).

Figure 1(b) shows the proportion of each kind of gateways used when a global broadcasting operation is initiated. As we can see, most of them are the ones which add only one transmitter node. This is true even for low intensities of node when the number of Leaf↔Leaf gateways elected was the highest. This shows that our algorithm can adapt and favor internal nodes naturally. As the mean number of gateways used is low and that for each gateway, we add only one transmitter node, the induced cost is low as well.

5.2 Broadcasting Performances

In order to evaluate our algorithm, we chose to compare it to representative broadcasting schemes seen in Section 2: blind flooding, LCA [8] (cluster-based schemes), Multi-Point Relay [13] (source-dependent dominating set) and Neighbors Elimination-Based [16] (source-independent dominating set). The significant characteristics we note are the proportion of nodes which need to re-emit the message (size of the dominating set), the mean number of copies of the broadcast message that a node receives (useless receptions) and the latency (time needed for the last node to receive the broadcast packet initiated at the source). As the main goal is to limit energy consumption and bandwidth occupation in order to maximize the network lifetime, all these values have to be as low as possible.

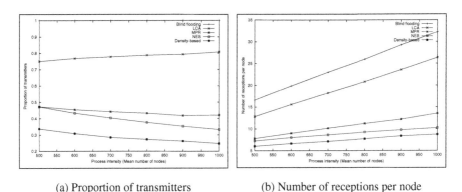

(a) Proportion of transmitters   (b) Number of receptions per node

Fig. 2. Proportion of transmitters *(a)* and Mean number of receptions per node *(b)*w.r.t the different algorithms and the mean number of nodes.(+: Blind Flooding; ×: LCA; *: MPR; □: NES; ■ Density-based;)

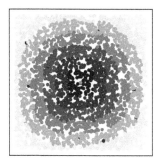 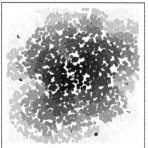

(a) Propagation with MPR   (b) Propagation with the density metric

Fig. 3. Propagation time of a general broadcasting initiated at a centered source (a) using MPR (b) using density-based clustering trees

Table 3. Mean and Max time for receiving the message. *"MAX" values represent the time needed for the last node to receive the packet. "MEAN" values represent the mean time a node needed for a node to receive the broadcast packet.*

	500 nodes		700 nodes		800 nodes		900 nodes		1000 nodes	
	MEAN	MAX	MEAN	MAX	MEAN	MAX	MEAN	MAX	MEAN	MAX
MPR	5.13	8.97	4.88	8.40	4.88	8.40	4.81	8.23	4.78	8.07
Density	6.31	11.05	6.22	10.78	6.24	10.95	6.15	10.66	6.19	10.74

Broadcasting over the whole network: global broadcasting. The broadcasting task is initiated at a randomly-chosen source over the whole network.

As Figure 2 plots, when a global broadcasting task is initiated, our algorithm induces less re-transmissions (Figure 2(a)) and less receptions(Figure 2(b)) than other algorithms. Thus, it spends less energy and resources.

Since in the MPR selection, the relays are selected in order to reach the 2-neighborhood after two hops, the k-neighborhood of the source is reached within k hops. Under the assumption of an ideal MAC layer, MPR gives the optimal results in terms of latency (Number of hops). We thus compare our heuristic to the MPR one to measure how far we are from the optimal solution. We consider a time unit as a transmission step (*i.e.*, 1 hop). Table 3 presents the results. Yet, we can note that, even if our algorithm is not optimal regarding the latency, results are very close to it. Figure 3 represents the propagation in time for a broadcast packet initiated at the centered source at time 0. Cluster-heads appear in blue and the source in green. The color of other nodes depends on the time they receive the broadcast packet. The darker the color, the shorter the time.

Broadcasting within clusters: cluster broadcasting. We now suppose that the broadcasting task is performed in each cluster, initiated at the cluster-heads. We thus have as many broadcastings as clusters.

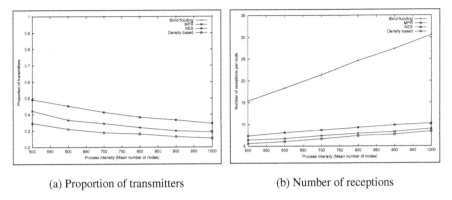

(a) Proportion of transmitters   (b) Number of receptions

Fig. 4. Mean number of receptions per node (a) and Proportion of transmitter nodes (b) for a cluster broadcasting scheme *w.r.t.* the process intensity and the metric used(+: Blind Flooding; ×: MPR; *: NES; □: Density-based;)

Table 4. Mean and Max time for receiving a cluster broadcast message. *"MAX" values represent the time needed for every node to receive the packet at least once, "MEAN" values the mean time a node has to wait till the first reception of the packet.*

	500 nodes		700 nodes		800 nodes		900 nodes		1000 nodes	
	MEAN	MAX	MEAN	MAX	MEAN	MAX	MEAN	MAX	MEAN	MAX
MPR	1.76	4.71	1.78	4.85	1.81	4.83	1.81	4.80	1.82	5.00
Density	1.80	5.08	1.83	5.38	1.87	5.29	1.87	5.50	1.88	5.30

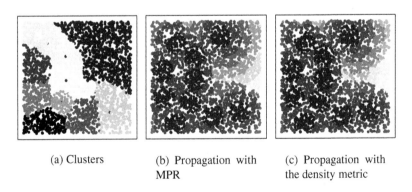

(a) Clusters   (b) Propagation with   (c) Propagation with
                          MPR   the density metric

Fig. 5. Propagation time of a cluster broadcast packet over the topology plotted in *(a)* using MPR *(b)* or density-based trees *(c)*

We can see on Figure 4, that our broadcasting algorithm still obtains the best results regarding the number of receptions (Figure 4(a)) and the transmitter ratio (Figure 4(b)). Table 4 and Figure 5 present the results regarding the latency. Once again, we can

Table 5. Proportion of nodes which receive the first copy of the packet by their parent or by one of their children

	500nodes	600nodes	700nodes	800nodes	900nodes	1000nodes
% by parent	78.74%	76.81%	74.57%	73.21%	71.31%	70.13%
% by a child	0.00%	0.00%	0.00%	0.00%	0.00%	0.00%

observe that, even if our algorithm is not optimal, results are very close as, in average, a node in our algorithm needs only 0.5 steps more than the optimal value to receive the packet. This also shows that, even if the routes in the trees from the cluster-heads to other nodes are not always the shortest ones, they are very close to them.

In Table 5, we give the proportion of nodes which receive the first copy of the packet by their parent or by one of their children. This feature shows whether the message which is sent by the cluster-head follows the branches of the trees. Indeed, a node u always receives the message by its parent (as all non-leaf nodes forward the message) but it could have received it before from another way as paths are not always optimal. In this case, the shorter route between u and its cluster-head is not found by following the route in the tree. We can thus see that routes are the shortest ones in number of hops for more than 70% of the cases. We can also observe that, as the message progresses down the tree, none of the nodes receives it the first by one of their children.

6   Conclusion and Perspectives

We have proposed a broadcasting scheme over a clustered topology for multi-hop wireless networks. We can thus obtain a double-use structure with low cost, as the maintenance is local and quasi-local. The cost is bounded by the tree depth which is a low constant. Moreover, two kinds of broadcasting task may be performed over it since we can perform global broadcastings in the whole network as well as broadcastings confined inside a cluster. Our proposed algorithm offers better results than some existing broadcasting schemes. More precisely, it saves more retransmissions since the number of internal nodes selected is lower (for both global and cluster-confined broadcastings). Moreover, the number of duplicated packets received is also lower. Note that reducing both emissions and receptions is an important factor when designing energy aware broadcasting protocols. Future works will be dedicated to investigate robustness of clustering-based broadcasting protocol in presence of node and link failures. Preliminary results tend to show that the confined broadcasting is more robust that the global one and thus we investigate more deeply the impact of the choice and number of gateways between clusters.

References

1. A. Amis, R. Prakash, T. Vuong, and D. Huynh. Max-Min D-cluster formation in wireless ad hoc networks. In Proceedings of the IEEE INFOCOM, Tel-Aviv, Isral, March 2000. IEEE.
2. J. Cartigny, F. Ingelrest, and D. Simplot-Ryl. RNG relay subset flooding protocol in mobile ad-hoc networks. IJFCS, pages 253–265, 2003.

3. G. Chen, F. Garcia, J. Solano, and I. Stojmenovic. Connectivity-based k-hop clustering in wireless networks. In HICSS02, Hawaii, USA, January 2002.

4. C. Chiang, H. Wu, W. Liu, and M. Gerla. Routing in clustered multihop, mobile wireless networks with fading channel. In ICCS/ISPACS96, Singapore, November 1996.

5. A. Ephremides, J. Wieselthier, and D. Baker. A design concept for reliable mobile radio networks with frequency hoping signaling. In IEEE 75, pages 56–73, 1987.

6. Y. Fernandess and D.Malkhi. k-clustering in wireless ad hoc networks. In POMC, Toulouse, France, 2002.

7. H. Lim and C. Kim. Multicast tree construction and flooding in wireless ad hoc networks. In ACM MSWiM Workshop at MobiCom 2000, Boston, MA, USA, August 2000.

8. C. Lin and M. Gerla. Adaptive clustering for mobile wireless networks. IEEE JSAC, 15(7):1265–1275, 1997.

9. N. Mitton, A. Busson, and E. Fleury. Self-organization in large scale ad hoc networks. In MED-HOC-NET 04, Bodrum, Turkey, June 2004.

10. N. Mitton, A. Busson, and E. Fleury. Broadcast in self-organizing wireless multi-hop network. Research report RR-5487, INRIA, February 2005.

11. N. Mitton, E. Fleury, I. Guerin-Lassous, and S. Tixeuil. Self-stabilization in self-organized multihop wireless networks. In WWAN05, Columbus, Ohio, USA, June 2005.

12. N. Nikaein, H. Labiod, and C. Bonnet. DDR-distributed dynamic routing algorithm for mobile ad hoc networks. In MobiHoc, Boston, MA, USA, November, 20th 2000.

13. A. Qayyum, L. Viennot, and A. Laouiti. Multipoint relaying: An efficient technique for flooding in mobile wireless networks. In HICSS02, Hawaii, USA, January 2002.

14. I. Stojmenovic, M. Seddigh, and J. Zunic. Dominating sets and neighbor elimination-based broadcasting algortithms in wireless networks. IEEE TPDS, 13(1), January 2002.

15. I. Stojmenovic and J. Wu. Broadcasting and activity scheduling in ad hoc networks. IEEE Mobile Ad Hoc Networking, pages 205–229, 2004.

16. J. Wu and H. Li. A dominating set based routing scheme in ad hoc wireless networks. Telecommunication Systems, pages 13–36, 2001.

17. J. Wu and W. Lou. Forward node set based broadcast in clustered mobile ad hoc networks. Wireless Communications and Mobile Computing, 3(2):141–154, 2003.

MIMOMAN: A MIMO MAC Protocol for Ad Hoc Networks*

Joon-Sang Park and Mario Gerla

Computer Science Department, University of California,
Los Angeles, CA 90095-1596, USA
{jspark, gerla}@cs.ucla.edu

Abstract. Multiple-Input Multiple-Output (MIMO) antenna systems
present a radical way to improve the performance of wireless communications. Such systems can be utilized in wireless ad hoc networks for
improved throughput in several ways. A straightforward way is to regard
the MIMO system just as a new link layer technology that provides a
higher data rate. In this way, MIMO systems can be integrated with
legacy Media Access Control (MAC) protocols such as 802.11 DCF. Recently, a few studies have proposed MAC protocols leveraging the advantages of MIMO systems in a different way. Those new MAC protocols enhance the network throughput by allowing simultaneous multiple
communications at a lower rate rather than a single communication at a
higher rate in a single collision domain. In this paper, we present a new
MIMO MAC protocol for Ad hoc Networks namely MIMOMAN. MI-
MOMAN tries to further increase the network throughput by combining
the two approaches, i.e., allowing simultaneous multiple communications
at a higher data rate. It is beneficial especially in a heterogeneous setting
where the number of antennas installed on each node varies. As a part
of the protocol, MIMOMAN also suggests a new beamforming algorithm
that allows spatial reuse among spatial multiplexing MIMO links. We
investigate the performance of MIMOMAN through a simulation study.

1 Introduction

One of the most interesting trends in wireless communications in recent times
is the proposed use of multiple-input multiple-output (MIMO) systems [2]. A
MIMO link employs an array of multiple antennas at both ends of the link. The
use of multiple antennas at both transmitted and receiver provides enhanced performance over diversity schemes where either the transmitter or receiver, but not
both have multiple antennas. MIMO systems can be utilized in ad hoc networks
for improved network throughput in several ways. The most straightforward approach is to regard the MIMO link just as a link with higher data rates. MIMO

---

* The work is sponsored in part by MINUTEMAN project. Contract/grant sponsor:
  Office of Naval Research. Contract/grant number: N00014-01-C-0016.

V.R. Sirotiuk and E. Chávez (Eds.): ADHOC-NOW 2005, LNCS 3738, pp. 207–220, 2005.

systems can increase the data rate (or capacity) of a wireless link by the factor of N where N is the number of antennas without increasing transmit power or bandwidth. The most well-known technique is called spatial multiplexing (SM). In a SM-MIMO configuration, the incoming data is demultiplexed into N distinct streams and transmitted out of N antennas at the same frequency, with the same modulation, and with the same signal constellation. The approach, viewing MIMO systems as a new link-layer technology, does not require any support from upper layers thus conventional protocols such as 802.11 DCF MAC can run atop them without any modification. The network performance enhancement comes only from the faster links. Proposals for the emerging standard IEEE 802.11n follow this approach [14].

Recently, a few studies [8, 7] have proposed another way of exploiting MIMO systems in ad hoc networks. The MAC protocols proposed in [8, 7] exploit the beamforming [6] capability of MIMO systems and improve the network throughput by allowing multiple simultaneous communications at a lower rate instead of having one higher rate communication at a time. This approach is similar to that taken by MAC protocols developed for directional antennas [1, 4, 10] in that they facilitate spatial reuse, whereby multiple transmissions can take place simultaneously in the same collision domain [3]. The main difference is their operating environment. Directional transmission requires line of sight propagation. Indoor and urban outdoor environments are typically rich in scattering and possible line of sight blocking, multi-path conditions are prevalent. MIMO systems exploit multi-path propagation to provide higher total capacity. The rich scattering environment in fact is a much more realistic condition in indoor and urban outdoor networks.

The two approaches increase the network performance in very different ways. The first approach is increasing the link capacity using techniques such as spatial multiplexing and the second is enhancing spatial reuse among nodes. In this paper, we attempt to further increase network throughput by harnessing both techniques: beamforming to enhance spatial reuse among nodes and spatial multiplexing to increase link capacity using spatial multiplexing. We present a new MIMO Mac protocol for Ad hoc Networks, MIMOMAN, at the heart of which is a new beamforming algorithm that allows spatial reuse among spatial multiplexing MIMO links. This is beneficial especially in a heterogeneous setting where the number of antennas installed on each node varies. Antenna heterogeneity itself is an issue in MIMO based wireless networks. Beside the fact that commercial MIMO products are expected to vary in the antenna setting, considering the applications of ad hoc networks it is always better in general for protocols to tolerate as many kinds of heterogeneity as possible without compromising other important facts. Our cross-layer, joint PHY/MAC solution to the antenna heterogeneity problem benefits from it rather than tolerates it.

The main contribution of this work is a fully distributed MAC protocol that exploits SM-MIMO links and cross-layer techniques to enable spatial reuse among them. MIMOMAN provides a methodology with which nodes can identify interferers and acquire channel information at each transmitter and receiver pair to implement transmit and receive antenna beamforming that achieves spatial multiplexing while at the same time nulling of co-existing, potentially interfering transmitter and receiver pairs. From a communications theoretic point of view, the contribution of MIMOMAN is the first proposal of the beamforming algorithm that enables spatial reuse among spatial multiplexing links in ad hoc networks. The main advantage of MIMOMAN with respect to conventional 802.11 DCF style MAC is that it allows high-powered nodes to utilize residual capacity of the wireless channel thereby increasing overall capacity of a network.

Another recent study [13] proposed a framework for utilizing MIMO systems in ad hoc networks. While our focus is a realistic MAC protocol design, its focus is rather a general theory in that it formulates the MIMO antenna utilization problem in ad hoc networks as a network-wide optimization problem and graph-coloring based solution is proposed.

The rest of the paper is organized as follows. We describe MIMOMAN in Section 2. Section 3 details our evaluation of the performance of MIMOMAN using simulation. Finally, Section 4 concludes the paper. We assume that readers are familiar with MAC protocols such as 802.11 DCF. This paper can be easily read with basic knowledge of communications theory but we expect that some knowledge of Linear Algebra suffices for understanding the protocol.

2   Protocol Description

In this section we describe details of the MIMOMAN protocol. We first consider a restricted case of the protocol termed bMIMOMAN which runs on the special type of MIMO system called beamforming MIMO. bMIMOMAN allows multiple simultaneous communications at a lower rate (i.e., only one stream per link). Then we extend it toward the general SM-MIMO system and describe the full version of MIMOMAN.

2.1   Preliminaries

The MIMO system can be thought of as a linear system where the input and output are vectors as shown in Figure 1. In the system, signals, channel coefficients, and antenna weights are all represented as complex numbers. Assuming N antennas at both the transmitter and receiver, the MIMO channel between them is represented with an NxN matrix of channel coefficients $\mathbf{H}$ where the h_{ij} element represents the individual channel gain between the ith transmitter antenna and j^{th} receiver antenna as shown in Figure 3. In rich scattering environments such as indoor or urban outdoor, $\mathbf{H}$ will be full-rank and can be characterized as a random matrix. The input $\mathbf{x}$ is the transmitted symbol vector and the output $\mathbf{y}$ is a received signal vector. The input $\mathbf{x}$ is linearly transformed by the channel $\mathbf{H}$ and appended with noise $\mathbf{n}$ resulting in the output $\mathbf{y}$. Given $\mathbf{H}$ and disregarding $\mathbf{n}$, the transmitted symbol vector $\mathbf{x}$ can be retrieved from the received signal vector $\mathbf{y}$ by $\mathbf{x} = \mathbf{H}^{-1}\mathbf{y}$, which is one of the simplest SM-MIMO decoders called zero-forcing decoder. A channel estimation procedure

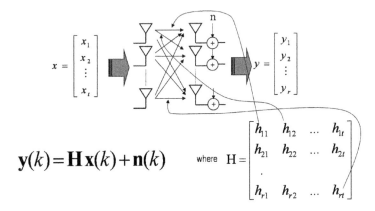

$$y(k) = \mathbf{H}\mathbf{x}(k) + \mathbf{n}(k) \qquad \text{where} \quad H = \begin{bmatrix} h_{11} & h_{12} & \cdots & h_{1t} \\ h_{21} & h_{22} & \cdots & h_{2t} \\ \vdots & & & \\ h_{r1} & h_{r2} & \cdots & h_{rt} \end{bmatrix}$$

Fig. 1. MIMO channel

precedes the real data exchange to get the channel $\mathbf{H}$. For better performance, additional signal processing operations can be applied to the transmitted and received signal vector, which can be represented as $\mathbf{y} = \mathbf{VHUx}$ where $\mathbf{U}$ and $\mathbf{V}$ are transmitter and receiver signal processing operations respectively. $\mathbf{U}$ and $\mathbf{V}$ are sometimes called transmitter/receiver beamforming. Through out this paper, we use uppercase and lowercase boldface letters to denote matrices and vectors, respectively, and superscripts T and H to denote the transpose operation and Hermitian operation, respectively. In our MAC designs, we maintain two logical channels, Control Channel and Data Channel, among which the total available bandwidth is divided. This division may be either frequency or code based. Only unicast DATA frames are sent over Data Channel. All other frames, RTS, CTS, ACK, and broadcast DATA frames, are sent over Control Channel. We assume the transceiver installed on each node is capable of listening to both channels concurrently but able to transmit over only one channel at a time. We assume flat fading meaning the two channels undergo the same channel characteristics. Our MAC protocols are based on IEEE 802.11 DCF MAC. Unless otherwise specified, our MAC protocols obey IEEE 802.11 DCF rules such as the exponential backoff.

2.2 bMIMOMAN: A Restricted Version of MIMOMAN

Figure 3 illustrates the abstract model of beamforming MIMO that bMIMOMAN is targeting. This type of MIMO transmits only one symbol at a time forming only one stream between a transmitter-receiver pair. By a stream we mean a time series of signals. At the transmitter side, each antenna transmits the same symbol or signal after applying its own weight to the signal (e.g., in the figure Antenna 1 transmits $s(t)w_{T1}$). At the receiver, received signals from all antennas are individually weighted and summed allowing the output $r(t) = s(t)\, \mathbf{w_T^H H w_R}$ where $\mathbf{w_T}$ is the weight vector $(w_{T1}\ w_{T2}\ w_{T3}\ w_{T4})^T$ of the transmitter, $\mathbf{H}$ is the MIMO channel matrix between the two nodes, and $\mathbf{w_R}$ is the weight vector

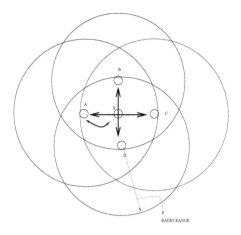

Fig. 2. Network Topology Example

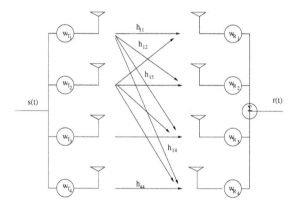

Fig. 3. Beamforming MIMO schematic with 4 antennas at both ends

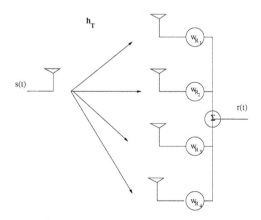

Fig. 4. Effective channel $\mathbf{h_T} = \mathbf{w_T^H H}$

of the receiver. The MIMO channel can be thought of as SIMO channel as shown in Figure 2.1. The effective channel then is $h_T = w_T^H H$. We assume that coefficients of the effective channel can be estimated using standard channel estimations methods. RTS and CTS MAC control frames shall carry training symbols for that purpose.

To allow simultaneous multiple transmissions, both ends beamform, i.e., adjust antenna weights, as follows:

1. Transmitter selects w_T such that it does not affect any ongoing communications.
2. Receiver selects w_R such that it receives only signal of interest and is not interfered by any other ongoing communications.

Let us consider a small network shown in Figure 2. Say at some point Node B wants to transmit a packet to Node D while Node C is sending a packet to A. If D can null the signal being transmitted from A to C and B transmits a signal which is nulled from A and C's perspective, we are enabling the two simultaneous transmissions. A node can null a signal easily if it knows the weight vector being used to transmit the signal. For example, D can null the signal from C by setting w_D such that $w_C^H H_{CD} w_D = 0$ where H_{CD} is the channel matrix between C and D. Also, a node can generate a signal that is nulled from the perspective of a certain receiver if it knows the weight vector of the receiver. For example, by adjusting w_D such that $w_D^H H_{DA} w_A = 0$, D can transmit a signal that means nothing to A. In our scenario, to accomplish the two simultaneous communications, A should know w_B and w_D where w_B and w_D are the weight vectors of B and D respectively and D should know w_A and w_C where w_A and w_C are the weight vectors of A and C respectively. This exchange of weight vectors is enabled by the procedure described subsequently.

Consider the topology shown in Figure 2 again. Assume at the beginning all the nodes are silent, i.e., there is no on-going communication. A node which wants to transmit data to another node, say Node A, transmits a RTS using the default weight vector, $[1\ 1\ 1\ 1]/\sqrt{4}$ in our 4-antenna example, or a random vector. The vector is normalized to have equal signal power regardless of the number of antennas. Note that different signal power results in different link performance. The weight vector used to transmit the RTS will be reused to transmit the following data packet and to receive the corresponding CTS. Once the designated receiver of the RTS, say Node C, receives the RTS, it responds with a CTS packet using the current weight vector. The weight vector used for transmitting CTS will be used to receive following data packet. The receiver estimates the SIMO channel vector $h_{AC} = w_A^H H_{AC}$. Since there is no on-going communication, Node C can switch its weight vector to $w_C = h_{AC}^T$ which maximizes the combined channel and array gain before it transmits CTS. When a node other than the designated receiver receives RTS, say Node D in our example, it estimates the effective channel h and adjusts the weight vector such that the signal from the sender of RTS is nullified (i.e., $h_{AD} w_D = 0$) for the duration of time specified in the RTS duration field. When a node other than the sender of the RTS receives the CTS, say Node S, it estimates the effective channel and stores the weight vector for the duration specified in the CTS duration filed. After the RTS/CTS handshaking Node A sends and C receives a data frame using $\mathbf{w_A}$ and $\mathbf{w_C}$ respectively. For the physical carrier sensing, a node listens to the Control Channel. If the Control Channel is free, it assume the wireless channel is free as in 802.11 case.

Now let us say B wants to initiate a data transmission toward D. Since it should ensure that C's signal reception is not disturbed, it picks a $\mathbf{w_B}$ such that $\mathbf{w_C^H H_{CB} w_B} = 0$ for a RTS transmission. The RTS itself does not interfere with data transmission but for the estimation purpose the beamformed version of the RTS has to be transmitted. Note that B already obtained $\mathbf{w_C^H H_{CB}}$ when it overheard C's CTS. B's counterpart, Node D, has to pick its weight vector such that the signal from A is nullified. Otherwise, D's decoding of the signal from B will be hindered by the interference from A. In fact, D already has done it when it overheard A's RTS so it either can use its current weight vector or select a new $\mathbf{w_D}$ such that the effective channel gain from B is maximized while the signal from A is nullified. This problem can be formulated as an optimization problem and using null-space projection it can be reduced to an eigenvalue problem in our case. The detailed algorithm is to be described in the next section. Note that any additional new transmission is only possible if both the sender and receiver have enough degrees of freedom. A node with N antennas can null out at most $N - 1$ stations in rich scattering environments. N is also known as the Degree of Freedom (DOF). Every time a node nulls out another node, it consumes one DOF. For example, a node with 4 antennas can null out the maximum of 3 other transmitters while transmitting its own stream.

A node should not start a new transmission if it cannot finish its transmission in the communication period (time to exchange RTS, CTS, DATA and ACK frames) of any ongoing communication. The reason is that during the course of communication nodes are unable to track neighbors' activities. So if a new transaction has commenced during their communication, there is high chance of either disturbing the new transmission or loosing the opportunity to start another transaction. Imagine a situation where A and B are in the range of D but they are hidden terminals to each other and A just finished its communication session and D is receiving B's signal. A sees a clear channel but it doesn't mean there is no activity. A would have known D's activity if it was not engaged in its communication beforehand. In ad hoc networks with conventional radios and MACs, this kind of situation does not happen.

The new restriction can be harsh for some nodes. If all the packets are the same size, then there cannot be any simultaneous transmissions. We get around this problem using the common packet aggregation technique. A node reserves the channel for a long period and send out multiple packets back-to-back. In 802.11 terms, we make PHY frames carry multiple MAC frames. In this way the session duration is made variable regardless of traffic so that several simultaneous sessions can be stack up. The packet aggregation technique is known to enhance channel utilization since it amortizes the fixed RTS/CTS overhead among multiple MAC frames.

When a node receives a RTS or CTS, the node stores all the information, the effective channel coefficient and the duration of the session, delivered by it. This is necessary because a node may not have enough degree of freedom to null out interfering signal. When a node receives RTS or CTS and it is out of DOF, it sets NAV to the closest finish time of on-going session. When NAV expires we recalculate the weightvector.

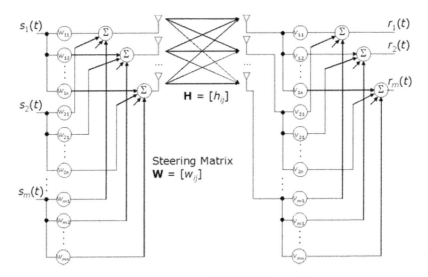

Fig. 5. MIMOMAN link model

2.3 MIMOMAN

In this section, we generalize bMIMOMAN toward spatial multiplexing MIMO and develop the full version of MIMOMAN. Figure 5 illustrates the SM-MIMO link model. As mentioned previously, SM-MIMO can transmit multiple signal/streams concurrently. At the sender, each stream/signal is applied its own weightvector and then all of them are summed and transmitted out of N antennas. At the receiver, multiple streams are generated by applying different weighvectors to the received signal vector and those are fed to MIMO decoder. For sending and receiving, each node maintains a weight matrix (or steering matrix) which is a collection of vectors since we need one weightvector for each stream. The relationship between the input signal vector at the transmitter and output signal vector at the receiver can be represented as $\mathbf{r}(t) = \mathbf{s}(t)\,\mathbf{W}^{\mathbf{H}}\mathbf{H}\mathbf{V}$ where $\mathbf{W}$, $\mathbf{H}$, and $\mathbf{V}$ are the weight matrix of the transmitter, the channel matrix, and the weight matrix of the receiver respectively. In fact, multiple streams in a communication link can be thought of as multiple senders and receivers each transmitting and receiving one stream sharing a channel. Following this track, bMIMOMAN can be naturally extended to spatial multiplexing MIMO. We use the story plot similar to the one used in the previous section for easier understanding by comparison. We assume that the total transmission power is constrained by the number of antennas.

At the beginning, Node A having data designated to another node, say Node C, transmits a RTS frame. In SM-MIMO networks with heterogeneous antenna settings, there is an issue in sending and receiving frames regarding the number of antennas and streams. If a node with 4 antennas sends a frame with 4 streams then only nodes with more than four antennas can decode the frame. Since the RTS sender does not know how many antennas the neighbor nodes including designated and non-designated receivers have or what channel condition they are experiencing, every RTS frame is sent using one stream. In the RTS frame, it is indicated that how many antennas the sender has and how many streams can be employed when sending following DATA. The number of available streams depends on the channel condition. As noted earlier, A node with N antennas can suppress up to N - 1 streams, i.e., it can transmit without interfering nodes receiving total N - 1 streams. How to set this value will be elaborated shortly.

For a receiver to decode multiple streams, it should aware of the effective channel for each stream. In the beamforming MIMO case, since each stream generates one RTS, the receiver can estimate the effective channels of multiple streams one by one. But in this case one RTS is for multiple streams thus RTS should be designed such that the effective channel estimation can performed for each stream. It is indicated that a block of training symbols is in the RTS such that the effective channel can be estimated. To do the channel estimation for multiple streams, the block is replicated as many time as the number of streams. Each block serves for each streams. The weightvector associated to a specific stream is applied when sending out the block corresponding to the stream.

This way, the neighbors of Node A learn $\mathbf{W_A^H H}$. In response to the RTS, Node C, the designated receiver, transmits a CTS frame towards Node A. Similarly, its neighbors learn their $\mathbf{W_B^H H}$'s. We set $\mathbf{W_B}$ as the right singular vectors of $\mathbf{W_A^H H}$, i.e., $\mathbf{W_B} = \mathbf{V}$ where $\mathbf{W_A^H H} = \mathbf{U^H D V}$ is a singular vector decomposition. This is to maximize SINR. For the same reason as the RTS case, a CTS should be sent out using one stream and carry blocks of training symbols. In the CTS frame, it is indicated that how many antennas the sender of the frame has and how many streams can be accepted for the following DATA communication. The number of acceptable streams depends on the channel condition and should be equal to or less than the number of available streams of the corresponding RTS sender. Once Node A receives the CTS, Node A starts transmission of DATA using the number of streams indicated in the CTS.

Now let's say Node B has data to send out. When transmitting frames, B should not interfere C's reception of data. To that end, B has to control the number of streams and set the weight matrix, i.e., beamform, appropriately. As indicated earlier, the total number of streams to be suppressed and to be transmitted at the same time is the same as the number of antennas installed on the node.

We can solve the stream control and beamforming problem jointly for the general case at Node S as follows. Let

k = the number of receiving neighbors, Node 1 to k, not to interfere,
$\mathbf{H_{iS}}$ = channel between Node i and S,
$\mathbf{S_i}$ = steering matrix of Node i,
$\mathbf{Z_{iS}} = \mathbf{H_{iS}^H S_i S_i^H H_{iS}}$, and
$\mathbf{Q} = \sum_{i=1}^{k} \mathbf{Z}_{iS} = \mathbf{PDP^T}$ which is an eigenvalue decomposition.

We first find the largest m such that $\lambda_{n-m+1} \leq \alpha$, 0 in our current design, where λ_p is $\mathbf{D}(p, p)$ (p^{th} largest eigenvalue of Q) and n is the number of antennas of Node S. And we set S's steering matrix $\mathbf{W_S} = [\mathbf{P}_n \ ... \ \mathbf{P}_{n-m+1} \ \mathbf{z} \ ... \ \mathbf{z}]$ where $\mathbf{P}_i$ is i^{th} column of $\mathbf{P}$ and $\mathbf{z}$ is zero column vector. Last (n - m) columns of $\mathbf{W_S}$ are zero columns. If there is no such λ, the transmission should not commence. From their CTSs transmission neighboring receivers are identified and corresponding $\mathbf{S_i^H H_i S}$ is estimated. Referring to the scenario depicted in Figure 2, once Node S starts a new transmission, Node 1, 2, 3, and 4 will undergo the interference with energy $\mathrm{tr}(\mathbf{W_S^H H_{1S}^H W_1 W_1^H H_{1S} W_S})$, $\mathrm{tr}(\mathbf{W_S^H H_{2S}^H W_2 W_2^H H_{2S} W_S})$, $\mathrm{tr}(\mathbf{W_S^H H_{3S}^H W_3 W_3^H H_{3S} W_S})$, and $\mathrm{tr}(\mathbf{W_S^H H_{4S}^H W_4 W_4^H H_{4S} W_S})$ respectively where tr() denotes the trace operation. What we do is choose such $\mathbf{W_S}$ that those interferences are nulled.

The beamforming problem at the receiver can be solved similarly. We maximize channel gain from the transmitter while nullify other known interfering signals. Let there be Node T and R, the transmitter and receiver, as well as k interferers, Node 1 to k. We define

$\mathbf{H_{iR}}$ = channel between Node i and R,
$\mathbf{S_i}$ = steering matrix of Node i,
$\mathbf{Z_{iR}} = \mathbf{H_{iR}^H S_i S_i^H H_{iR}}$, and
$\mathbf{Q} = \sum_{i=1}^{k} \mathbf{Z}_{iR} = \mathbf{PDP}^T$.

We find the largest m such that $\lambda_{n-m+1} = \alpha \ (= 0$ in our current design) where λ_p is $\mathbf{D}(p, p)$ (p^{th} largest eigenvalue of Q). If there is no such λ, it means that a node cannot suppress known interferers thus the reception should not commence. Otherwise m becomes the number of available streams at the receiver. And let

$\mathbf{U} = [\mathbf{P}_n...\mathbf{P}_{n-m+1}]$,
$\mathbf{N} = \mathbf{UU}^H$, and
$\mathbf{M} = \mathbf{NZ_{RT}N} = \mathbf{XAX}^T$.

Set R's steering matrix $\mathbf{W_R} = [\mathbf{X}_n \ ... \ \mathbf{X}_{(n-m+1)} \ \mathbf{z} \ ... \ \mathbf{z}]$ where $\mathbf{X}_i$ is i^{th} column of $\mathbf{X}$. What we are doing here is optimizing the combined channel and antenna array gain between the the transmitter while nullifying known interferes. We formulate the problem as an optimization problem and using null-space projection it is reduced to an eigenvalue problem.

The main advantage of MIMOMAN compared to legacy 802.11 style MACs comes from the fact that it gives more *power* to the nodes with larger number of antennas. From a node with larger number of antennas point of view, the wireless channel captured by another node with fewer number of antennas is underutilized. Once the underutilization is detected, MIMOMAN allows high power nodes to kick in. By overhearing RTSs and CTSs, a node can identify the underutilization of the channel as described above.

The sender and receiver negotiate the number of streams to be used for the data transmission. This means that the data rate and thus transmission time of a frame will vary. This causes several problems including setting the *duration* field in the RTS. We get around the problem again with the packet aggregation technique described in the previous section.

When preparing a RTS frame, a sender collects as many packets as possible from the application layer and calculates the time to transmit all the packets at the highest possible rate. The highest possible rate depends on the channel condition. It is (base rate) * (number of antennas - number of known receiving neighbors). If there is no interferer, the sender sets the duration field as the smaller value of the channel coherence time which is predefined and the expected transmission time. The channel coherence time is expected time during which the channel doesn't change. When node are moving with the speed of 10 m/s, it is about 10ms according to [12]. If there is any known on-going communication, the sender sets the duration field as the earliest termination time of any on-going communication. But a sender should be able to send at least one packet so if the earliest termination time of any on-going communications is less than the transmission time of the first packet at the base rate then, the sender should not initiate any data transmission procedure.

On the reception of CTS, the data rate is known to the sender since it is indicated in the CTS how many streams can be employed transmitting data. The sender creates an aggregated DATA frame by packing as many packets that can be transmitted in the duration specified in the CTS at the rate specified in the CTS.

3 Performance Evaluation

In this section we describe our simulation setup and present the results characterizing the performance of MIMOMAN.

We implemented MIMOMAN in Qualnet [9]. Data structures holding antenna/stream weights and the channel information are added and the gain due to the MIMO channel **H** which is instantiated whenever needed and antenna weights is calculated on the fly and applied to the computation of the Signal-to-Noise ratio (SNR) for each packet. We assume a quasi-static Rayleigh fading environment [11]. The channel coefficient for each transmit-receive antenna pair is an i.i.d. complex Gaussian random variable with zero mean unit variance and the channel is invariant during a complete session including RTS, CTS, and DATA frames. We assume that the channel estimation can be done with no error. Beamforming algorithms in the simulator are implemented using LAPACK [5]. We use the SNR threshold based reception model and the threshold value of 10dB which is the usual number for single antenna based network simulations. In fact, the more antennas the less the bit error rate is. That is, the bit error ratio of the 4x4 sys-

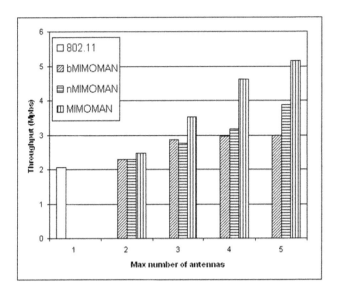

Fig. 6. Comparison of throughput of MIMOMAN protocols (20 node randomly deployed on $700\text{x}100m^2$ field)

tem is less than 2x2 or 1x1 system for the same SNR. We use the same threshold regardless of the number of antennas/streams employed, which is a rather conservative approach. More realistic reception error model is BER based which we plan to use for future work. It requires two step procedure: First, generate a BER table by performing separate link-level simulation. Then integrate the BER table into our simulation environment. We use the default values as Qualnet provides for the configurable parameters (e.g., 15dBm transmission power) unless otherwise specified. The radio range is about 250m.

We compare four protocols: 802.11, bMIMOMAN, nMIMOMAN, and MIMOMAN. nMIMOMAN is another restricted version of MIMOMAN. Contrary to bMIMOMAN, it does not allow simultaneous communicating node pairs but a node can use multiples streams to transmit data. Thus beamforming in nMIMOMAN only can increase SNR. And like MIMOMAN, stream number negotiation and packet aggregation are performed. Out of total 2Mbps link bandwidth 1.6Mbps is allocated to Data Channel. 802.11 uses only one channel of 2Mbbs. We set the channel coherence time to 10ms.

We consider a saturated network scenario with 20 static nodes randomly distributed over the $700\text{x}100m^2$ field. Each node varies in the number of antennas installed. Each node is assigned a random number between 1 and the maximum value of antennas. In the 802.11 case the maximum number can only be 1. Every node generates constant bit rate (CBR) traffic of 1000 packets/sec and 1KB/packet. Not to be influence by higher layer protocols static routes are used and each node sends packets only to one of its 1-hop neighbors. We measure

Throughput which is defined as the total number of data bits received by all the nodes divided by the simulation time. Figure 6 illustrates the performance of the protocols as a function of the maximum number of antennas. The numbers are averaged over 10 independent simulation runs. Typical hardware limitations might not allow the manufacture of a laptop or a mobile device with more than 5 antennas, hence we limit our simulations to the max 5-antenna case. The most notable result demonstrated in the figure is that MIMOMAN's performance scales well with the maximum number of antennas compared to others. MIMO-MAN shows about 35% throughput enhancement per antenna over 802.11. In the max 5-antenna case, the improvement is around 150% over 802.11, 75% over bMIMOMAN, and 35% over nMIMOMAN. The reason why bMIMOMAN cannot keep up with MIMOMAN is that for each initiation of a communication session, the network incurs overhead due to RTS/CTS handshaking and collision of RTSs. Since bMIMOMAN allows only one stream per communicating node pair, to saturate the wireless channel, i.e., nodes completely consume their DOFs, in a certain period of time bMIMOMAN should initiate several communications in the time frame. Thus the overhead can be quite high and the saturation may not be possible in some cases. Whereas MIMOMAN saturates the channel with fewer number of communications since multiples streams per communication are permitted. bMIMOMAN exhibits only about 45% improvement over 802.11 even in the max 5-antenna case. Since only one communication is permitted in a collision domain, the channel has to be underutilized unless nodes with the highest number of antennas capture the channel always which is prevented by contention scheme used in 802.11 from which nMIMOMAN is driven. This guarantees some level fairness, that situation is prevented. Whereas in MIMOMAN case, if a node detects underutilization of the channel it will kick in to take advantage of it, which gives full exploitation the channel thereby higher throughput than nMIMOMAN. Note that 802.11's achieved throughput higher than the channel bandwidth is due to enough spatial separation between nodes. Intuitively, the throughout should increase with the number of antennas since the nodes will have more degrees of freedom to take advantage of the underutilized channel. However the increase is not N-fold, where N is the maximum number of antennas. MIMOMAN does not achieve the expected maximum factor improvement of throughput for several reasons such as channel division, MAC control frame overheads, packet aggregation overhead, and aforementioned session initiation overhead.

4 Conclusion

In this paper, we presented MIMOMAN, a MIMO Mac protocol for Ad hoc Networks. MIMOMAN provides a methodology with which nodes can identify interferers and acquire channel information to implement transmit and receive antenna beamforming that achieves spatial multiplexing while at the same time nulling of co-existing, potentially interfering transmitter/receiver pairs. Our simulation results confirmed the advantages of MIMOMAN.

References

1. R.R. Choudhary, X. Yang, R. Ramanathan and N.H. Vaidya, *Using Directional Antennas for Medium Access Control in Ad Hoc Networks,* In Proceedings of ACM MOBICOM 2002.
2. D.Gesbert, M.Shafi, D.Shiu, P.J Smith and A. Naguib, *From Theory to Practice: An Overview of MIMO Space-time Coded Wireless Systems,* IEEE Journal of Selected Areas in Communications Vol 21, No.3 April 2003
3. Y.B. Ko, V. Shankarkumar and N.H. Vaidya, *Medium Access control protocols using directional antennas in ad hoc networks,* In Proceedings of INFOCOM 2000.
4. T. Korakis, G. Jakllari, and L. Tassiulas, *A MAC protocol for full exploitation of Directional Antennas in Ad-hoc Wireless Networks,* In Proceedings of ACM MOBIHOC 2003.
5. LAPACK - Linear Algebra PACKage, http://www.netlib.org/lapack
6. J. C. Liberti, Jr. and T. S. Rappaport, *Smart Antennas for Wireless Communications: IS-95 and Third Generation CDMA Applications,* Prentice Hall, 1999.
7. J.-S. Park, A. Nandan, M. Gerla, and H. Lee, *SPACE-MAC: Enabling Spatial Reuse using MIMO channel-aware MAC,* In Proceedings of IEEE ICC 2005.
8. J.C Mundarath, P.Ramanathan and B.D Van Veen *NULLHOC: A MAC Protocol for Adaptive Antenna Array Based Wireless Ad Hoc NEtworks in Multipath Enviroments,* In Proceedings of IEEE GLOBECOM 2004.
9. *QualNet user manual* www.scalable-networks.com
10. R. Ramanathan, *On the Performance of Ad Hoc Networks with Beamforming Antennas,* In Proceedings of ACM MOBIHOC 2001.
11. T. S. Rappaport, *Wireless Communications Principles and Practice,* Prentice-Hall, 1996.
12. S. Sadeghi, V. Kanodia, A. Sabharwal, and E. Knightly, *OAR: An opportunistic auto-rate media access protocol for ad hoc networks,* In Proceedings of ACM MOBICOM 2002.
13. K. Sundaresan and R. Sivakumar, *A unified MAC layer framework for ad-hoc networks with smart antennas,* In Proceedings of ACM MOBIHOC 2004.
14. World-Wide Spectrum Efficiency (WWiSE) Proposal for 802.11n, http://www.wwise.org
15. IEEE 802.11 standard, http://standards.ieee.org/getieee802/download/802.11-1999.pdf, 1999.

Reed-Solomon and Hermitian Code-Based Scheduling Protocols for Wireless Ad Hoc Networks

Carlos H. Rentel and Thomas Kunz

Department of Systems and Computer Engineering,
Carleton University,
Ottawa, Ontario, K1V 9R9. Canada
{crentel, tkunz}@sce.carleton.ca

Abstract. In this work we investigate bounds on throughput and delay perform-ance of a scheduling protocol that derives its decisions from codes traditionally used to correct or detect errors in the information carried over a noisy channel. In this paper we study the particular cases in which the Reed-Solomon and Hermitian code constructions are used. It is found that Hermitian codes outper-form Reed-Solomon codes in minimum throughput guarantee and delay metrics when the number of nodes is in the order of thousands. The *relative minimum distance* of the code used to schedule the transmissions is identified as an im-portant property that can be used to identify codes that can enable scheduling patterns with better minimum performance guarantees. Furthermore, the termi-nology of error control coding is used to present a more general and construc-tive framework for the study of code-based scheduling protocols.

1 Introduction

The efficient schedule of the transmissions in a Wireless Ad Hoc network is a chal-lenging task. The most popular schemes are those derived from the ALOHA protocol, such as the CSMA/CA protocol used in the IEEE 802.11 standard. The family of ALOHA protocols offer good performance for a few number of nodes transmitting bursty traffic over a shared channel. However, ALOHA-like protocols suffer from in-stability and lack of any performance guarantee. It is necessary to have more structure in the scheduling approach if real-time applications are to be supported with some minimum performance guarantees. A way to add structure is by allowing the trans-missions to happen during known boundaries or slots. On the other hand, a highly structured scheduling protocol usually suffers from being inflexible and difficult to manage in a distributed way. An example of a structured scheduling protocol is clas-sic time division multiple access (TDMA). In classic TDMA every node is assigned a unique time-slot in which to transmit; in this way collisions are completely avoided and the packet delay becomes highly deterministic. However, the throughput and de-lay efficiency of such a scheme is questionable due to the lack of a slot re-utilization technique that takes advantage of the spatial separation between the nodes.

We define a code-based scheduling protocol as one that utilizes the code-words of a structured code to select the slots over which the nodes in the network will transmit. A structured code is constructed using mathematical techniques aimed at creating a

V.R. Sirotiuk and E. Chávez (Eds.): ADHOC-NOW 2005, LNCS 3738, pp. 221 – 234, 2005.

set of code-words with some desirable properties. We exclude in this definition the use of code-words formed by elements drawn from a finite and discrete random distribution. Code-based scheduling protocols might also be referred to as topology-transparent [1], [2]. However, protocols that do not use structured code-words can also be topology-transparent (e.g., slotted-ALOHA), therefore we prefer the term code-based scheduling when referring to protocols that utilize code-words to perform the scheduling task. The transparency of a scheduling protocol implies that the procedure's performance does not depend on the relative position of the nodes in the network. However, some dependency on general network parameters, such as the number of nodes or the maximum density of the network is generally accepted. Examples of structured codes are numerous and a possible way to classify them is based on their method of construction. Codes constructed using algebraic principles include the popular Reed-Solomon (RS) codes and the Hermitian codes; whereas codes constructed with more empirical methods include the convolutional and turbo codes.

A code is denoted as $C(n, k, q)$, where n is the length of the code-words, k is the rank of the code, and q is the dimension of the Galois field over which the code is defined [3]. A $C(n, k, q)$ code has q^k code-words of length n in GF (q). The code-words of *any* code can be used to schedule channel-slots (e.g., time or frequency slots) in a multiple access system. . However, a proper motivation for the use of certain code constructions over others should be given and is the main subject of this paper.

The remainder of this paper is as follows. Section 2 presents a discussion on some related work. Section 3 presents the model and code constructions used in this work. Section 4 presents the results of the comparative performance evaluation between RS and Hermitian code-based scheduling. Finally, Section 5 concludes the paper and highlights future research directions.

2 Related Work

The use of code-words for scheduling purposes is not new. Code-based scheduling has been used in the past to assign the time-frequency slots of a frequency-hopping multiple access (FHMA) system. [4] and [5] proposed the use of code-words to avoid *hits* in a FHMA system. Solomon [5] proposed the use of the code-words of a RS code to assign the time-frequency slots to the transmitters of a FHMA system in order to minimize collisions at a common receiver. The main rationale behind this choice is that RS codes are maximum distance separable (MDS) [3]. An MDS code has the largest possible minimum Hamming distance d_{min} (i.e., Any linear block code satisfies $d_{min} \leq n - k + 1$, and an MDS code satisfies it with equality). Intuitively, among the code-based scheduling protocols, one could think that the one that utilizes the code-words of an MDS code is the best candidate to achieve an optimum minimum guaranteed performance. However, as shown in this paper, this is not the case.

The authors in [1] were the first to propose a framework for the application of code-based scheduling in the context of wireless Ad Hoc networks. However, the method is restricted to the code-words of a singly-extended RS code without any apparent connection to other code constructions (i.e., variations of the same code, or even different codes can be used for scheduling). The method proposed in [1] is based on polynomials over a Galois field. Each user is assigned a different polynomial P characterized by a degree k and coefficients in $GF(q)$. The evaluation of the polynomial is used to select the slots in a frame. The frame is sub-divided in q sub-frames of q slots each, and the ith user is assigned a slot y in a sub-frame x according to the following evaluation $y = P_i(x) \bmod q$. The difference between two different polynomials also results in a polynomial of degree less than or equal to k, therefore the number of roots of the difference between two polynomials will be bounded by k. The latter translates into the fact that the number of common points between any two different polynomials will be bounded by k as well (this arguments is also used in [5]). Additionally, it is argued that if information on the maximum number of inter-ferers that any node could have is known, it could be possible to guarantee a mini-mum performance as long as the scheduling is performed following the unique poly-nomial evaluation. Note that polynomial evaluation in $GF(q)$ is one of the possible methods used to construct RS codes. [2] uses the same construction as [1] (i.e., sin-gly-extended RS codes) to find an optimum value for the parameters of the code that maximize a lower-bound on the minimum throughput. However, this optimality is only valid when using RS codes that are singly-extended. To the best of our knowl-edge, we are the first to analytically show that a code-based scheduling protocol achieves better lower-bound minimum throughput when using a doubly-extended RS code rather than a singly extended RS code. Additionally, we compute the optimum code dimension that maximizes the minimum throughput of a scheduling protocol that uses the code-words of a doubly-extended RS code. The important fact that a code-based scheduling protocol is able to guarantee a minimum performance is first men-tioned in [1] and expanded in [2].

A more recent work in [6] identifies a generalization to the procedure in [1] based on Orthogonal Arrays (OAs). Additionally, they identified certain OA constructions that are analogous to the doubly-extended RS code which can provide better mini-mum performance guarantee than the method proposed in [2]. However, no analytical result is presented. Note that a large number of different code constructions produce OAs, including RS, BCH, Reed-Muller codes among others [7]. However, there are still a large number of code constructions that are not encapsulated in the concept of an OA. Additionally, different code constructions may have different performance if analyzed out of the scope of the OA theory. Part of our goal is to highlight the fact that generalization of the procedure in [1] is trivial since in principle any code-words derived from a linear or non-linear non-binary code can be used as a scheduling pat-tern, therefore a great deal of coding theory knowledge is available for us to experi-ment with. Additionally, all the possible modifications a code can undergo (e.g., trun-cation, augmentation, extension, shortening) represent a higher degree of generalization. The latter does not try to diminish the generalization efforts, but we believe that fewer attention has been given to the task of isolating the specific code constructions and techniques that can offer improved performance and practical im-plementations.

An average throughput performance comparison is presented in [8] between the family of code-based scheduling protocols derived from OAs and slotted-ALOHA. It is found that, in all the cases studied, slotted-ALOHA outperforms the given code-based scheduling approach. A code-selection algorithm is presented that can solve this problem and potentially make the average performance of a code-based scheduling protocol larger than the one achieved by slotted-ALOHA.

To the best of our knowledge we are the first to present an evaluation of two important and specific code constructions in terms of their bounded scheduling performance in the context of Wireless Ad Hoc networks.

3 Problem Model and Constructions

Assume a code $C(n, k, q)$ of length n, rank k in GF(q). $C(n, k, q)$ has q^k code-words of length n in an alphabet form by the elements of GF(q). The code-words of any such code can be utilized as scheduling patterns for the nodes of a wireless Ad Hoc network. For the sake of brevity we refer here to time-slots in a TDMA-like scheme. Every node access the medium assuming a time-slotted structure as the one shown in Figure 1.

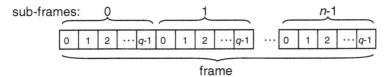

Fig. 1. Time-slotted structure of a code-based scheduling protocol

A lower-bound throughput of a node can be written as,

$$G_{min} = \frac{n - (n - d_{min})I_{max}}{nq} \tag{1}$$

Where I_{max} is the estimated maximum number of interferers a node can have at any given time, d_{min} is the minimum Hamming distance of the code, q is the dimension of the finite field in which the code is defined, and n is the length of the code. The maximum degree D_{max}, used in previous works (e.g., [1], [2]), has been changed for the more appropriate I_{max}, in this way nodes more than two-hops away could be considered when modeling the interferers of a given node (this is a more realistic situation in a wireless medium). A bound I_{max} could be achieved with techniques such as power control or any other topology control mechanism. However, this is a topic out of the scope of this paper. The numerator in (1) is the minimum possible number of successful slots in a frame of nq slots. G_{min} is a lower bound since the code-words of any good code will be separated by a Hamming distance larger than or equal to d_{min}. The following inequalities are imposed on the parameters of a code in order to guarantee a positive value of G_{min},

$$q^k \geq N$$

$$n \geq (n - d_{\min})I_{\max} + 1 \Rightarrow n \leq \frac{1 - d_{\min}I_{\max}}{1 - I_{\max}} \tag{2}$$

$$q \geq I_{\max} + 1$$

N is the number of nodes in the network. The first inequality in (2) guarantees a unique code-word for every node in the network, and the second inequality ensures that a node will have at least some successful transmissions (greater than zero) in a frame. That is, a node should be assigned a number of opportunities to transmit in a frame greater than the maximum number of collisions possible (i.e., $n > (n - d_{\min})I_{\max}$) if one wishes to have a minimum of performance greater than zero. The third inequality is explained in Section 4.

One important design goal is to find the dimension and rank of the specific code for which (1) is maximized constrained to (2). It is necessary however, to know the minimum Hamming distance and length of the code. Additionally the parameters $I_{\max}$ and N of the network must also be available (i.e., at least upper bounds to these parameters are necessary if a minimum throughput is to be guaranteed). Equation (1) can be re-written as,

$$G_{\min} = \frac{1}{q} - \left(1 - \frac{d_{\min}}{n}\right)\frac{I_{\max}}{q} \tag{3}$$

We note that code constructions that produce codes with larger $d_{\min}/n$ ratio for a given dimension q and $I_{\max}$ will have larger $G_{\min}$ guarantees. In other words, larger $G_{\min}$ are possible for codes with more *relative minimum distance*. This is the reason an extended RS code will guarantee a larger minimum throughput as will be proved shortly. Hermitian codes are an example of long codes (i.e., longer than RS codes) that possess a good $d_{\min}/n$ ratio property. Hermitian codes are constructed using algebraic geometry (AG) principles. In particular, Hermitian codes are derived from a Hermitian curve in a finite field. In fact, RS codes can be seen as AG codes over a straight line in a finite field, and therefore form a specific case of the more general set of AG codes. Codes can be constructed that are derived from many algebraic curves, including Elliptic and Hyper-elliptic curves. However, Hermitian curves have more points per given dimension; this is one of the reasons that, for some code parameters, the Hermitian codes possess a larger relative minimum distance than the RS codes of the same dimension. Next, however, we prove that a doubly extended RS code offers larger minimum throughput guarantee than a non-extended or a singly-extended RS code, and compute the optimal value of q that maximizes (1).

An RS code can always be doubly extended without loosing its MDS property (the method of constructing this extension can be found in [9]). A triple extension of a RS code is also possible without loosing the MDS property, however, only when the code has the following parameters $(n, k, q) = (2^m + 2, 3, 2^m)$ or $(2^m + 2, 2^m - 1, 4)$ [9]. The minimum distance of a RS code is given by $d_{\min} = n - k + 1$, substituting in (1) yields,

$$G_{\min}^{RS} = \frac{n - (k-1)I_{\max}}{nq} \tag{4}$$

Non-extended, singly-extended, and doubly extended RS codes have $n = q-1$, q, and $q+1$ respectively by definition. Substituting the latter values of n in (4) we obtain the minimum throughput for each code version,

$$G_{\min}^{neRS} = \frac{q-1-(k-1)I_{\max}}{(q-1)q}$$

$$G_{\min}^{seRS} = \frac{q-(k-1)I_{\max}}{q^2} \tag{5}$$

$$G_{\min}^{deRS} = \frac{q+1-(k-1)I_{\max}}{(q+1)q}$$

We take the ratios of the minimum throughputs in (5) and expand the factors to get,

$$\frac{G_{\min}^{neRS}}{G_{\min}^{deRS}} = \frac{q^2 - qKI_{\max} - 1 - KI_{\max}}{q^2 - qKI_{\max} - 1 + KI_{\max}} < 1 \Rightarrow G_{\min}^{neRS} < G_{\min}^{deRS}$$

$$\frac{G_{\min}^{neRS}}{G_{\min}^{seRS}} = \frac{q^2 - (1 + KI_{\max})q}{q^2 - (1 + KI_{\max})q + KI_{\max}} < 1 \Rightarrow G_{\min}^{neRS} < G_{\min}^{seRS} \tag{6}$$

Where $K = k-1$. Therefore $G_{\min}^{deRS} > G_{\min}^{seRS} > G_{\min}^{neRS}$. The latter implies the fact that the results found in [2] are not optimum since larger $G_{\min}$ are possible with the same values of k and q. Note that k in this and the following sections represents the rank of the code as it is frequently used in the error-control coding literature. This is different to the k used in [2], which, in that case, represents the maximum degree of the polynomial used to construct the codes. The relationship between both is $k_{[2]} = k - 1$. The value of q that maximizes $G_{\min}^{deRS}$ can be found by solving,

$$\frac{\partial}{\partial q}\left(\frac{q+1-KI_{\max}}{(q+1)q}\right) = 0 \Rightarrow$$

$$q^* = KI_{\max} - 1 \pm \sqrt{KI_{\max}(KI_{\max}-1)} \tag{7}$$

In order to satisfy the second inequality in (2), namely $n \geq (n-d_{\min})I_{\max} + 1 \Rightarrow q+1 \geq KI_{\max} + 1$, we must choose,

$$q^* = KI_{\max} - 1 + \sqrt{KI_{\max}(KI_{\max}-1)}, \tag{8}$$

which after substitution in $G_{\min}^{deRS}$ results in,

$$G^*_{de_{min}} =$$

$$\begin{cases} \dfrac{\sqrt{KI_{max}(KI_{max}-1)}}{\left(KI_{max}+\sqrt{KI_{max}(KI_{max}-1)}\right)\left(KI_{max}-1+\sqrt{KI_{max}(KI_{max}-1)}\right)}, & q \geq KI_{max} \\[3ex] \dfrac{N^{1/(K+2)}+1-KI_{max}}{\left(N^{1/(K+2)}+1\right)N^{1/(K+2)}}, & otherwise \ (q = N^{1/K+2}) \end{cases}$$

(9)

Figure 2 plots $G^*_{de_{min}}$ versus the result obtained in [2] for the case when $q \geq KI_{max}$ and assuming $D_{max} = I_{max}$ to make the comparison meaningful. Note that a doubly extended RS code outperforms the proposed scheme in [2], which is actually the performance of an optimal singly extended RS code. However, the difference becomes negligible for $KI_{max} > 10$. That is, as the node density increases, the difference between both codes becomes negligible.

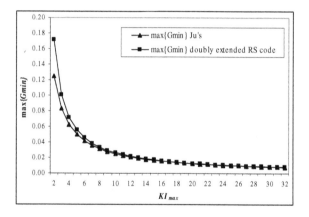

Fig. 2. max$\{G_{min}\}$ using result in [2] and a doubly-extended RS code

Next we provide a definition of Hermitian codes and their construction following the description in [10]; the construction of RS codes is well documented in the literature and will not be presented here.

Definition: Let q be a power of some prime p. The Hermitian curve over $GF(q^2)$ is given by the following equation,

$$x^{q+1} = y^q + y$$

(10)

The number of points (x, y) satisfying (10) over $GF(q^2)$ are $n = q^3$ not counting the point at infinity.

Construction of Hermitian codes: Let the points on a Hermitian curve over $GF(q^2)$, except for the point at infinity, be $\{P_1, P_2, \cdots, P_n\}$. Given an arbitrary positive integer u, construct a set of functions, denoted $L(uP_\infty)$, as follows

$$L(uP_\infty) = span\left\{ x^r y^s \mid rq + s(q+1) \leq u \right\} \tag{11}$$

The Hermitian code-words $H(q^3, k, q^2)$ can be obtained by evaluating all the functions $f \in L(uP_\infty)$ at the points $\{P_1, P_2, \cdots, P_n\}$, where k is the total number of functions f found using (11). The term $span\{f_1, f_2, \cdots f_n\}$ represents the set of functions that can be written as a linear combination of the functions in $\{f_1, f_2, \cdots f_n\}$, and u is an arbitrary but positive integer number that determines the rank of the code.

Note that this construction is very similar to the construction of an RS code if $L(uP_\infty)$ is substituted by $P = span\left\{1, x, x^2, \cdots, x^k\right\}$, which is nothing more than all polynomials of degree less than or equal to k. Every such polynomial is evaluated over the points on a straight line in $GF(q)$ (i.e., all the different elements in $GF(q)$). The minimum distance of a Hermitian code satisfies, in general, the following inequality [11],

$$d_{min}^H \geq q^3 - u \tag{12}$$

4   Performance Evaluation of RS and Hermitian Codes

For the following comparisons we numerically find the Hermitian and RS codes that maximize the minimum throughput in (1) subject to (2). Note that the only dimensions allowed in a Hermitian code are of the form q^2, where q is the power of a prime. The minimum distance of the Hermitian code is computed according to the following expression,

$$d_{min}^H = \min_{H_{cw} \in H, H_{cw} \neq 0} weight\{H_{cw}\} \tag{13}$$

Where H_{cw} is a code-word different to the all zero code-word, and $weight\{\}$ is the weight of a code-word [3].

Minimum and maximum delay (i.e., DT_{min}, DT_{max}) are defined as,

$$DT_{max} = (\text{frame length})/(\text{min. number of successful transmisions in a frame})$$
$$DT_{min} = (\text{frame length})/(\text{max. number of transmisions in a frame}) \tag{14}$$

For a doubly-extended RS code, the parameters in (14) take the following form for any value of q,

$$DT_{max} = (q+1)q/(q+1-(k-1)I_{max})$$
$$DT_{min} = q$$

Figures 3 and 4 show $\max\{G_{min}\}$ DT_{min}, and DT_{max} for $N = 100$ nodes. Note that both constructions have roughly the same performance, except at lower densities (i.e., smaller I_{max}), in which the DE-RS code has a better performance, particularly in terms of minimum throughput, than the Hermitian code.

In classic TDMA the throughput of a node is guaranteed to be $1/N$ and the delay N slots. The performance of code-based scheduling approaches fall below classic TDMA as the number of possible interferers increases beyond a certain threshold. In general, the minimum performance guarantee of code-based approaches will degrade with respect to classic TDMA as the node density increases (assuming that node density will increase the number of interferers of a node). In the limit, when all the nodes are one-hop from one another the classic TDMA approach will have better than or equal *minimum* performance than any code-based approach. To see this, note that in a code-based scheduling protocol the sub-frame size must satisfy $q \geq I_{max} + 1$ when a minimum throughput guarantee is required; this is because a node transmits once in every sub-frame, and in order to guarantee at least one successful transmission in any sub-frame of a frame, the sub-frame size must be larger than the maximum number of interferers a given node may have in a locality. Substituting $I_{max} = N - 1$ (i.e., the maximum number of interferers in a single-hop network of N nodes) into the previous inequality and equation (1) yields,

$$G_{min}^{s-hop} = \frac{n - (n - d_{min})(N - 1)}{nq}$$

$$q \geq N \tag{15}$$

Equation (15) is maximized for a given n, d_{min}, and N when q attains the smallest value constrained to $q \geq N$, that is $q^* = N$. Re-arranging (15) after substitution yields,

$$\max\{G_{min}^{s-hop}\} = \frac{1}{N} - \frac{(n - d_{min})(N - 1)}{nN} \tag{16}$$

Since $n \geq d_{min}$ and $N > 1$ for non-trivial cases, then, $\max\{G_{min}^{s-hop}\} \leq G_{TDMA}$, where $G_{TDMA} = 1/N$. Figures 5 through 10 show $\max\{G_{min}\}$, DT_{min}, and DT_{max} when N = 500, 2000, and 10000 nodes. As the number of nodes increases, the Hermitian code offers higher minimum throughput guarantees than the RS code, particularly at lower node densities. As the number of nodes increases, the I_{max} threshold below which Hermitian codes offer better performance than RS codes increases as well. The minimum and maximum delay of Hermitian codes tend to be less than the corresponding ones for RS codes as the number of nodes increases. The previous characteristic makes Hermitian codes attractive for large cooperative sensor networks in which thousands of wirelessly connected sensors are spread over large geographical areas. However, the maximum number of interferers of a given node must be controlled if some performance guarantee is desired. The latter applies equally to both constructions, and to the majority of wireless medium access control protocols.

Table I shows the parameters of some Hermitian and RS codes. Note that the minimum Hamming distance for codes of rank one will be the same as the RS codes of the same dimension. However, as the rank of the code increases beyond two, the Hermi-

Hermitian codes have larger d_{min}/n ratios. Take for instance, the RS and Hermitian codes of dimension $q = 16$ and rank $k = 3$. This doubly extended RS code has $n = 17$, and $d_{min} = 17 - 3 + 1 = 15$. The Hermitian code has $d_{min} = 59$ with $n = 64$. Note that the Hermitian code is not maximum distance separable (MDS) with these parameters since $d_{min} < n - k + 1$. However, the ratio d_{min}/n is larger.

Between two codes that have the same dimension, the code that has the larger d_{min}/n ratio will have larger G_{min} and hence it will be a preferred choice (i.e., regardless of it being a MDS code or not). When the number of nodes in the network is considerably large, the rank of the codes will be larger in order to satisfy the first inequality in (2) while maintaining a value of q considerably smaller than N. This is observed in the codes that maximize G_{min} in Figures 3 through 10.

Table 1. Parameters of some Hermitian and DE-RS codes (Both codes have same dimension)

q^2	n (Her)	k	dmin (Her)	dmin/n (Her)	dmin (RS)	dmin/n (RS)
4	8	1	8	1.00	5	1.00
4	8	2	6	0.75	4	0.80
4	8	3	5	0.63	3	0.60
4	8	4	4	0.50	2	0.40
9	27	1	27	1.00	10	1.00
9	27	2	24	0.89	9	0.90
9	27	3	23	0.85	8	0.80
9	27	4	21	0.78	7	0.70
16	64	1	64	1.00	17	1.00
16	64	2	60	0.94	16	0.94
16	64	3	59	0.92	15	0.88
16	64	4	56	0.88	14	0.82

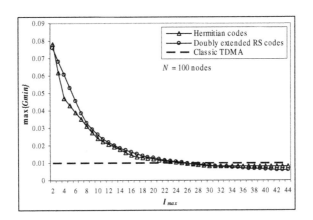

Fig. 3. $\max\{G_{min}\}$ with DE-RS, Hermitian codes, and classic TDMA, $N = 100$ nodes

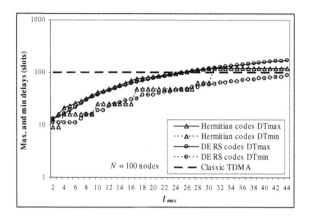

Fig. 4. $DT_{\max}$ and $DT_{\min}$ with DE-RS, Hermitian codes and classic TDMA, $N = 100$ nodes

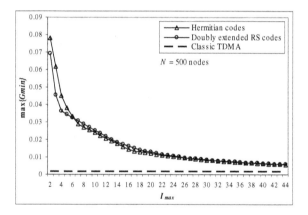

Fig. 5. $\max\{G_{\min}\}$ with DE-RS, Hermitian codes, and classic TDMA, $N = 500$ nodes

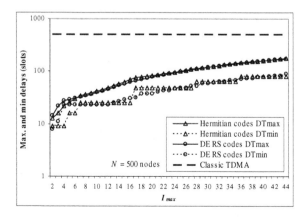

Fig. 6. $DT_{\max}$ and $DT_{\min}$ with DE-RS, Hermitian codes and classic TDMA, $N = 500$ nodes

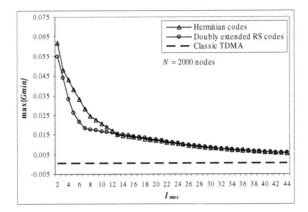

Fig. 7. $\max\{G_{\min}\}$ with DE-RS, Hermitian codes, and classic TDMA, $N = 2000$ nodes

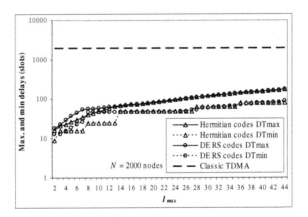

Fig. 8. $DT_{\max}$ and $DT_{\min}$ with DE-RS, Hermitian and classic TDMA, $N = 2000$ nodes

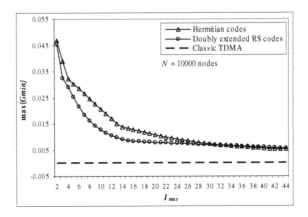

Fig. 9. $\max\{G_{\min}\}$ with DE-RS, Hermitian codes, and classic TDMA, $N = 10000$ nodes

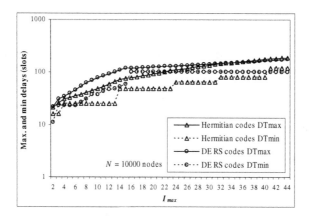

Fig. 10. $DT_{\max}$ and $DT_{\min}$ with DE-RS, Hermitian and classic TDMA, $N = 10000$ nodes

5 Conclusion

The advantage of Hermitian codes is exposed through a minimum throughput and delay performance comparison with a scheduling protocol that utilizes the code-words of a Reed-Solomon code. Hermitian codes can guarantee more minimum throughput and less maximum and minimum delay than the RS codes when the rank of the code is increased beyond two. Therefore Hermitian codes are good candidates for scheduling in large networks. The minimum relative distance of a code is identified as an important metric for the selection of good scheduling codes, regardless of the code being MDS.

Hermitian codes could also be utilized to increase the number of code-words available when the number of nodes increases beyond q^k without the need to increase the dimension of the code (increase k rather than q). Note that changing the dimension of the code translates into a different frame size, which means a need to re-distribute all the code-word assignments in the entire network. If, however, a higher-rank code with an unchanged dimension is used, the original code-words assigned to the old nodes can still be used since they form a sub-set of the new code-set, and therefore, only additional code-word assignments will be needed for the newly arriving nodes.

Another potential advantage of Hermitian codes is that higher-rank codes could be used to generate an extremely large set of code-words (e.g., billions), which in turn means that a node would randomly select a code-word from this large set. The possibility of two nodes picking the same code-word would be almost zero assuming the number of code-words is much larger than the number of nodes. This would represent a reduction in the overhead created by the assignment of unique code-words to nodes. As part of our future work we are investigating techniques that combine contention and code-based scheduling.

References

1. I. Chlamtac and A. Farago, "Making transmission schedules immune to topology changes in multi-hop packet radio networks," IEEE/ACM Trans. Networking, vol. 2, No. 1, Feb. 1994, pp. 23-29
2. Ji-Her Ju and Victor O. K. Li, "An optimal topology-transparent scheduling method in multi-hop packet radio networks," IEEE/ACM Trans. Networking, vol. 6, No. 3, June 1998, pp. 298-306
3. F.J. MacWilliams and N. J. A. Sloane, The theory of error-correcting codes, North Holland Pub. 1977. ISBN: 0444850090
4. G. Einarsson, "Address assignment for a time-frequency-coded, spread-spectrum system," The Bell Systems Tech. Journal, vol. 59, No. 7, Sept. 1980, pp. 1241-1255
5. G. Solomon, "Optimal frequency hopping sequences for multiple-access," Proceedings of the Symposium of Spread Spectrum Communications, vol. 1, AD-915 852, 1973, pp. 33-35
6. V.R. Syrotiuk, C. J. Colbourn and A. C. H. Ling, "Topology-transparent scheduling for MANETs using orthogonal arrays," International Conference on Mobile Computing and Networking, San Diego, CA, 2003, pp. 43-49
7. A.S. Hedayat, N. J. A. Sloane, J. Stufken, Orthogonal Arrays: Theory and Applications, Springer; 1st edition. June 22, 1999. ISBN: 0387987665
8. Carlos H. Rentel and Thomas Kunz, "On the average throughput performance of code-based scheduling protocols for Wireless Ad Hoc networks," ACM MobiHoc poster, Urbana-Champaign, Il, 2005.
9. S. B. Wicker, Error Control Systems for Digital Communication and Storage, Prentice Hall; 1995. ISBN: 0132008092
10. O. Pretzel, Codes and algebraic curves, Oxford science publications. 1998. ISBN: 0198500394
11. I. Blake, C. Heegard, T. Hoholdt, V. Wei, "Algebraic-Geometry Codes," IEEE Transactions on Information Theory, vol. 44, No. 6, Oct. 1998, pp. 2596-2618

An Intelligent Sensor Network for Oceanographic Data Acquisition

Chris Roadknight[1], Antonio Gonzalez[2], Laura Parrot[3], Steve Boult[4],
and Ian Marshall[3]

[1] BT, Adastral Park, Suffolk. UK IP5 3RE
christopher.roadknight@bt.com
[2] UCL, Ross Building, Adastral Park, IP5 3RE
ae.gonzalez@ee.ucl.ac.uk
[3] The University of Kent, UK
Lvp3@kent.ac.uk, i.w.marshall@kent.ac.uk
[4] Intelisys Ltd, Campus Ventures Centre, Oxford Road, Manchester M13 9PG
s.boult@man.ac.uk

Abstract. In this paper we describe the deployment of an offshore wireless sensor network and the lightweight intelligence that was integrated into the data acquisition and forwarding software. Although the conditions were harsh and the hardware readily available, the deployment managed to characterise an extended period due the some unique algorithms. In-situ adaptation to differences in data types, hardware condition and user requirements are demonstrated and analysis of performance carried out. Statistical tests, local feedback and global genetic style material exchange ensure limited resources such as battery and bandwidth are used efficiently by manipulating data at the source.

1 Introduction

Research into wireless sensor networks is not new [1,2,3] and networks of devices are already being deployed [4], but robust and efficient deployment of these networks will require hands-off configuration and management as the size of these networks increase.

Efficient and effective parameter measurement of oceanographic environments is an ongoing challenge in itself but also serves as a useful test case for applying sensor networks and assessing the performance of on-board intelligence. For dynamic, real world scenarios it is preferable for devices to make decisions in real time that adapts to limiting parameters (battery life, network conditions etc) and conditions (temperature, wave height etc). Enabling a device to make good decisions in real time is one type of artificial intelligence. This paper proposes a hybrid algorithm that is shown to fulfil some of the important requirements.

Monitoring of environmental conditions also serves as a useful test case for more heterogeneous pervasive networks [5]. Mobile phones, laptops, PDAs all have limited network, hardware and battery power but extensive functional capabilities, getting the most out of these devices in an increasingly networked society is a primary goal for ICT research.

V.R. Sirotiuk and E. Chávez (Eds.): ADHOC-NOW 2005, LNCS 3738, pp. 235–243, 2005.
© Springer-Verlag Berlin Heidelberg 2005

2 Experimental Design

The remit of this project was to replace single expensive monitoring stations with cheaper, connected devices that operated as an intelligent network. There is some evidence that many less accurate devices can characterise an area better than a single accurate device.

2.1 SECOAS

The Self-Organising Collegiate Sensor (SECOAS) Network Project [3, 7] is a collaborative effort that is providing original solutions to implementation of offshore wireless sensor networks [8]. Simple, 'off-the-shelf' hardware has been used that works together to characterise the conditions of an area of coastline, the test site is an off-shore wind farm [9] where the impact of the structure is largely unknown. This is made possible by a range of onboard AI that simultaneously aims to conserve battery power, optimise bandwidth usage, compress data and assess the usefulness of data. Most types of behaviour can be pre-programmed, a device can be told in advance what to sense, when to forward, when to compress etc but this type of pre-programmed behaviour relies on accurate knowledge of the system being studied, the hardware used and the conditions of the trial. This knowledge is seldom accurately known which leads to inefficient use of resources. This could be expressed as oversampling of relatively uninteresting periods and undersampling of interesting periods or, completing the experiment with unused battery power or the sampling period being cut short to overuse of the battery. A truly adaptive approach could make best use of resources and reconfigure even when catastrophic events such as the loss of a node occurs.

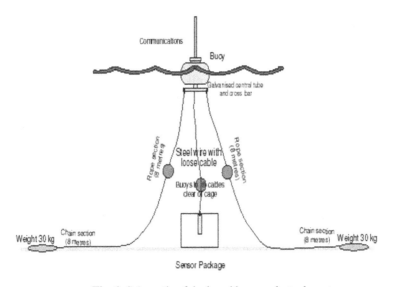

Fig. 1. Schematic of deployed buoy and attachments

Like all wireless sensor network systems, resources such as battery power will be limited in the SECOAS project [10,11]. Analogue to digital converters, microchips and radio equipment all use battery power, so the usefulness of every reading needs to be assessed on the node, to save resource expensive transmission.

6 buoys were deployed for 2 weeks in October 2004. One of which is shown in figure 1. The buoy was anchored in position and had also had a tethered sensing module (figure 1). An RS232 interface was used between the network module and the sensing module with the purpose of collecting sampling data. In the initial deployment only single hop radio transmission was used.

2.2 ARNEO (Autonomous Resilient Network for Environmental Observation)

Wireless propagation can be seriously affected during periods when exceptional events of interest might be happening (e.g. storms or intense wave activity). Since they can last for long periods and are very difficult to predict, activity-aware mechanisms for data communication which avoid energy starvation are required. The opportunities for transmissions are scarce and shorter in range, therefore the expected error rate is high. Devices can be temporarily or permanently obstructed from performing due to a multitude of reasons that cannot be reasonably avoided. The causes are extremely varied, thus aiming to predict or manage individual cases would make decision taking inefficient. The system must dynamically self-reconfigure functional communication paths as needed. Multi-hop communication is required for covering vaster areas than can not be covered using other wireless configurations, in addition of providing resilience to individual failure.

Deployments must adaptively handle situations ranging from small to large numbers of heterogeneous devices, from sporadic small areas of intermittent obstructions to complete communication blackouts lasting many hours, from operating with the entire device population to a large proportion temporally buried, blocked or damaged. Whereas disruptive factors cannot be eradicated, it is possible to reduce the negative impact they produce on message recovery.

Autonomous Resilient Network for Environmental Observation Systems (ARNEO) provides network services for interconnecting sensors and consequently providing access to scientific models in line with the requirements of Environmental observation. ARNEO is an autonomous wireless ad hoc sensor network providing resilient transport across dense sensor deployments and scientific models of environmental phenomena.

It is expected that the collection of data produced at large by the sensor population possess strong correlation and redundancy. It is clear that isolated samplings may provide little information towards the understanding of the phenomena in comparison to the information provided by the majority of the deployment. Nevertheless, it also indicates that collecting a high percentage of generated samplings can still produce sensible depiction of the phenomena depending of the scientific model. Under certain considerations, scientific models may evaluate the possibility of having their demands satisfied with lower recover rates.

Our longer term intention for ARNEO is to create a large-scale multi-hop network, however during the described trial it was decided to experiment with a simple single cluster of six nodes as a representative of the target system. Wireless devices based on a microcontroller of the series PIC 16XXX and radio modules at 173.25 MHz were deployed using a STDMA hierarchy as in [12]. Tools, instruments and programs were deployed to record relevant parameters and collection of sampling data and synchronisation parameters

2.3 Hybrid Data Handling Algorithm

In the context of a sensor network data handling can be seen as anything involved with the acquisition, mathematical manipulation and forwarding on of environmental measurements. The 'algorithm' we use for this involves the data passing through several rounds on analysis and manipulation, versions of which have been used for other ad-hoc and active network tasks [13,14]. Firstly, before the readings are even stored on the data logger some preprocessing occurs but this occurs on the monitoring device so can be ignored. The major decisions are what to do with the data on the logger. Given enough bandwidth and enough battery every single reading would be sent back to the base station but this is not feasible.

The first decision involves passing the data through a 3 entry sliding window looking for sufficient deletion conditions. Given a time-series of sensor readings at t0, t1, t2 a simple analysis of the reading at t1 can decide how useful it is. If the reading at t1 is the average of the readings at t0 and t2 then its deletion will make no effect on the characterisation of a time series, given that it's value can be interpolated from readings at t0 and t2. A deviation from the average by a small amount may also be acceptable if improved compression is required. A trade off between loss of information and compression must be made.

Next we use internal condition monitoring that affects the frequency of some actions, using negative feedback to obtain a homeostatic behaviour. A node may carry out none, one or many actions during a specific time period. Actions such as sensing, forwarding and queue management. Each action has a cost in terms of queue occupancy, battery usage and bandwidth usage. By monitoring the condition of these resources the probability of carrying out these actions can be modified. For instance, if the queue length is near it's maximum it would be prudent to take fewer readings and/or to do more forwarding or if the battery is being used at an unsustainable rate higher battery usage behaviours should be reduced and lower usage ones increased. We term this 'local learning' as changes are affected using only local knowledge.

Finally, a genetic style transfer and fitness based evaluation of internal performance parameters can enable nodes that are performing well to share their configuration with nodes that are performing less well. Methods 1 and 2 both involve several parameters, values that effect the performance (e.g. Reading at T1 is deleted if + or − Z% of the average of Reading T0,T2. Sensing probability is reduced by X if queue is above Y). Effective values for these parameters are discovered in advance using multi-parameter optimisation on a simulated environment. But this can only be as good as the simulated environment. By encoding these parameters in a genetic fashion the performance of the nodes can be evaluated and the genetic material for the 'fittest' nodes can be spread, while the genetic make up of the less fit nodes is modified or dies out.

3   Results

While 6 nodes were deployed (see figure 2 for example) only 2 gathered data for the whole 2 week period, 2 more gathered data for part of the period.

Data for electrical conductance, temperature, pressure (from which wave height can be inferred) and turbidity were collected on the data logger ant minutely intervals [figures 3-6 show data over various time ranges].

Fig. 2. Deployed Buoys off Scroby Sands

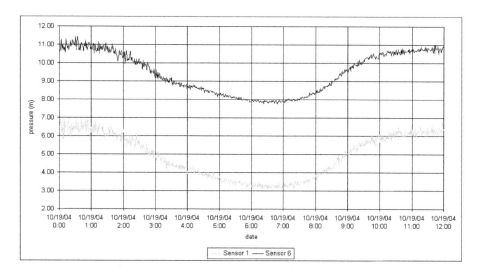

Fig. 3. Readings depth (derived from pressure) at 2 buoys over 12 hours

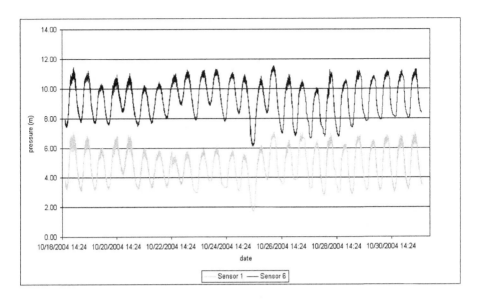

Fig. 4. Readings for depth (derived from pressure) at 2 different buoys over 3 weeks

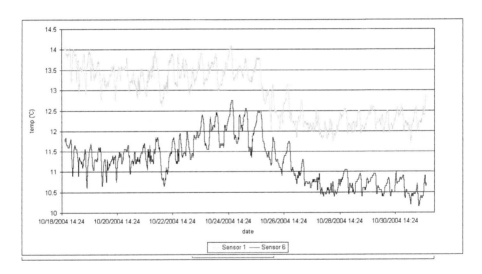

Fig. 5. Readings for temperature from 2 different buoys over 3 weeks

More details on the deployment specifics are presented elsewhere [6,8,15], as are details of the radio based data transmission [12]. For the purposes of this paper we were interested in how the data gathered was handled and how this compared to othermethods that could have been applied. Firstly, if we look at the data logged at 2 nodes we will see that the profile differs for each reading type (figs 3-6) and differs

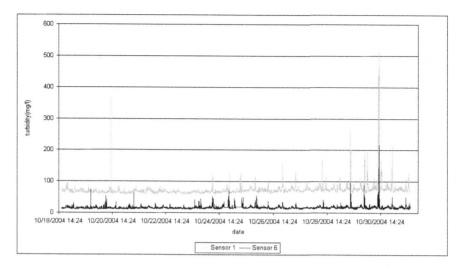

Fig. 6. Readings for turbidity from 2 different buoys over 3 weeks

from node to node (fig 3-6).   For example temperature varies less wildly thanturbidity, pressure has a strong periodic behaviour (tidal), temperature is different at different nodes. There are other less obvious characteristics in the dataset. Any real time data handling software must be able to adapt to different data profiles in real time regardless of how many underlying trends and characteristics are present.

This variation in behaviour has an effect on optimal compression. Table 1 shows how different methods of compression and forwarding (sliding window, round robin etc) perform on different datasets (pressure, temperature, turbidity). Random sampling just takes the desired compression and probabilistically deletes samples to affect this. Similarly, round robin deletes samples sequentially to cause the desired compression, for instance 75% compression would be made by deleting 3 samples then leaving 1 and so on. The variables were tuned so that all compression techniques over all sensor datasets compressed the data by 50% with the exception of 'sliding window' for turbidity and pressure. This is because the sliding window approach requires the middle value of three to vary by less than X% before it can be deleted, both turbidity and pressure we too variable for a sensible value for X to be found.

It is apparent that for less volatile datasets (temperature) a simple, non-adaptive approach like round robin sampling performs well but for more complex, volatile datasets a more adaptive approach performs better. Over the long term evolvable parameters should produce a more stable and robust solution but due to the inevitable introduction of mutated variables performance over the short term may be inferior. Evolvable parameters appear to improve performance for turbidity and temperature but incur an overall cost for pressure readings, possibly due to their volatile nature.

While at this stage different algorithms seem to be optimal, the aim is to modify the algorithm so that it can adapt to be optimal in all situations. This will be explored in experiments later in 2005.

Table 1. Performance of different algorithms at in-situ data compression (*less than 50% compression achieved)

	Absolute error (bits)		
	Sensor (average)		
	turbidity	pressure	temperature
Random sampling	7837	12005	543
Round robin sampling	6277	8573	274
Sliding Window (SW)	4236*	15015*	221
SW + local learning (LL)	4491	7120	645
SW + LL + evolvable parameters	4340	7559	347

4 Discussion

Embodying Artificially Intelligent (AI) techniques developed in simulations and models often prove to be an exercise in pragmatism. This is a result of hardware limitations and transparency concerns. Our experience is no different, but our real world deployment of a real-time intelligent sensor array still contained a significant amount of decision making power. The goal, of a system that made the right decisions all of the time, was only partly achieved. Tests against comparative techniques showed in some cases the deployed approach works best, in others it does not. For simple, slow moving time series data, the probabilistic nature of an evolvable, adaptive, layered approach is outperformed by a simple round robin sampling approach. For very bursty, fast moving data it proves better to apply a rule based approach. But if a 'good' solution is required that works over a range of data types then the implemented approach seems appropriate. This is due to the robustness of the approach but also the inherent ability to learn and evolve the longer the experiments run means that these systems will perform better the longer they are in place. We hope future longer-term experiments will demonstrate this.

References

1. Edgar H. Callaway. Wireless Sensor Networks: Architectures and Protocols. 2003. ISBN: 0849318238
2. I. Akyildiz, W. Su, Y. Sankarasubramaniam, E. Cayirci, "A Survey On Sensor Networks", IEEE Communications, August 2002, pp. 102-114.
3. L. Sacks, M. Britton, I. Wokoma, A. Marbini, T. Adebutu, I. Marshall, C. Roadknight, J. Tateson, D. Robinson and A. Gonzalez-Velazquez, "The development of a robust, autonomous sensor network platform for environmental monitoring," *in Sensors and their Applications XXII*, Limerick, Ireland, 2nd-4th September, 2003
4. A. Mainwaring, J. Polastre, R. Szewczyk D. Culler, J. Anderson,"Wireless Sensor Networks for Habitat Monitoring", First ACM Workshop on Wireless Sensor Networks and Applications (WSNA) 2002, pp. 88-97
5. M. Shackleton, F. Saffre, R. Tateson, E. Bonsma and C Roadknight. Autonomic computing for pervasive ICT – a whole system approach. BTTJ Vol 22, No3. 2004.

6. C Roadknight, L Parrott, N Boyd and I W Marshall "Real–Time Data Management on a Wireless Sensor Network" International Journal of Distributed Sensor Networks. Taylor & Francis. Volume 1, Number 2. April-June 2005 Pages: 215 - 225

7. SECOAS web site. www.adastral.ucl.ac.uk/sensornets/secoas

8. C Roadknight, L Parrott, N Boyd and I W Marshall. "A Layered Approach to in situ Data Management on a Wireless Sensor Network". Second Interna-tional Conference on Intelligent Sensors, Sensor Networks and Information Processing. 5-8 December 2005. Melbourne, Australia

9. Scroby sands wind farm homepage: http://www.offshorewindfarms.co.uk/sites/scroby-sands.html

10. A. Gonzalez-Velazquez, M. Britton, L. Sacks and I. Marshall, "Energy Savings in Wireless Ad-Hoc Sensor Networks as a Results of Network Synchro-nisation," in the London Communications Symposium, University College London, 8th-9th Septem-ber, 2003.

11. M. Brittan and L. Sacks. The SECOAS Project: Development of a Self-Organising, Wireless Sensor Network for Environmental Monitoring. SANPA 2004

12. A. Gonzalez-Velazquez, I.W. Marshall and L. Sacks. A self-synchronised scheme for automated communication in wireless sensor networks, Proceedings Intelligent Sensors, Sensor Networks and Information Processing Conference, 2004.

13. C.Roadknight and I.W.Marshall, Adaptive management of an active services network, BTTJ 18, 3, Oct 2000

14. I W. Marshall, C. Roadknight. Emergent Organisation in Colonies of Simple Automata. ECAL 2001: 349-356

15. C. Roadknight and I W. Marshall "Sensor Networks of Intelligent Devices," 1st European Workshop on Wireless Sensor Networks (EWSN '04), Berlin, 2004

Performance Analysis of the Hierarchical Layer Graph for Wireless Networks

Stefan Rührup*, Christian Schindelhauer**, and Klaus Volbert**

Heinz Nixdorf Institute, University of Paderborn, Germany
{sr, schindel, kvolbert}@uni-paderborn.de

Abstract. The Hierarchical Layer Graph (HL graph) is a promising network topology for wireless networks with variable transmission ranges. It was introduced and analyzed by Meyer auf der Heide et al. 2004. In this paper we present a distributed, localized and resource-efficient algorithm for constructing this graph. The qualtiy of the HL graph depends on the domination radius and the publication radius, which affect the amount of interference in the network. These parameters also determine whether the HL graph is a c-spanner, which implies an energy-efficient topology. We investigate the performance on randomly distributed node sets and show that the restrictions on these parameters derived from a worst case analysis are not so tight using realistic settings. Here, we present the results of our extensive experimental evaluation, measuring congestion, dilation and energy. Congestion includes the load that is induced by interfering edges. We distinguish between congestion and realistic congestion where we also take the signal-to-interference ratio into account. Our experiments show that the HL graph contains energy-efficient paths as well as paths with a few number of hops while preserving a low congestion.

1 Introduction and Overview

Topology control is an important issue in the field of wireless networks. Excellent surveys are presented by X.-Y. Li [7,6,8] and R. Rajaraman [11]. The general goal is to select certain connections to neighboring nodes that may be used for the network communication. On the one hand each node should have many connections to neighboring nodes to achieve fault-tolerance. On the other hand, if a node has many links, i.e. a high in-degree, then the probability of interference among these links is high and maintaining such a high number of links is not practicable. Note, that low in-degree alone does not imply low interference (cf. Figure 3). It is the task of the topology control algorithm – as part of the link layer – to select suitable links to neighboring nodes such that connectivity is

---

* DFG Graduiertenkolleg 776 "Automatic Configuration in Open Systems".
** Supported by the DFG Sonderforschungsbereich 376: "Massive Parallelität: Algorithmen, Entwurfsmethoden, Anwendungen." and by the EU within the 6th Framework Programme under contract 001907 "Dynamically Evolving, Large Scale Information Systems" (DELIS).

V.R. Sirotiuk and E. Chávez (Eds.): ADHOC-NOW 2005, LNCS 3738, pp. 244–257, 2005.

guaranteed and the number of links per node as well as the amount of interference is small.

If the nodes can adjust their transmission power, then energy can be saved and interference can be reduced when using short links. Nevertheless, connectivity of the network must be guaranteed. The problem of assigning transmission ranges to the nodes of a wireless network such that the total power consumption is minimized under the constraint that network is strongly connected (i.e., each node can communicate along some path in the network to every other node) is called the *minimum range assignment problem*. This problem is known to be NP hard [4,1]. Note, that we consider another model: We assume that every node is allowed to adjust its transmission range at any time. In this model connectivity is no longer a problem, but power consumption and the higher amount of interference caused by larger transmission ranges are the important aspects for designing a topology control algorithm.

Meyer auf der Heide et al. [10] proposed and analyzed a promising network topology, called the Hierarchical Layer Graph (HL graph), for power-variable wireless networks. In this paper we present a distributed, localized and resource-efficient algorithm for constructing the HL graph. We have implemented this topology control algorithm in our simulation environment SAHNE [14,12] and we present the results of our extensive experimental evaluations concerning the HL graph using realistic settings. We compare our results with the results using the *unit disk graph* as the network topology. The unit disk graph contains an edge between two nodes if and only if the Euclidean distance between these nodes is at most 1. The topology of a wireless network with fixed transmission ranges (normalized to 1) corresponds to the unit disk graph. The results of our simulations show the impact of the topology on a hop-minimal and an energy-minimal routing.

The remainder of this paper is organized as follows. In Section 2 we review the formal definition of the HL graph introduced in [10]. In this paper we concentrate on the resources routing time, defined in terms of congestion and dilation, and energy. In Section 3 we present the measures that we use to compare wireless network topologies. In Section 4 we describe the main functionality of our topology control algorithm for the HL graph. In Section 5 we give the settings of our simulations before we present the results of our simulations in Section 6. Finally, in Section 7 we conclude this work and discuss further directions.

2   The Hierarchical Layer Graph

The Hierarchical Layer Graph (HL graph) was first introduced in [2]. The set of nodes V is divided into several layers. The idea is to establish many short edges on the lower layers, that constitute an energy-optimal path system, and create only a small number of long edges on the upper layers, that ensure connectivity and allow short paths, i.e. paths with a small number of hops. The HL graph consists of the layers $L_0, L_1, \ldots, L_w$. The lowest layer L_0 contains all the nodes. The next layers contain fewer and fewer nodes. Finally, in the uppermost layer

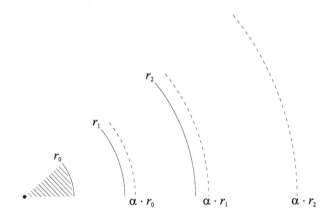

Fig. 1. The radii of the HL graph ($\beta < \alpha < \beta^2$)

only one node remains. If a node v belongs to the set of layer-i nodes, i.e. $v \in V(L_i)$, then it belongs also to each layer $V(L_j)$ with $0 \leq j < i$. In each layer a minimal distance between the nodes is required: $\forall u, v \in V(L_i) : |u - v| \geq r_i$. All the nodes in the layers below must be located within the radius r_i around a node in $V(L_i)$: $\forall u \in V(L_{i-1}) \exists v \in V(L_i) : |u - v| \geq r_i$. The distance constraints are defined by the parameters $\alpha \geq \beta > 1$. The smallest radius r_0 is chosen such that $r_0 < \min_{u,v \in V}\{|u - v|\}$. The other radii are defined by $r_i := \beta^i \cdot r_0$. These radii also define the edge set of each layer (see Figure 1). An edge (u, v) with $u, v \in V(L_i)$ belongs to the edge set of the i-th layer $E(L_i)$, if its length does not exceed the minimal distance r_i by the factor α: $E(L_i) := \{(u, v) | u, v \in V \land |u - v| \leq \alpha \cdot r_i\}$. The HL graph contains only symmetric edges. In this paper we assume that all nodes have random and unique IDs which serve for breaking the symmetry in leader election. A node belongs to the i-th layer, if there is no other node within distance r_i with higher priority, i.e. higher ID: $u \in V(L_i) \Leftrightarrow \neg\exists v : |u - v| \leq r_i \land ID(u) < ID(v)$. A node in $V(L_i)$ is a leader in $V(L_{i-1})$ and all the layers below. The **rank** of a node v denotes the number of the highest layer it belongs to: $R = \text{Rank}(v) \Leftrightarrow v \in \bigcup_{i=0}^{R} V(L_i) \land v \notin V(L_{i+1})$. In the following we refer to r_i as **domination radius** (β-radius) and to $\alpha \cdot r_i$ as **publication radius** (α-radius). According to the definition of the HL graph, the domination radii determine which nodes belong to a layer and the publication radii determine which edges are established (see Figure 2).

3   Measures for Network Topologies

The quality of a routing scheme depends on the quality of the topology of the network. Instead of focussing on one specific routing algorithm we consider path systems that use the topology and investigate the quality of the paths systems using the measures congestion, dilation, and energy which have been proposed in [9,2]: Given a path system $\mathcal{P}$. The **dilation** is given by the maximum of the

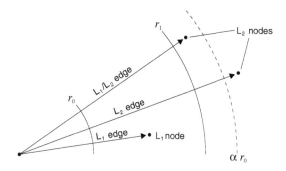

Fig. 2. The edge sets of the HL graph ($\beta < \alpha < \beta^2$)

lengths of all paths in $\mathcal{P}$. Regarding energy we distinguish between unit energy, which reflects the power consumption of maintaining the edges, and flow energy, which reflects the power consumption of using the paths for communication. The energy used in maintaining a communication link e is proportional to $|e|^2$, where $|e|$ denotes the Euclidean length of e. Then, the **unit energy** is defined by $\sum_{e \in E(\mathcal{P})} |e|^2$. If we take the **load of an edge** $\ell(e)$ (i.e. the number of paths in $\mathcal{P}$ using this edge) into account we obtain the **flow energy**, which is defined by $\sum_{e \in E(\mathcal{P})} \ell(e)|e|^2$.

The **load** of a path system is the maximum load of an edge: $L(\mathcal{P}) :=$ $\max_{e \in E(\mathcal{P})} \ell(e)$ (for wired networks this is often called congestion, see [5]). The definition of congestion for wireless networks contains also the load which is induced by interfering edges. The **congestion of an edge** e is given by $C(e) := \ell(e) + \sum_{e' \in \text{Int}(e)} \ell(e')$ where $\text{Int}(e)$ is the set of edges e' that interfere with e. The **congestion** of a path system is defined by $C(\mathcal{P}) := \max_{e \in E(\mathcal{P})} C(e)$.

The set of interfering edges $\text{Int}(e)$ can also be described according to the model used in [9]: An edge $e' = (v_1, v_2)$ interferes with $e = (u_1, u_2)$ if $\exists v \in e', \exists u \in e : |v - u| \leq r$, where r is the transmission range used by node v. The motivation for modeling interference this way is that communication in networks usually includes the exchange of acknowledgements, so that interference is also a problem for the sender who expects to receive an acknowledgement. Beside this model, we use a more realistic model in the simulation, including the power attenuation according to the free space propagation model and the signal-to-interference ratio. In particular, we consider the **realistic congestion $C_r(\mathcal{P})$** that we introduced in [12]. The realistic congestion includes load, interference, and properties of the propagation model. The definition is the same as for congestion, but for counting the interfering edges, we take the power attenuation and the signal-to-interference ratio (SIR) into account. Let us assume, that transmissions take place on all edges. An edge $e' = (v_1, v_2)$ interferes with an edge $e = (u_1, u_2)$ if the transmission on e' causes a received power p' at u_1 or u_2 that is higher than the received power p caused by a transmission on e divided by the SIR (i.e. $p/p' < \text{SIR}$).

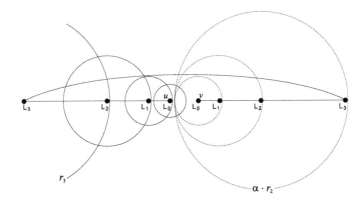

Fig. 3. Worst case construction for the HL graph

For the computation of congestion, we consider two path systems which are constructed by solving an all-pairs-shortest-paths problem w.r.t. hop-optimal and energy-optimal paths. The hop-optimal path system $\mathcal{P}_d$ optimizes dilation, whereas the energy-optimal path system $\mathcal{P}_e$ optimizes flow energy.

In [10] it is shown that a c-spanner allows us to approximate an energy-optimal path system within a constant factor and a congestion-optimal path system within a factor of $\mathcal{O}(\log n)$ for n nodes in general positions. Note, that in a c-spanner there is at least one path between two arbitrary nodes u and v of length at most $c \cdot |u - v|$ (for more details we refer to [13]). If $\alpha = \beta$ then $\alpha \cdot r_i = r_{i+1}$, i.e. the publication radius of layer i and the domination radius of layer $i + 1$ are congruent. In this case the HL graph is not a c-spanner for any constant c (see Figure 3). However, due to the definition, the graph is still connected, unless the maximum transmission radius is limited. For $\alpha > 2\frac{\beta}{\beta-1}$ the HL graph is a c-spanner with $c = \beta \frac{\alpha(\beta-1)+2\beta}{\alpha(\beta-1)-2\beta}$ [10]. It is known that the amount of interference in the HL graph can be upper bounded by $\mathcal{O}(\log n)$. Hence, the HL graph allows an approximation of a congestion-optimal path system by a factor of $\mathcal{O}(\log^2 n)$ for n nodes in general positions.

4   The HL Topology Control Algorithm

The hierarchical layer topology control algorithm basically consists of two parts (see Figure 4): Leader election and link establishment. As the topology control is meant to be a small part of the protocol stack we use a simple leader election mechanism: We assume that the nodes have unique IDs which are regarded as priorities. Initially, each node is in layer 0. Then each node tries to become leader in layer 0 which means that it becomes member of layer 1. For that, it sends a *claim for leadership* (CFL) message with the transmission range that is covering the domination-radius and increases its rank by one. If other nodes with a higher priority receive this message, then they respond with a *disagreement*

Preliminaries
RANDOM$(x_0, \ldots, x_i)$ is a random number uniformly chosen from $\{x_0, \ldots, x_i\} \subset \mathbb{N}$.
CREATEPACKET(type, source, target, layer, attempt) creates a packet
SEND(packet, p) sends a packet with transmission power p.
r is the rank of the node, the ID of the node determines its priority
$1/\rho$ is the expected length of an interval between updating a neighbor.
M is a list of messages a node receives.
$p_\alpha[i]$ is the power for the publication radius of layer i
$p_\beta[i]$ is the power for the domination radius of layer i

HLTC()
```
 1    r ← 0
 2    for t ← 1 to ∞
 3    do with Probability ρ
 4         if RANDOM(0, 1) = 0
 5         then ℓ ← RANDOM(0, ..., r)
 6                  P ← CREATEPACKET(NNP, ID, undefined, ℓ, 1)
 7                  SEND(P, p_α[ℓ])
 8         else   a ← 1
 9              repeat
10                        M ← ∅
11                        interference ← false
12                        P ← CREATEPACKET(CFL, ID, undefined, r + 1, a)
13                        SEND(P, p_β[r + 1])
14                        wait 2^{a−1} time steps
15                        and add received DIS-Messages to M
16                        update interference
17                        a ← a + 1
18              until interference = false
19              if M = ∅
20              then r ← r + 1
```

Fig. 4. The hierarchical layer topology control algorithm (HLTC)

(DIS). A node that receives a disagreement has to decrease its rank. If it receives no disagreement it can try to become leader on the next upper layer.

The second part of the protocol, the link establishment, is done the following way: A node sends neighbor notification packets (NNP) for a certain layer ℓ. The layer is chosen randomly between 0 and the rank of the node. If a message for layer ℓ is received by a node with a rank less than ℓ, then it is ignored. Each node belonging to layer ℓ that receives the NNP message includes the sender in the list of neighbors[1]. Figure 4 shows the pro-active part of the algorithm. The nodes respond to the control messages in the following way:

Response to CFL packet for layer ℓ: If the rank of the receiver of the CFL packet is at least ℓ, then the following action is performed: If the priority of the sender is higher, then the receiver has to decrease its rank to $\ell - 1$. If the priority of the receiver is higher, the receiver answers with a DIS packet. The transmission of the DIS packet is delayed for a random time according to an exponential backoff scheme: Before sending the DIS packet the receiver waits for for a number of time steps that is randomly chosen from the interval $\{1, \ldots, 2^{a-1}\}$,

---

[1] In practical environments the antennae often have deformed radiation patterns. In this case the receiver of a message cannot assume to reach the sender with the same transmission power. For the link establishment one would use additional acknowledgements for the NNP packets.

where a is the number of transmission attempts by the sender of the CFL packet. If the rank of the receiver of the CFL packet is smaller than ℓ, then no further action is needed.

Response to NNP packet for layer ℓ: If the rank of the receiver of the NNP packet is at least ℓ, then the sender is included in the list of neighbors. The transmission power for using this link is given by the pre-defined power level for layer ℓ.

5   Simulation Settings

In this section we present our experimental settings. We want to study the quality of the resulting topology in large-scale networks with high node density. One would assume, that such networks can be characterized by high interference and high congestion. We use our simulator for mobile ad hoc networks, SAHNE [14,12], that enables us to perform simulations of wireless networks with a large number of nodes. SAHNE has been designed for this purpose and is characterized by a lean model of the physical layer. It's data structures enable efficient range queries in order to determine the nodes that possibly receive a packet.

We consider the following scenario (cf. Table 1): From 50 up to 300 nodes are deployed (randomly and identically distributed) over an area of size 50 m × 50 m. The nodes communicate with a low power radio transceiver which supports variable transmission ranges, e.g., the CC1000 from ChipCon[2]. We assume 256 transmission power levels within this range and a maximum transmission range of 50 m. We use the signal propagation model proposed in [3] for flat rural environments, which is similar to the free space propagation model but with a path loss exponent of 3. All the transceivers use one frequency. Instead of a stochastic error model, we use a stronger condition for the successful reception of a packet: A packet can be successfully received only if the signal-to-interference ratio is at least 10.

We assume synchronized transmission and simulate the exchange of packets by the topology control algorithm. We do not simulate multi-hop data transmission because we do not want to focus on a specific routing algorithm. Instead, we assess the quality of the constructed topology as a whole using the measures defined in Section 3. Therefore, the network simulator calculates shortest paths for all pairs of nodes based on the current network topology in order to determine the load on the edges of the network graph. According to the simplified theoretical interference model and the more realistic interference model, the interference of the edges is calculated, which is needed to determine the congestion. As distance metric we use either the hop-distance or the energy-consumption of the links. So we can see whether the topology is suitable for shortest path routing in general.

At the beginning of a simulation the nodes are placed uniformly at random in the fixed simulation area, i.e. with an increasing number of nodes the density of

---

[2] ChipCon SA, www.chipcon.com. The CC1000 has an programmable output ranging from -20 dBm to 5 dBm.

Table 1. Simulation parameters

simulation area	50 m × 50 m
number of nodes	up to 300
max. transmission power	5 dBm
min. transmission power	-20 dBm
min. reception power	-107 dBm
path loss exponent	3
signal-to-interference ratio	10
max. transmission range	50 m

the nodes also increases, and with a higher density the probability of interference grows. The maximum transmission is also fixed such that it is unlikely that nodes are not connected. This is important for the all-pairs-shortest-paths routing: Missing connections would reduce the load and thus affect the calculation of congestion. Values for unit energy and flow energy are normalized, such that the transmission at maximum power requires the amount of energy of one. We call this unit of measurement **standard energy**. All the simulations are done for 20 node sets. In the tables and diagrams the average values (together with the standard error) are given.

6   Experimental Results

We performed numerous experiments with different combinations of the parameters α and β that determine the publication radius and the domination radius. In general, the HL graph is a c-spanner for a constant c if $\alpha > 2\frac{\beta}{\beta-1}$. In our simulations sometimes we deliberately chose these parameters such that this inequality is not fulfilled, in order to see whether this restriction is critical for random node placements. A comparison between different settings of α and β is difficult, so we decided to choose a fixed number of layers w and a factor $\delta = \alpha/\beta$, which is the ratio of publication radius and domination radius (α and β are derived from w and δ). We performed experiments with a fixed number of layers ranging from four to ten. With more than ten layers, the radii cannot be assigned to different power levels.

For comparison we constructed a unit disk graph, see Section 1, with a fixed transmission range of 25 m (UDG_{25}) and 50 m (UDG_{50}). With a lower transmission range most of the resulting networks are not connected (if 100 nodes are equally distributed over an area of 50 m × 50 m). The results of these experiments are shown in Table 2 and Table 3 in the Appendix.

The HL graph yields a small load for a high number of layers and $\delta \geq 2$. In this case the hop minimization of the path system results in a lower congestion than the energy minimization. One would expect that in an energy-optimal path system the short edges on the lower layers would be preferred. These edges usually cause fewer interferences. But paths using these edges are longer and (larger hop-count). This causes the relative high load values of the energy-optimal path system. A comparison between the load and the congestion shows the great im-

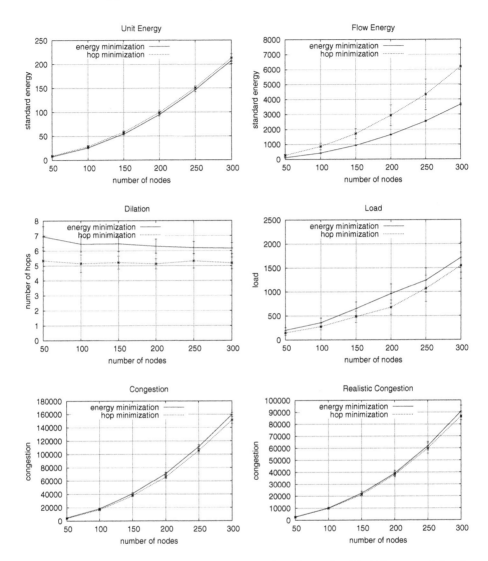

Fig. 5. Energy, dilation, load and congestion of the HL graph with 10 layers and $\delta = 2$ for up to 300 nodes (average values of 20 node sets)

pact of interferences. Congestion includes the load of interfering edges, which is significantly higher than the original load.

Note, that for constructing the HL graph no range adaption or assignment is performed. Choosing different power levels for the different layers is only a coarse-grained approximation of the optimal transmission range. Thus, the transmission power is not optimal for reaching a certain neighboring node. This makes the comparison of scenarios with a different number of layers difficult. But it can explain why a choice of ten layers yields better results than fewer layers, because with more layers a better approximation of the optimal transmission ranges can be achieved.

The UDG$_{25}$ contains path systems with a higher load, but with a smaller congestion than the HL graph for $w = 10$ and $\delta = 2$, because in the HL graph longer edges are allowed. If an edge on the top layer of the HL graph has a high load, then its load is assigned to most of the other edges. This effect is not so strong in the UDG$_{25}$, because the transmission range is limited. The UDG$_{50}$ is nearly a complete graph; so, a node can reach nearly all other nodes within one or two hops and, therefore, the load is very small. This results in a low congestion, which is about 30% lower as in the HL graph (realistic congestion for hop-optimal path systems). But the high price for this positive effect is the huge energy consumption: The flow energy in the UDG$_{50}$ (even without energy minimization) is over 20 times larger than in the HL graph with ten layers and $\delta = 2$. Note, that the path systems are not constructed with respect to congestion optimization.

For ten layers and $\delta = 2$ we performed further experiments for 50 to 300 nodes. The results are shown in Figure 5. We observe a super-linear increase of congestion and energy, due to the fact that the area is fixed and thus the node density increases. With increasing node density, more energy-efficient paths become available, which results in a decrease of the dilation in energy-optimal path systems. Regarding unit energy, we observe no significant difference between energy minimization and hop minimization, in contrast to flow energy: With energy minimization the use of long edges is avoided, which results in a smaller flow energy. With hop minimization it is possible that few long edges are used by most of the paths. The large deviation from the average indicates that this is not always the case. Regarding congestion, the difference between hop optimization and energy optimization is small. On the one hand, short paths can reduce the load; on the other hand, the long edges thereby used can cause more interference. It seems that these effects cancel out each other or that they are not so dominant.

A high number of layers yields also a small dilation in combination with $\delta \geq 2$. Especially with hop minimization a very small dilation can be achieved. Here, energy minimization can reduce unit energy and flow energy. Also a small δ, that reduces the publication radius and causes shorter edges on the lower layers, causes a significant reduction of unit energy and flow energy. But even with $\delta = 2$ the HL graph yields a flow energy that is about six times smaller than that of an energy minimized path system in the UDG. Yet, the unit energy of the HL graph is about four times smaller.

This demonstrates the advantages of the HL graph: It contains energy-saving paths as well as paths with a small number of hops. In comparison to a unit disk graph with a smaller restricted transmission range (UDG$_{25}$) the HL graph has a smaller load but a higher congestion; however, unit energy and flow energy are significantly smaller in the HL graph. So the routing strategy can decide, if a packet has to be delivered quickly or with low power consumption – the HL graph contains suitable paths for both cases.

7 Conclusion

In this paper we presented the results of our extensive experimental evaluations concerning the hierarchical layer graph (HL graph) for power-variable wireless networks. For this purpose we have developed and implemented the hierarchical topology control algorithm that allows us to construct the HL graph in a local and distributed way. We could show by simulations that the restrictions on the domination radius and the publication radius are not so tight using realistic settings, i.e. the HL graph gives a well-suited topology concerning congestion, dilation and energy also for values for α and β that do not fulfill the inequality $\alpha > 2\frac{\beta}{\beta-1}$ for nodes in general position. For practical considerations the simulation results show how the power levels of a radio transceiver should be set to achieve a good network topology. With a high number of layers the benefit of range adaption becomes visible: With ten layers and $\delta = 2$ we could achieve moderate congestion and small flow energy. Finally, the decision which paths to choose is left to the routing algorithm – the HL graph contains short paths (small hop-count) as well as energy efficient paths.

References

1. A. E. F. Clementi, P. Penna, and R. Silvestri. On the power assignment problem in radio networks. *Electronic Colloquium on Computational Complexity (ECCC'00)*, 2000.
2. M. Grünewald, T. Lukovszki, C. Schindelhauer, and K. Volbert. Distributed Maintenance of Resource Efficient Wireless Network Topologies (Ext. Abstract). In *8th European Conference on Parallel Computing (EURO-PAR'02)*, pages 935–946, 2002.
3. Simon Haykin and Michael Moher. *Modern Wireless Communications*. Prentice Hall, 2004.
4. L. M. Kirousis, E. Kranakis, D. Krizanc, and A. Pelc. Power consumption in packet radio networks. *Theoretical Computer Science*, 243(1-2):289–305, 2000.
5. F. T. Leighton. *Introduction to Parallel Algorithms and Architectures Arrays, Trees, Hypercubes*. Morgan Kaufmann Publishers, Inc., San Mateo, California, 1992.
6. X.-Y. Li. Applications of Computational Geometry in Wireless Ad Hoc Networks, Book Chapter of Ad Hoc Wireless Networking, Kluwer, edited by X. Z. Cheng, X. Huang, and D.-Z. Du. 2003.
7. X.-Y. Li. Topology Control in Wireless Ad Hoc Networks, Book Chapter of Ad Hoc Networking, IEEE Press, edited by S. Basagni, M. Conti, S. Giordano, and I. Stojmenovic. 2003.
8. X.-Y. Li. Wireless Sensor Networks and Computational Geometry, Book Chapter of Handbook of Sensor Networks, edited by Mohammad Ilyas et al. CRC Press. 2003.
9. F. Meyer auf der Heide, C. Schindelhauer, K. Volbert, and M. Grünewald. Energy, Congestion and Dilation in Radio Networks. In *14th ACM Symposium on Parallel Algorithms and Architectures (SPAA'02)*, pages 230–237, 2002.
10. F. Meyer auf der Heide, C. Schindelhauer, K. Volbert, and M. Grünewald. Congestion, Dilation, and Energy in Radio Networks. *Theory of Computing Systems (TOCS'04)*, 37(3):343–370, 2004.

11. R. Rajaraman. Topology Control and Routing in Ad hoc Networks: A Survey. In *SIGACT News, June 2002*, pages 60–73, 2002.
12. S. Rührup, C. Schindelhauer, K. Volbert, and M. Grünewald. Performance of Distributed Algorithms for Topology Control in Wireless Networks. In *17th Int. Parallel and Distributed Processing Symposium (IPDPS'03)*, page 28(2), 2003.
13. C. Schindelhauer, K. Volbert, and M. Ziegler. Spanners, Weak Spanners, and Power Spanners for Wireless Networks. In *Proc. of the 15th International Symposium on Algorithms and Computation (ISAAC'04)*, pages 805–821, 2004.
14. K. Volbert. A Simulation Environment for Ad Hoc Networks Using Sector Subdivision. In *10th Euromicro Workshop on Parallel, Distributed and Network-based Processing (PDP'02)*, pages 419–426, 2002.

Appendix

Table 2 and Table 3 contain the simulation results for the HL graph with different parameters α and β using a set of 100 randomly distributed nodes and the simulation settings shown in Table 1. The parameters α (publication radius) and β (domination radius) are derived from the number of layers w and $\delta := \alpha/\beta$ (see Section 5). The results are compared with the results using the unit disk graph (UDG) as topology with fixed transmission ranges $r = 25\,\mathrm{m}$ and $r = 50\,\mathrm{m}$.

Table 2. Load and congestion of the HL graph with 100 nodes (average values of 20 node sets with standard error)

energy-optimal path system					
w δ	α	β	load	congestion C	realistic congestion C_r
4 1	2.52	2.52	1376.29 +/- 411.12	20023.40 +/- 1997.95	11240.70 +/- 976.70
4 2	4.00	2.00	461.15 +/- 139.10	18797.30 +/- 1078.45	10329.70 +/- 885.49
4 3	5.25	1.75	299.40 +/- 64.19	18849.80 +/- 537.63	11628.20 +/- 694.02
6 1	1.74	1.74	1573.00 +/- 517.08	22202.50 +/- 4310.91	12258.50 +/- 2387.25
6 2	3.03	1.52	429.05 +/- 82.04	18656.00 +/- 830.34	10104.50 +/- 1095.90
6 3	4.20	1.40	269.95 +/- 74.30	18748.00 +/- 517.35	11373.80 +/- 821.95
8 1	1.49	1.49	1612.38 +/- 433.30	20119.50 +/- 3588.20	11183.20 +/- 1696.22
8 2	2.69	1.35	409.15 +/- 146.06	17913.20 +/- 979.74	9913.30 +/- 731.78
8 3	3.81	1.27	244.105 +/- 59.27	18244.10 +/- 406.21	11294.10 +/- 556.85
10 1	1.36	1.36	1165.80 +/- 532.90	17270.00 +/- 3298.22	9001.60 +/- 1612.68
10 2	2.52	1.26	358.50 +/- 99.13	18357.40 +/- 835.89	10120.30 +/- 701.61
10 3	3.61	1.21	256.63 +/- 42.27	18148.40 +/- 469.53	11554.60 +/- 907.49
UDG $r = 25\,\mathrm{m}$			731.83 +/- 177.87	9545.28 +/- 1458.02	5904.89 +/- 989.62
UDG $r = 50\,\mathrm{m}$			251.82 +/- 65.68	11830.70 +/- 701.20	6967.59 +/- 1012.87
hop-optimal path system					
w δ	α	β	load	congestion C	realistic congestion C_r
4 1	2.52	2.52	686.00 +/- 202.31	16859.90 +/- 687.142	11176.30 +/- 926.78
4 2	4.00	2.00	242.00 +/- 51.81	16187.40 +/- 688.529	10433.90 +/- 333.77
4 3	5.25	1.75	141.95 +/- 46.02	16524.80 +/- 476.963	11140.60 +/- 584.62
6 1	1.74	1.74	1148.46 +/- 314.28	18343.10 +/- 3316.36	11786.90 +/- 2302.61
6 2	3.03	1.52	234.20 +/- 58.96	16364.90 +/- 627.88	9943.85 +/- 755.67
6 3	4.20	1.40	198.32 +/- 99.78	17174.30 +/- 355.95	11053.60 +/- 629.66
8 1	1.49	1.49	1099.75 +/- 196.14	17992.80 +/- 978.47	10103.60 +/- 1261.27
8 2	2.69	1.35	281.05 +/- 72.37	16635.20 +/- 633.17	9986.75 +/- 470.35
8 3	3.81	1.27	194.95 +/- 77.85	17242.50 +/- 588.09	10922.10 +/- 864.51
10 1	1.36	1.36	1137.27 +/- 252.81	17346.50 +/- 1735.72	9618.55 +/- 1556.63
10 2	2.52	1.26	278.05 +/- 69.68	16907.20 +/- 653.95	9925.10 +/- 713.20
10 3	3.61	1.21	195.00 +/- 49.55	17493.70 +/- 348.91	11340.00 +/- 585.41
UDG $r = 25\,\mathrm{m}$			808.79 +/- 369.47	10104.10 +/- 1530.98	6177.32 +/- 1088.30
UDG $r = 50\,\mathrm{m}$			258.80 +/- 63.82	12265.10 +/- 797.52	7041.10 +/- 822.51

Table 3. Dilation, unit energy and flow energy of the HL graph with 100 nodes (average values of 20 node sets with standard error)

w	δ	α	β	dilation	unit energy	flow energy
energy-optimal path system						
4	1	2.52	2.52	14.93 +/- 2.53	3.50 +/- 0.59	209.58 +/- 40.48
4	2	4.00	2.00	6.90 +/- 0.45	33.02 +/- 2.73	432.44 +/- 16.78
4	3	5.25	1.75	5.00 +/- 0.00	118.86 +/- 8.88	756.46 +/- 23.64
6	1	1.74	1.74	16.82 +/- 4.17	2.64 +/- 0.36	221.77 +/- 56.69
6	2	3.03	1.52	6.85 +/- 0.49	28.42 +/- 1.38	431.52 +/- 12.46
6	3	4.20	1.40	5.00 +/- 0.00	108.18 +/- 8.55	749.98 +/- 23.77
8	1	1.49	1.49	17.25 +/- 1.83	2.35 +/- 0.10	227.50 +/- 38.76
8	2	2.69	1.35	6.60 +/- 0.50	25.76 +/- 1.54	437.28 +/- 21.97
8	3	3.81	1.27	4.79 +/- 0.42	97.12 +/- 5.62	751.62 +/- 29.17
10	1	1.36	1.36	13.20 +/- 0.84	2.26 +/- 0.22	181.25 +/- 19.43
10	2	2.52	1.26	6.45 +/- 0.51	25.67 +/- 1.37	422.47 +/- 15.48
10	3	3.61	1.21	4.74 +/- 0.45	95.68 +/- 7.08	743.09 +/- 24.96
UDG r = 25 m				9.50 +/- 0.79	116.86 +/- 8.33	5118.71 +/- 161.63
UDG r = 50 m				4.06 +/- 0.24	2996.24 +/- 116.84	19672.20 +/- 433.35
hop-optimal path system						
4	1	2.52	2.52	7.75 +/- 0.93	8.84 +/- 2.46	1139.20 +/- 238.72
4	2	4.00	2.00	4.45 +/- 0.51	59.26 +/- 10.99	1126.91 +/- 154.73
4	3	5.25	1.75	3.95 +/- 0.23	161.93 +/- 23.37	1249.81 +/- 129.22
6	1	1.74	1.74	10.62 +/- 2.33	4.12 +/- 0.73	929.19 +/- 210.77
6	2	3.03	1.52	5.00 +/- 0.32	34.85 +/- 4.30	919.84 +/- 188.72
6	3	4.20	1.40	4.00 +/- 0.33	117.10 +/- 10.99	1051.85 +/- 159.39
8	1	1.49	1.49	12.13 +/- 2.03	3.15 +/- 0.61	582.44 +/- 280.00
8	2	2.69	1.35	5.10 +/- 0.45	30.89 +/- 1.97	884.97 +/- 136.52
8	3	3.81	1.27	4.05 +/- 0.22	108.05 +/- 7.54	943.38 +/- 101.19
10	1	1.36	1.36	12.09 +/- 1.51	2.83 +/- 0.56	663.70 +/- 456.18
10	2	2.52	1.26	5.15 +/- 0.59	28.26 +/- 2.60	857.53 +/- 217.14
10	3	3.61	1.21	4.00 +/- 0.00	102.35 +/- 6.53	946.29 +/- 62.57
UDG r = 25 m				9.84 +/- 1.61	116.39 +/- 5.47	5216.28 +/- 552.05
UDG r = 50 m				4.00 +/- 0.00	3113.90 +/- 236.34	19338.20 +/- 757.30

Heuristic Algorithms for Minimum Bandwith Consumption Multicast Routing in Wireless Mesh Networks

Pedro M. Ruiz and Antonio F. Gomez-Skarmeta

University of Murcia, Murcia, E-30071, Spain

Abstract. We study the problem of computing multicast trees with minimal bandwidth consumption in multi-hop wireless mesh networks. For wired networks, this problem is known as the Steiner tree problem, and it has been widely studied before. We demonstrate in this paper, that for multihop wireless mesh networks, a Steiner tree does not offer the minimal bandwidth consumption, because it neglects the wireless multicat advantage. Thus, we re-formulate the problem in terms of minimizing the numbrer of transmissions, rather than the edge cost of multicast trees. We show that the new problem is also NP-complete and we propose heuristics to compute good approximations for such bandwidth-optimal trees. Our simulation results show that the proposed heuristics offer a lower bandwidth consumption compared with Steiner trees.

1 Introduction and Motivation

A wireless multihop network consists of a set of nodes which are equipped with wireless interfaces. Nodes which are not able to communicate directly, use multihop paths using other intermediate nodes in the network as relays. When the nodes are free to move, these networks are usually known as "mobile ad hoc networks". We focus on this paper in static multihop wireless networks, also known as "mesh networks". These networks have recently received a lot of attention in the research community, and they are also gaining momentum as a cheap and easy way for mobile operators to expand their coverage and quickly react to temporary demands.

In addition, IP multicast is one of the areas which are expected to play a key role in future mobile and wireless scenarios. Key to this is the fact that many of the future services that operators and service providers forsee are bandwidth-avid, and they are strongly based on many-to-many interactions. These services require an efficient underlying support of multicast communications when deployed over multihop extensions where bandwidth may become a scarce resource.

The problem of the efficient distribution of traffic from a set of senders to a group of receivers in a datagram network was already studied by Deering [1] in the late 80's. Several multicast routing protocols like DVMRP [2], MOSPF [3], CBT [4] and PIM [5]) have been proposed for IP multicast routing in fixed networks. These protocols have not been usually considered in mobile ad hoc

V.R. Sirotiuk and E. Chávez (Eds.): ADHOC-NOW 2005, LNCS 3738, pp. 258–270, 2005.

networks because they do not properly support mobility. In the case of mesh networks, one may think that they can be a proper solution. However, they were not designed to operate on wireless links, and they lead to sub-optimal routing solutions which are not able to take advantage of the broadcast nature of the wireless medium (i.e. sending a single message to forward a multicast message to all the next hops rather than replicating the message for each neighbor). Moreover, their routing metrics do not aim at minimizing the cost of the multicast tree, which limits the overall capacity of the mesh network.

The problem of finding a minimum cost multicast tree is well-known as the minimum Steiner tree problem. Karp [7] demonstrated that this problem is NP-complete even when every link has the same cost, by a transformation from the exact cover by 3-sets. There are some heuristic algorithms [8] to compute minimal Steiner trees. For instance, the MST heuristic ([9,10]) provides a 2-approximation, and Zelikovsky [11] proposed an algorithm which obtains a 11/6-approximation. Recently, Rajagopalan and Vazirani [12] proposed a 3/2-approximation algorithm. However, given the complexity of computing this kind of trees in a distributed way, most of the existing multicast routing protocols use shortest path trees or sub-optimal shared trees, which can be easily computed in polynomial time.

Similarly, multicast ad hoc routing protocols proposed in the literature [6] do not approximate a minimal cost multicast tree either. For ad hoc networks, most of the works in the literature are devoted to the improvement of multipoint forwarding efficiency have been related to the particular case of flooding (i.e. the broadcast storm problem). Only a few papers like Lim and Kim [14] analyzed the problem of minimal multicast trees in ad hoc networks, but they only defined several heuristics for the case of flooding, based on the minimum connected dominating set (MCDS).

Although it is widely assumed that a Steiner tree is the minimal cost multicast tree, we show in this paper that it is not generally true in wireless multihop networks (see Fig. 1). The problem of minimizing the bandwidth consumption of a multicast tree in an ad hoc network needs to be re-formulated in terms of minimizing the number of data transmissions. By assigning a cost to each link of the graph computing the tree which minimizes the sum of the cost of its edges, existing formulations have implicitly assumed that a given node v, needs k transmissions to send a multicast data packet to k of its neighbors. However, in a broadcast medium, the transmission of a multicast data packet from a given node v to any number of its neighbors can be done with a single data transmission. Thus, in ad hoc networks the minimum cost tree is the one which connects sources and receivers by issuing a minimum number of transmissions, rather than having a minimal edge cost.

In this paper we show that the Steiner tree does not always give an optimal solution. Additional contributions of this papers are the demonstration that the problem of minimizing the cost of a multicast tree in a wireless mesh network is also NP-complete, and the proposal of enhanced heuristics to approximate such optimal trees, which we call minimal data overhead trees. Our simulation

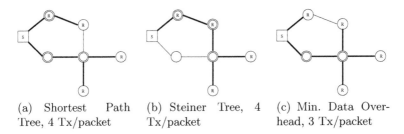

(a) Shortest Path Tree, 4 Tx/packet

(b) Steiner Tree, 4 Tx/packet

(c) Min. Data Overhead, 3 Tx/packet

Fig. 1. Differences in cost for several multicast trees over the same ad hoc network

results show that the proposed heuristics produce multicast trees with a lower bandwidth consumption that previous heuristics for Steiner trees over a variety of scenarios. In addition, they offer a huge reduction in the cost compared to the shortest path trees used by most of the ad hoc multicast routing protocols proposed so far.

The remainder of the paper is organized as follows: section 2 describes our network model, formulates the problem and shows that it is NP-complete. The description of the proposed algorithm is given in section 3. In section 4 we explain our simulation results. Finally, section 5 provides some discussion and conclusions.

2   Network Model and Problem Formulation

2.1   Network Model

We represent the ad hoc network as an undirected graph $G(V, E)$ where V is the set of vertices and E is the set of edges. We assume that the network is two dimensional (every node $v \in V$ is embedded in the plane) and mobile nodes are represented by vertices of the graph. Each node $v \in V$ has a transmission range r. Let $dist(v_1, v_2)$ be the distance between two vertices $v_1, v_2 \in V$. An edge between two nodes $v_1, v_2 \in V$ exists iff $dist(v_1, v_2) \leq r$ (i.e. v_1 and v_2 are able to communicate directly). In wireless mobile ad hoc networks some links may be unidirectional due to different transmission ranges. However, given that lower layers can detect and hide those unidirectional links to the network layer, we only consider bidirectional links. That is, $(v_1, v_2) \in E$ iff $(v_2, v_1) \in E$.

2.2   Problem Formulation

Given a multicast source s and a set of receivers R in a network represented by a undirected graph, we are interested in finding the multicast tree with the minimal cost in terms of the total number of transmissions required to deliver a packet from s to every receiver. To formulate the problem, we need some previous definitions.

Definition 1. Given a graph $G = (V, E)$, a source $s \in V$ and a set of receivers $R \subset V$, we define the set T as the set of the possible multicast trees in G which connect the source s to every receiver $r_i \in R$. We denote by F_t, the set of relay nodes in the tree $t \in T$, consisting of every non-leaf node, which relays the message sent out by the multicast source. We can define a function $C_t : T \to \mathbb{Z}^+$ so that given a tree $t \in T$, $C_t(t)$ is the number of transmissions required to deliver a message from the source to every receiver induced by that tree.

Lemma 1. Given a tree $t \in T$ as defined above, then $C_t(t) = 1 + |F_t|$.

Proof. By definition relay nodes forward the message sent out by s only once. In addition, leaf nodes do not forward the message. Thus, the total number of transmissions is one from the source, and one from each relay node. Making a total of $1 + |F_t|$. ∎

So, as we can see from lemma 1, the to minimize $C_t(t)$ we must somehow reduce the number of forwarding nodes $|F_t|$.

Definition 2. Under the conditions of definition 1, let $t^* \in T$ be the multicast tree such that $C_t(t^*) \leq C_t(t)$ for any possible $t \in T$. We define the data overhead of a tree $t \in T$, as $w_d(t) = C_t(t) - C_t(t^*)$. Obviously, with this definition $w_d(t^*) = 0$.

Based on the previous definitions, the problem can be formulated as follows. Given a graph $G = (V, E)$, a source node $s \in V$, a set of receivers $R \subset V$, and given $V' \subseteq V$ defined as $V' = R \cup \{s\}$, find a tree $T^* \subset G$ such that the following conditions are satisfied:

1. $T^* \supseteq V'$
2. $C_t(T^*)$ is minimum

From the condition of T^* being a tree it is obvious that it is connected, which combined with condition 1) establishes that T^* is a multicast tree. Condition 2) is equivalent to say that $w_d(T^*)$ is minimum, and establishes the optimality of the tree. As we show in the next theorem, this problem is NP-complete.

Theorem 1. Given a graph $G = (V, E)$, a multicast source $s \in V$ and a set of receivers R, the problem of finding a tree $T^* \supseteq R \cup \{s\}$ so that $C_t(T^*)$ is minimum is NP-complete.

Proof. According to lemma 1, minimizing $C_t(T^*)$ is equivalent to minimize the number of relay nodes $F \subseteq T^*$. So, the problem is finding the smallest set of forwarding nodes F that connects s to every $r \in R$. If we consider the particular case in which $R = V - \{s\}$, the goal is finding the smallest $F \subseteq T^*$ which covers the rest of nodes in the graph $(V - \{s\})$. This problem is the well-known vertex cover problem [13], which is NP-complete. So, by including a particular case which is NP-complete, our problem is also NP-complete. ∎

In the next theorem we show that in general the tree with the minimal edge-cost is not the one with the minimal data-overhead. Before presenting the theorem we give some definitions used within the proof of the theorem.

Definition 3. Under the same conditions of definition 1, and provided that each edge $e \in E$ has an associated cost $w(e) > 0$, we can define a function $C_e : T \rightarrow \mathbb{Z}^+$ so that given a tree $t \in T$, $C_e(t)$ is the edge cost of t defined as:

$$C_e(t) = \sum_{e \in E} w(e) \tag{1}$$

For the particular case of ad hoc networks, we can consider every edge to have the same cost. For simplicity in the calculations we assume that $w(e) = 1, \forall e \in E$. Even in that particular case, the problem of finding the multicast tree T^* so that $C_e(T^*)$ is minimum (also called Steiner tree) is NP-complete as R. Karp showed in [7]. In this particular case of unitary edge cost, $C_e(T^*)$ equals to the number of edges, which is $|V| - 1$ by a definition of tree.

Theorem 2. Let $G = (V, E)$ be an undirected graph. Let $s \in V$ be a multicast source and $R \subseteq V$ be the set of receivers. The Steiner multicast tree $T^* \subseteq G$ so that $C_e(T^*)$ is minimal may not be the minimal data-overhead multicast tree.

Proof. We will show that given an Steiner tree T^*, it is possible to find a tree T' such that $C_e(T') \geq C_e(T^*)$ and $C_t(T') \leq C_t(T^*)$. Let's denote by F' and F^* the number of forwarding nodes in each of the trees. For T' to offer a lower data overhead, the following condition must hold:

$$1 + |F'| \leq 1 + |F^*|$$

In a multicast tree, the number of forwarding nodes can be divided into those which are also receivers and those which are not. The latter are usually called Steiner nodes, and we will denote the set of such nodes as $\mathbb{S}$. The number of forwarding nodes which are also receivers can be easily computed as $|R - \mathbb{L}|$ being $\mathbb{L}$ the set of leaf nodes. Of course, every leaf node is also a receiver. Thus, the previous inequality, is equivalent to the following one:

$$1 + |\mathbb{S}'| + (|R| - |\mathbb{L}'|) \leq 1 + |\mathbb{S}^*| + (|R| - |\mathbb{L}^*|) \Rightarrow$$

$$|\mathbb{S}'| - |\mathbb{L}'| \leq |\mathbb{S}^*| - |\mathbb{L}^*| \Rightarrow$$

$$|\mathbb{S}'| - |\mathbb{S}^*| \leq |\mathbb{L}'| - |\mathbb{L}^*| \tag{2}$$

In addition, by the definition of C_e, $C_e(T') \geq C_e(T^*) \Rightarrow |V'| - 1 \geq |V^*| - 1$. Given that the number of vertices is exactly the sender plus the number of Steiner nodes plus the number of receivers, we can derive the following inequality:

$$|\mathbb{S}'| + |R| \geq |\mathbb{S}^*| + |R| \Rightarrow |\mathbb{S}'| \geq |\mathbb{S}^*|$$

So, according to Eq. 2 it is possible to build a tree T' so that $C_t(T') \leq C_t(T^*)$ provided that the number of additional Steiner nodes added ($|\mathbb{S}'| - |\mathbb{S}^*|$), is lower than the additional number of leaf nodes ($|\mathbb{L}'| - |\mathbb{L}^*|$). Such an example is shown in Fig. 1. ∎

In general, Steiner tree heuristics try to reduce the cost by minimizing the number of Steiner nodes $|\mathbb{S}^*|$. In the next section we will present our heuristics being able to reduce the bandwidth consumption of multicast trees, by just making receivers be leaf nodes in a cost effective way, according to Eq. 2.

3 Proposed Algorithms

Given the NP-completeness of the problem, within the next subsections we describe two heuristic algorithms to approximate minimal data-overhead multicast trees. As we learned from the demonstration of theorem 2, the best approach to reduce the data overhead is reducing the number of forwarding nodes, while increasing the number of leaf nodes. The two heuristics presented below try to achieve that trade-off.

3.1 Greedy-Based Heuristic Algorithm

The first proposed algorithm is suited for centralized wireless mesh networks, in which the topology can be known by a single node, which computes the multicast tree.

Inspired on the results from theorem 2, this algorithm systematically builds different cost-effective subtrees. The cost-effectiveness refers to the fact that a node v is selected to be a forwarding node only if it covers two or more nodes. The algorithm shown in algorithm 1, starts by initializing the nodes to cover ('aux') to all the sources except those already covered by the source s. Initially the set of forwarding nodes ('MF') is empty. After the initialization, the algorithm repeats the process of building a cost-effective tree, starting with the node v which covers more nodes in 'aux'. Then, v is inserted into the set of forwarding nodes (MF) and it becomes a node to cover. In addition, the receivers covered by v ($Cov(v)$) are removed from the list of nodes to cover denoted by 'aux'. This process is repeated until all the nodes are covered, or it is not possible to find more cost-effective subtrees. In the latter case, the different subtrees are connected by an Steiner tree among their roots, which are in the list 'aux' (i.e. among the nodes which are not covered yet). For doing that one can use any Steiner tree heuristic. In our simulations we use the MST heuristic for simplicity.

Theorem 3. The proposed algorithm results in a tree whose data-overhead is upper-bounded by that of the one obtained with a Steiner tree heuristic.

Proof. Let's consider the worse case in which no cost-effective tree can be formed. There are two possible cases:

1. There are no receivers in the range of the source. Then $Cov(s)=\varnothing$ and the resulting tree (T_1) is exactly a Steiner tree among s and all the receivers computed using any Steiner tree heuristic (T_2). Thus, $C_t(T_1) = C_t(T_2)$.
2. There are receivers in the range of the source. Then, the resulting tree (T_1) is a Steiner tree among the source s and all the receivers except $Cov(s)$. This tree is a subtree of the Steiner tree from s to every receiver (T_2), so $C_t(T_1) \leq C_t(T_2)$. ∎

Algorithm 1. Greedy minimal data overhead algorithm

```
1: MF ← ∅ / * mcast − forwarders * /
2: V ← V - {s}
3: aux ← R-Cov(s) + {s} / * nodes − to − cover * /
4: repeat
5:     node ← argmax_{v∈V}(|Cov(v)|) s.t. Cov(v)≥2
6:     aux ← aux-Cov(v)+{v}
7:     V ← V-{v}
8:     MF ← MF + {v}
9: until aux = ∅ or node = null
10: if V!=∅ then
11:     Build Steiner tree among nodes in aux
12: end if
```

3.2 Distributed Version of the Algorithm

The previous algorithm may be useful for some kind of networks, however a distributed approach is much more appealing for the vast majority of scenarios. In this section we present a slightly different version of the previous algorithm, being able to be run in a distributed way.

The previous protocol consists of two different parts: (i) construction of cost-efficient subtrees, and (ii) building a Steiner tree among the roots of the subtrees.

To build a Steiner tree among the roots of the subtrees, we assumed in the previous protocol the utilization of the MST heuristic. However, this is a centralized heuristic consisting of two different phases. Firstly, the algorithm builds the metric closure for the receivers on the whole graph, and then, a minimum spanning tree (MST) is computed on the metric closure. Finally, each edge in the MST is substituted by the shortest path tree (in the original graph) between the to nodes connected by that edge. Unfortunately, the metric closure of a graph is hard to build in a distributed way. However, we can approximate such an MST heuristic with the simple, yet powerful, algorithm presented in algorithm 2. The source, or the root of the subtree in which the source is (called source-root) will start flooding a route request message (RREQ). Intermediate nodes, when propagating that message will increase the hop count. When the RREQ is received by a root of a subtree, it sends a route reply (RREP) back through the path which reported the lowest hop count. Those nodes in that path are selected as multicast forwarders (MF). In addition, a root of a subtree, when propagating the RREQ will reset the hop count field. This is what makes the process very similar to the computation of the MST on the metric closure. In fact, we achieve the same effect, which is that each root of the subtrees, will add to the Steiner tree the path from itself to the source-root, or the nearest root of a subtree. The way in which the algorithm is executed from the source-root to the other nodes guarantees that the obtained tree is connected.

The second part of the algorithm to make distributed is the creation of the cost-effective subtrees. However, this part is much simpler and can be done locally with just a few messages. Receivers flood a Subtree_Join (ST_JOIN)

Algorithm 2. Distributed approximation of MST heuristic

```
 1: if thisnode.id = source − root then
 2:    Send RREQ with RREQ.hopcount=0
 3: end if
 4: if rcvd non duplicate RREQ with better hopcount then
 5:    prevhop ← RREQ.sender
 6:    RREP.nexthop ← prevhop
 7:    RREQ.sender ← thisnode.id
 8:    if thisnode.isroot then
 9:       send(RREP)
10:       RREQ.hopcount ← 0
11:    else
12:       RREQ.hopcount++;
13:    end if
14:    send(RREQ)
15: end if
16: if received RREP and RREP.nexthop = thisnode.id then
17:    Activate MF_FLAG
18:    RREP.nexthop ← prevhop
19:    send(RREP)
20: end if
```

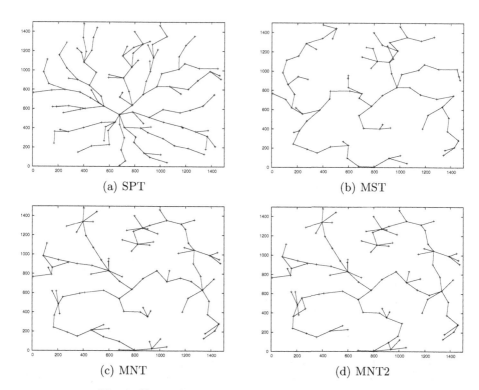

(a) SPT   (b) MST

(c) MNT   (d) MNT2

Fig. 2. Shape of example trees produced by heuristics

message only to its 1-hop neighbors indicating the multicast group to join. These neighbors answer with a Subtree_Join_Ack (ST_ACK) indicating the number of receivers they cover. This information is known locally by just counting the number of (ST_JOIN) messages received. Finally, receivers send again a Subtree_Join_Activation (ST_JOIN_ACT) message including their selected root, which is the neighbor which covers a higher number of receivers. This is also known locally from the information in the (ST_ACK). Those nodes which are selected by any receiver, repeat the process acting as receivers. Nodes which already selected a root do not answer this time to ST_JOIN messages.

As we can see in Fig. 2, the multicast trees produced by the proposed heuristics (MNT and MNT2) are very leafy, as suggested by theorem 2, to take advantage of the broadcast nature of the medium. Unlike our approaches, the Steiner tree heuristic (MST) tries to reduce the number of Steiner nodes, producing thus multicast trees with fewer leaves. As expected, the shortest path multicast tree (SPT) does not aim at minimizing the cost of the tree.

In the next section, we shall see that this distributed version of the algorithm is not as efficient as the centralized one, but offers a good approximation to the centralized scheme. This is because instead of really computing the metric closure in the graph, we just approximate it. However, the performance of the distributed approach is still better than the one offered by the Steiner tree.

4   Simulation Results

In order to assess the effectiveness of our proposed algorithms we have simulated them under different conditions. The algorithms that we have simulated are the two proposed approaches as well as the MST heuristic to approximate Steiner trees. In addition, we also simulated the shortest path tree algorithm, which is the one which is used by most multihop multicast routing protocols proposed to date.

As in many similar papers, we do not consider mobility in our simulations, because we are dealing with wireless mesh networks. In mobile ad hoc networks, changes in the topology make useless to approximate optimal solutions, which may become suboptimal within a few seconds.

4.1   Performance Metrics

We are interested in evaluating the optimality of the the topology of the multicast tree produced by the different algorithms. That is the reason why we use different metrics from the typical performance measurements (e.g. packet delivery ratio) which strongly depend upon the underlying wireless technology under consideration. In our particular case, the metrics under consideration are:

– Number of transmissions required. The total number of packet transmitted either by the source or by any relay node to deliver a data packet from the source to all the receivers.

– Mean number of hops. The number of multicast hops from a receiver to the source averaged over the total number of receivers.

So, by considering these metrics along with a perfect MAC layer (i.e. without collisions, retransmissions or interferences) we guarantee an unbiased comparison.

4.2   Simulation Methodology

All the approaches have been evaluated under a different number of receivers, and a varying density of the nodes in the network. In particular, the number of receivers considered was between 1 and 40% of the nodes, which corresponds to the range from 5 to 200 receivers. The density of the network varied between 100 and 500 $nodes/Km^2$.

For each combination of simulation parameters, a total of 91 simulation runs with different randomly-generated graphs were performed, making a total of more than 100000 simulations. The error in the graphs shown below are obtained using a 95% confidence level.

4.3   Performance Evaluation

In the figures below, SPT refers to the shortest path tree and MST corresponds to the MST heuristic to approximate Steiner trees. Finally, MNT and MNT2 correspond to the proposed centralized and distributed heuristics respectively.

In Fig. 3(a) we show for a network with an intermediate density how the number of transmissions required varies with respect to the number of receivers. As expected, when the number of receivers is lower than 20, the proposed schemes do not offer significant differences compared to the Steiner tree heuristic. This is clearly explained by the fact that the nodes tend to be very sparse and it is less likely that it is possible to build cost-effective trees. However, as the number of receivers increases, the creation of cost-effective trees is favored, making the the proposed schemes to achieve significant reductions in the number of transmissions required. In addition, given that the SPT approach doesn't aim at minimizing the cost of the trees, it shows a lower performance compared to any of the other approaches. Regarding the two proposed approaches the distributed approach, by avoiding the use of the metric closure, gets a slightly lower performance compared to the centralized approach. However, both of them have a very similar performance which allow them to offer substantial bandwidth savings compared to the Steiner tree (i.e. MST heuristic).

To evaluate the impact on the length of the paths, we performed the analysis shown in Fig. 3(b). As expected the SPT approach is the one offering the lowest mean path length. The Steiner tree heuristics as well as the proposed ones offer a higher mean path length. This is clearly due to the fact grouping paths for several receivers makes them not to use their shortest paths. As we can see, the this metric is much more variable to the number of receivers than the number of transmissions was for the heuristic approaches. This is why the error bars are reporting a larger confidence interval for MST, MNT and MNT2.

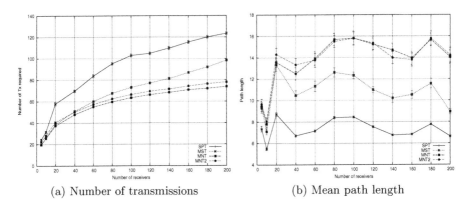

(a) Number of transmissions   (b) Mean path length

Fig. 3. Number of Tx and mean path length at increasing number of receivers

Another important aspect to consider is how the performance varies regarding the density of the network. This results are of paramount importance to determine under which scenarios the proposed approaches perform better. In particular we consider two different cases: a medium number of receivers represented by a 20% of the nodes, and a high number of receivers represented by a 36% of the nodes. As we showed before, the case of a very low number of receivers is not interesting because most of the approaches offer a similar performance.

In Fig. 4(a) and Fig. 4(b) we present the results for the medium number of receivers and high number of receivers respectively. As the figure depicts, the higher the density, the better is the performance of all the approaches. This makes sense, because the higher the density the lower the path lengths, so in general one can reach the receivers with less number of transmissions regardless of the routing scheme. However, if we compare the performance across approaches, we can see that the reduction in the number of transmissions that our proposed

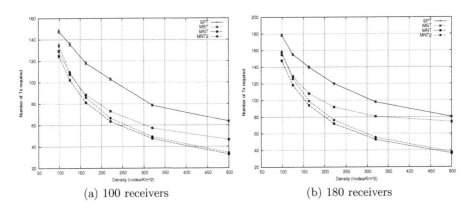

(a) 100 receivers   (b) 180 receivers

Fig. 4. Number of Tx with varying network density

heuristics achieve compare to the other approaches is higher as the density of the network increases.

This can be easily explained by the fact that for higher densities it is more likely that several receivers can be close to the same node, which facilitates the creation of cost-effective subtrees.

In addition, the higher the density, the closer in performance are the centralized and the distributed approaches. This is because in dense networks, the number of hops between any pair of nodes is also reduced. This makes the difference between metric closure and its approximation in number of hops to be reduced as well. This makes our approach very appealing for dense networks such as sensor networks in which the mean degree of a node is usually very high.

If we compare the two figures we can see that the difference in the number of receivers just varies a little bit the concrete performance differences among approaches. However, the density of the network has an strong effect on the overall performance of the solutions.

5   Conclusions and Discussion

As we have shown, the generally considered minimal cost multicast tree (Steiner tree) does not offer an optimal solution in multihop wireless networks. The problem is that the original Steiner tree problem formulation does not account for the reduction in bandwidth that can be achieved in a broadcast medium. Given those limitations we re-formulate the problem in terms of minimizing the number of transmissions required to send a packet from a multicast source to all the receivers in the group.

We have shown that this formulation is adequate for multihop wireless networks, and we have also demonstrated that this problem is NP-complete. So, we have introduced two new heuristic algorithms to deal with the problem of optimizing multicast trees in wireless mesh networks. Our simulation results show that the proposed heuristics manage to beat the Steiner tree MST heuristic over a variety of scenarios and network densities.

In particular, our results show that the higher the density of the network, the higher are the performance gains introduced by our heuristics compared to the other approaches. These results seem very promising as a possible future direction to address similar issues in sensor networks in which the network topology is generally very dense, and reverse multicast trees are very common as a mechanism to gather information from the sensor network.

Acknowledgment

This work has been partially funded by the Spanish Science and Technology Ministry by means of the "Ramon y Cajal" workprograme, the SAM (MCYT, TIC2002-04531-C04-03) project and the grant number TIC2002-12122-E. Part of this work was performed under a Visiting Fellowship of Pedro M. Ruiz at the International Computer Science Institute (www.icsi.berkeley.edu).

References

1. S. Deering, "Multicast Routing in a Datagram Internetwork," *Ph.D. Thesis, Electrical Engineering Dept., Stanford University*, Dec. 1991.
2. S.-E. Deering and D.-R. Cheriton, "Multicast Routing in datagram internetworks and extended LANs," *Transactions on Computer Systems*, vol.8, no.2, May 1990, pp. 85–110.
3. J. Moy, "Multicast routing extensions for OSPF," *Computer communications of the ACM*, vol.37, no.8, August 1994, pp.61–66.
4. T. Ballardie, P. Francis and J. Crowcroft, "Core Based Trees (CBT) – An architecture for scalable inter-domain multicast routing, " *Proc. of ACM SIGCOMM'93*, San Francisco, CA, October 1993, pp.85–95.
5. S. Deering, D.-L. Estrin, D. Farinacci, V. Jacobson, C.-G. Liu and L. Wei, "The PIM architecture for wide-area multicast routing," *IEEE/ACM Transactions on Networking*, vol.4, no.2, April 1996, pp. 153–162.
6. C. Cordeiro, H. Gossain and D. Agrawal, "Multicast over Wireless Mobile Ad Hoc Networks: Present and Future Directions" *IEEE Network*, no. 1, Jan 2003, pp. 52–59.
7. R.-M. Karp, "Reducibility among combinatorial problems," *In Complexity of computer computations*, Plenum Press, New York, 1975, pp.85–103.
8. B.-M. Waxman, "Routing of Multipoint Connections," *IEEE Journal on Selected Areas in Communications*, vol. 6, no. 9, December 1998, pp. 1617–1622.
9. L. Kou, G. Markowsky, and L. Berman, "A fast algorithm for Steiner trees," *Acta Informatica*, no. 15, vol. 2, 1981, pp. 141–145.
10. J. Plesnik, "The complexity of designing a network with minimum diameter," *Networks*, no. 11, 1981, pp. 77–85.
11. A. Zelikovsky, "An 11/6-approximation algorithm for the network Steiner problem," *Algorithmica*, no. 9, 1993, pp.463–470.
12. S. Rajagopalan and V. V. Vazirani. "On the bidirected cut relaxation for the metric Steiner tree problem," in Proceedings of the 10th Annual ACM-SIAM Symposium on Discrete Algorithms, 1999, pp. 742–751.
13. S. Even, "Graph Algorithms," *Computer Science Press*, 1979, pp. 204–209.
14. H. Lim and C. Kim, "Multicast Tree Construction and Flooding in Wireless Ad Hoc Networks," *Proceedings of the 3rd ACM international workshop on Modeling, analysis and simulation of wireless and mobile systems*, Boston, MA, USA. August, 2000, pp. 61–68.

Probability Distributions for Channel Utilisation

Christian Schindelhauer[1,*] and Kerstin Voß[2,**]

[1] Heinz Nixdorf Institute, Paderborn University
`schindel@uni-paderborn.de`
[2] Paderborn Center for Parallel Computing - PC[2], Paderborn University
`kerstinv@uni-paderborn.de`

Abstract. Sensor nets have many undisputed fields of application. A paradigm of communication is the use of **one** control channel in the MAC layer. We challenge this paradigm for nodes with very restricted hardware resources. In our model nodes support the use of different channels and use clock synchronisation.

We present a simple probabilistic synchronised channel utilisation scheme for wireless communication. The main features are its simplicity, robustness against radio interference, the high throughput caused by less interfering signals, and predictable energy consumption. For this, the channel selection is based on a carefully chosen probability distribution maximising the expected number of successfully delivered packets up to a constant factor.

Combined with a standard synchronisation scheme it provides a novel energy-efficient, robust, and fast message delivery service for sensor networks where data gathering is not available due to memory restrictions.

Topics: Access control, Sensor Networks.

Keywords: Sensor network, frequency selection, medium access control.

1 Introduction

The growing research interest in sensor nets is motivated by the broad applicability of such systems. The primary task of the sensor nodes is to send observed sensory attributes of their environment to a central station [1]. Sensor nodes consist of a processor (CPU), storage (RAM and FLASH), radio transceiver, sensor, and an autarkic energy source. They need to be inexpensive and small and due to this their hardware resources are sparse [2,3], i.e. low battery energy, tiny RAM, and small computational power. So, algorithms for sensor nets must be time and memory efficient. Due to the fact that nodes run on batteries, the range for sending and receiving is very limited if they communicate wirelessly. Consequently, in order to send data to the central station, nodes have to cooperate with each other by providing multi-hop-connections. In wireless communication radio interference occasionally occurs and connections can fail. Furthermore,

---

* Partially supported by the DFG-Sonderforschungsbereich 376 and by the EU within 6th Framework Programme under contract 001907 "Dynamically Evolving, Large Scale Information Systems" (DELIS).
** Contact author: Kerstin Voß, PC[2], Fuerstenalle 11, 33102 Paderborn, Germany, Phone. +49 5251 606321, Fax: +495251 60-6297.

V.R. Sirotiuk and E. Chávez (Eds.): ADHOC-NOW 2005, LNCS 3738, pp. 271–284, 2005.

nodes can move or drop out due to loss of energy. According to these characteristics, routing protocols have to manage communication with dynamic partners.

Some state of the art radio transceivers offer only a small number of frequencies. For example, the *Chipcon CC1000* supports at most 13 frequencies of 64 kHz, if the safety section is as large as the interval for a channel [4]. However, if the message transmission does not need much time, a higher number of frequencies can be simulated using internal rounds. Typically, most energy is consumed for sending and receiving [5]. According to this, during inactive rounds only little energy will be consumed. The energy consumption in such waiting periods should be neglected.

In case of high communication traffic, a greater number of frequencies will increase the throughput. However, nodes have to select carefully the channels in order to avoid as much interference as possible. Finally, if two nodes try to send on the same channel, they normally interfere with another. This problem leads to the following question: Which probability distribution in the frequency selection would attain best results? For many cases the uniform distribution is the best choice. However, for very few or a large number of senders and receivers there are better probability distributions. According to the Data Link Layer in the OSI Model (see IEEE P802.3ae) this is a typical problem of the Medium Access Layer because the question relates to the use of the communication medium.

This paper discusses which probability distribution should be used in the MAC layer to maximise the number of successful delivered messages within one parallel sending attempt. The structure of the paper is as following: At first, we take a look at related work. Afterwards the explicit problem and model are described. Section 4 presents the results of our research which includes analysis and empirical examples. A perspective of the future work follows and the conclusion forms the end.

2   Related Work

Only special routing algorithms can be used in sensor nets because of the nodes' sparse resources and the network dynamic. In the past several algorithms have been developed. Few protocols work with data aggregation but these cannot be used with every node specification (if the nodes have only little RAM, e.g. 256 Byte). We refer to [6] and mention the LEACH protocol [7] as examples for protocols gathering data. Additionally, the Data Collection Protocol (DCP) is notable [8].

Rudiments of routing in mobile ad-hoc networks (reactive and proactive) are rarely transferred for sensor networks. Indeed resources are limited for mobile ad-hoc networks but these resource-restrictions are certainly stronger for sensor nets. Anyhow, the Pulse Protocol [9,10] based on a reactive approach has been used for sensor nets.

Usually, different communication directions are considered in sensor networks. We assume that the clock synchronised sensor nodes deliver packets to and from a central station. This communication pattern is the standard communication mode for sensor nets. Sometimes the communication between several nodes in the net is also of interest as for example in the direct diffusion mechanism [7,11].

In sensor nets also different methods are proposed for the underlying MAC layer. Are collisions avoided or not? To avoid interference, nodes arrange the data transfer with several control messages (See PAMA [12] and MACA [13,14]). If the transmitted messages are very small, it can be advantageous that no control messages are used [15]. This method was already integrated in ALOHA [16]: In ALOHA nodes send their messages at any time. If collisions appear, the packages are repeated after a randomised waiting-time. A variant is the slotted ALOHA [17,18,19]: Sending is exclusively allowed in special defined slots.

Several MAC protocols with coordinated wake up have been developed. Very energy-efficient is the *time-slotted* PMAC [20] because the sleep time depends on the traffic in the net. If there nodes recognise no communication activities, they sleep longer to save energy. The sleep-wakeup pattern can be used in nearly every protocol and also with our new channel selection.

In the considered sensor network all nodes use the same medium. Following, multiplexing is needed in the MAC layer [14] which can be achieved in four dimensions: space, time, frequency, and code. Our work is based on frequency division multiplexing (FDM) because the used frequency band is divided in several channels. Additionally, time division multiplexing (TDM) can be integrated if the radio transceiver does not support as many channels as required in the protocol. TDM simulates channels with the aid of internal rounds. However, messages can collide without access control.

All these discussed sensor net protocols only work with one or a few number of channels. No protocol takes advantage of the number of supported frequencies. A clever utilisation of this multitude increases the throughput which further leads to energy savings and a longer lifetime of the nodes. This is our motivation to discuss the integration of this feature into the MAC layer.

As a basis principle we choose a probabilistic channel selection. Our mechanism needs no control channel on which the communication participants arrange a rendezvous on another channel and initialise the pseudo-random channel hopping. This approach is realised in Bluetooth for every Piconet: The clients follow the pattern of the master [21,22]. Bluetooth's method of frequency hopping was also transferred especially for smart devices with Bluetooth like special sensor nodes in [23].

3 The Problem

To increase throughput towards the central node, we use zoning which classifies nodes according to their hop distance. Then, multi-hop communication towards the central station can be reduced to series of single hops with an unknown number of possible senders n and receivers m. Senders and receivers can choose among C channels. A node is a sender if it carries a packet to be delivered. If a node is allowed and capable to accept such a packet, it is a receiver. Because of the tininess of our packets, we do not use a sophisticated medium access protocol. Instead we implement an easy three message hand-shake for a selected channel at some synchronised point of time (possibly simulated by a TDM).

- Every sender sends one packet, which includes the message and some control information, on a random channel.
- A receiver chooses a random channel for listening a time duration sufficient to receive one packet. If one sender sends on the same channel synchronously, the

receiver answers on the same channel with an acknowledgement packet containing the sender's ID.
- If a sender receives an acknowledgement, it erases the packet. Otherwise it tries to re-transmit the message in an subsequent round.

The numbers n and m of sending and receiving nodes are unknown to the protocol and to the nodes. Nodes are informed only about parts of their neighbourhood because of their memory restrictions. The communication consists of a single packet on a channel at a specific point of time. To reach a good throughput the nodes are synchronised, so that sending and listening take place at the same time. For the beginning we concentrate on one sending and receiving attempt, i.e. one point of time.

3.1   The Model

The significant measure for the throughput is the expected value of the successful transmitted messages in one sending phase.

Definition 1. *Given n senders, m receivers and C channels. Each sender chooses channel $i \in C$ independently with probability p_i and each receiver chooses this channel with probability p'_i. Define the random variable $M^{n,m}_{p_i,p'_i}$ as the number of forwarded messages in the considered round. This value corresponds to the number of channels chosen by exactly one sender and at least one receiver.*

The optimisation objective is to maximise the expected number of delivered messages during a single round for all numbers of senders from a given range $[n_0, N_0]$ and receivers from $[m_0, M_0]$:

$$\min_{n \in [n_0,N_0]} \min_{m \in [m_0,M_0]} \mathbf{E}\left[M^{n,m}_{p_i,p'_i}\right] .$$

Considering the expectation this is equivalent to minimise the following term:

Lemma 1.

$$\mathbf{E}\left[M^{n,m}_{p_i,p'_i}\right] = \sum_{i=1}^{C} n \cdot p_i \cdot (1 - p_i)^{(n-1)} \cdot \left(1 - (1 - p'_i)^m\right)$$

Proof. follows from the independence of the channel selection.

3.2   Candidates for Probability Distributions

In this paper we consider the following probability distributions:

Definition 2.
1. Uniform distribution: $p^{uni}_i := \dfrac{1}{C}$
2. Geometric distribution:

$$p^{geo}_i := \begin{cases} \frac{1}{2^i}, & \text{if } i < C \\ \frac{1}{2^{C-1}}, & \text{if } i = C \end{cases}$$

3. Factorised geometric distribution *with parameter $s \in \mathbb{N}$ uses the geometric distribution stretched with factor C/s, assuming C is a multiple of s.*

$$p^{s\text{-}geo}_i := \frac{1}{s} p^{geo}_{\lceil \frac{i}{s} \rceil} = \frac{1}{s \cdot 2^{\lceil \frac{i}{s} \rceil}}$$

4. Pareto distribution *with parameter* $\alpha \geq 1$: $\quad p_i^\alpha := \dfrac{k_{\alpha,C}}{i^\alpha}$,

with $k_{\alpha,C} = \left(\sum_{i=1}^{C} \frac{1}{i^\alpha}\right)^{-1}$. *Note that for* $\alpha > 1$: $k_{\alpha,C} = O\left(\frac{1}{\alpha-1}\right)$ *and for* $\alpha = 1$: $k_{\alpha,C} = \Theta(\ln C)$.

Both, senders and receivers, can use independently from another one of these probability distributions.

The *uniform distribution* chooses each channel with the same probability resulting in an expected number of $\frac{n}{C}$ senders and $\frac{m}{C}$ receivers on each channel. For many cases the uniform distribution seems to be a good choice. We will see that this intuition is correct if there are approximately as many receivers as senders and if these numbers fits approximately to the number of channels. If there are significantly less senders than available channels, other distributions outperform the uniform distribution.

The *geometric distribution* performs best for one sender. The expected number of messages decreases rapidly with more senders. Also it never reaches a higher value than one. So, it is very surprisingly that the closely related *factorised distribution* performs extraordinarily well in simulations. For our envisaged scenario of a range of $[10, 100]$ senders and receivers, a scaling factor of $s \in [12, 14]$ outperforms all other probability distributions. The paper we will explain this phenomenon. We conjecture that the factorised geometric distribution is optimal if the number of channels is large enough.

The *Pareto distribution* with factor one achieves better results than distributions based on factors $\alpha = 2, 3 \ldots$. It turns out that these distributions using any factor outperform all other probability distributions if the number of channels is smaller than the number of senders.

4 Results

In this section we give a mathematical analysis of the above defined distributions, present a lower bound, case differentations for upper bounds, and graph plots for the practical relevant scenarios.

The communication characteristics leads to three analysis cases: First, a sender chooses an empty channel with at least one receiver. This is the only case in which messages are assumed to be transmitted successfully. Second, a sender selects a channel without any receiver. Third, two senders collide on a channel, then we assume no message is delivered to a potential listening receiver.

For the message delivery the presence of multiple receivers is not negative. However, this may result in duplicating a message which may cause problems in possible subsequent rounds. Delimiting such effects is part of ongoing work.

4.1 Approximating the Expected Number of Messages

We start our analysis with a *master*-lemma giving asymptotic tight bounds for the expected number of messages.

Lemma 2. *For constants $c = (1 - \frac{1}{e}) \min_i \{1 - p_i\}$ and $c' = e^{\max_i \{p_i\}}$ we have*

$$c \sum_{i=1}^{C} H\left(\frac{(n-1) \cdot p_i}{1 - p_i}\right) \cdot \min\{1, mp_i'\} \leq \mathbf{E}\left[M_{p_i, p_i'}^{n,m}\right] \leq c' \sum_{i=1}^{C} H(n \cdot p_i) \cdot \min\{1, mp_i'\},$$

where $H(x) := xe^{-x}$.

Proof. First note that for all $p \in [0, 1]$ and $m > 1$:

$$\left(1 - \frac{1}{e}\right) \min\{1, pm\} \leq (1 - (1 - p)^m) \leq \min\{1, pm\} .$$

Further, observe that $(1 - p)^{n-1} \leq e^{-pn+p} \leq e^{-pn}e^p$ which implies the upper bound.

To proof the lower, note that $\forall n > 1 : (1 - \frac{1}{n})^{n-1} \geq \frac{1}{e}$ which implies $(1 - p) \geq e^{-\frac{p}{1-p}}$. Therefore: $np(1 - p)^{n-1} \geq npe^{-\frac{np}{1-p}} \geq (n - 1)pe^{-\frac{np}{1-p}}$. From this the lower bound follows straight-forward. □

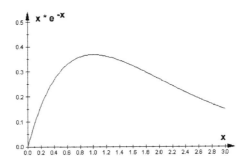

Fig. 1. The function $x \mapsto x \cdot e^{-x}$

We investigate probability distributions p with $0 \leq p_i \leq \frac{1}{2}$. So, $c \geq \frac{1}{2}\left(1 - \frac{1}{e}\right) \approx 0.316$ and $c' \leq e^{\frac{1}{2}} \approx 1.649$. For the analysis it is crucial to understand the meaning of the function $H(x) = xe^x$ visualised in Fig. 1. Additionally for the estimation of the performance of the probability distributions, we will use the following Lemma.

Lemma 3. *It holds*

$$\forall x \in [0, \tfrac{1}{2}] : \frac{1}{\sqrt{e}}x \leq xe^{-x} \leq x \quad , \quad \forall x \in [\tfrac{1}{2}, 2] : \frac{2}{e^2} \leq xe^{-x} \leq \frac{1}{e},$$

$$\forall x \geq 2 : 2 \cdot e^{-x} \leq xe^{-x} \leq 2 \cdot 2^{-x} , \quad \int_{x=0}^{\infty} xe^{-x}dx = 1$$

Proof. follows by applying straight-forward mathematical analysis. □

4.2   The Performance of the Distributions

In this section we assume senders and receivers to use the same probability distribution, i.e. $p'_i = p_i$.

Theorem 1. *Uniform distribution:*

$$
\mathbf{E}\left[M^{n,m}_{p^{uni},p^{uni}}\right] = \begin{cases} \Theta(\frac{nm}{C}), & \text{if } n \leq C \text{ and } m \leq C \\ O(m \cdot 2^{-\frac{n}{C}}), & \text{if } n \geq C \text{ and } m \leq C \\ \Theta(n), & \text{if } n \leq C \text{ and } m \geq C \\ O(C \cdot 2^{-\frac{n}{C}}), & \text{if } n \geq C \text{ and } m \geq C \end{cases}
$$

Proof. follows by applying Lemma 2 and Lemma 3 to an extensive case study.   □

This result underlines the aforementioned intuition: If the number of senders approximates the number of channels and sufficient receivers are available, then the uniform distribution is asymptotically optimal. It performs worse if less or more senders want to communicate.

Theorem 2. *Geometric distribution:*

$$
\mathbf{E}\left[M^{n,m}_{p^{geo},p^{geo}}\right] = O(1) .
$$

Proof. Consider the terms $p_i n = \frac{1}{2}n, \frac{1}{4}n, \ldots$. Using Lemma 3 there are three cases for the term $H(p_i n)$:

1. $p_i n \leq \frac{1}{2}$: Then $H(p_i n) \leq p_i n$ and this implies $\sum_{i:p_i n \leq \frac{1}{2}} H(p_i n) \leq 1$.
2. $\frac{1}{2} < p_i n \leq 2$: Then $H(p_i n) \leq \frac{1}{e}$. For those i's the following estimation has to be valid: $i - 1 < \log n \leq i + 1$. Here, we have $\sum_{i:\frac{1}{2} < p_i n \leq 2} H(p_i n) \leq \frac{2}{e}$.
3. $p_i n \geq 2$: Then $1 + n2^{-i} = 1 + p_i n \leq p_{i-1} n = n2^{-i+1}$. Therefore

$$
\sum_{i:p_i n \geq 2} H(p_i n) \leq \sum_{j=2}^{\infty} H(j) \leq 1 .
$$

Combining all cases gives an upper bound of $2 + \frac{2}{e}$.   □

Theorem 3. *Factorised geometric distribution: Let $s \leq n$ and $C \geq s \log n$:*

$$
\mathbf{E}\left[M^{n,m}_{p^{s\text{-}geo},p^{s\text{-}geo}}\right] = \begin{cases} \Theta(s\frac{m}{n}), & \text{if } m \leq n \\ \Theta(s), & \text{if } m \geq n \end{cases}
$$

Proof. From $s \leq \frac{C}{n}$ and $s \leq n$ follows that there exist $2s$ indices i with $\frac{1}{2} \leq np_i \leq 2$. These indices satisfy $\frac{1}{2} \leq \frac{n}{s2^{\lceil \frac{i}{s} \rceil}} \leq 2$ which is equivalent to

$$
s(\log n - \log s) - s \leq i \leq s(\log n - \log s) + s
$$

For these s indices the term $\min\{1, mp_i\}$ is $\Theta(1)$ if $m \geq n$, otherwise $\Theta(\frac{m}{n})$.
From Lemma 3 follows that for each of these indices $\frac{2}{e^2} \leq H(np_i) \leq \frac{1}{e}$ is valid.

This already implies a lower bound of $(1 - \frac{1}{e})\frac{2}{e^2}s\min\{1, mp_i\}$. Within this range of indices the sum is upperbounded by $\frac{1}{\sqrt{e}}s\min\{1, mp_i\}$.

Now consider $i \leq s(\log n - \log s) - s$ then $H(np_i) \leq np_i$ combined with $\sum_{i \leq s(\log n - \log s) - s} H(np_i) \leq s$. For the case $i \geq s(\log n - \log s) + s$ note that $np_{i+s} \leq np_i - 1$. Since $np_i \geq 2$, $H(np_i) \leq 2 \cdot 2^{-np_i}$ is valid (from Lemma 3) which also implies an upper bound of $\sum_{i \geq s(\log n - \log s) + s} H(np_i) \leq s$. For $m \geq n$, this implies the claim.

In the case $m \leq n$ for $i \leq s(\log n - \log s) + s$ we have $\min\{1, mp_i\} = O(\frac{m}{n})$. For $i \geq s(\log n - \log s) + s$ we have to consider the sum

$$\sum_{i \geq s(\log n - \log s) + s} H(np_i)\min\{1, mp_i\} \leq \sum_{i \geq s(\log n - \log s) + s} H(np_i)mp_i$$

$$\leq \frac{m}{n}\sum_{i \geq s(\log n - \log s) + s} 2 \cdot 2^{-np_i}np_i \leq 2s\frac{m}{n}\sum_{i=1}^{\infty} 2^i 2^{-2^i} = O\left(\frac{sm}{n}\right)$$

Then the claim follows by Lemma 2. $\square$

Theorem 4. *Pareto distribution: For $\alpha \geq 1$, $C \geq (nk_{\alpha,C})^{\frac{1}{\alpha}}$, and $k_{\alpha,C} = \frac{1}{\sum_{i=1}^{C} i^{-\alpha}}$ we observe*

$$\mathbf{E}\left[M_{p^\alpha,p^\alpha}^{n,m}\right] = \begin{cases} \Theta\left((nk_{\alpha,C})^{\frac{1}{\alpha}}\frac{m}{n}\right), & \text{if } m \leq n \\ \Theta\left((nk_{\alpha,C})^{\frac{1}{\alpha}}\right), & \text{if } m > n \end{cases}$$

Proof. Let $k := k_{\alpha,C}$. First we take a look at the case $m > n$. For the senders we consider three cases:

1. $np_i = \frac{nk}{i^\alpha} \leq \frac{1}{2} \Leftrightarrow i \geq (2nk)^{\frac{1}{\alpha}}$. From Lemma 3 it follows $H(np_i) \leq np_i$. This implies

$$\sum_{i \geq (2nk)^{\frac{1}{\alpha}}} \frac{nk}{i^\alpha} \leq \frac{1}{2}\sum_{i \geq 1}\frac{1}{i^\alpha} \leq \frac{1}{2}k$$

2. $\frac{1}{2} \leq np_i \leq 2 \Leftrightarrow \left(\frac{nk}{2}\right)^{\frac{1}{\alpha}} \leq i \leq (2nk)^{\frac{1}{\alpha}}$. Hence, we have from Lemma 3

$$\sum_{\left(\frac{nk}{2}\right)^{\frac{1}{\alpha}} \leq i \leq (2nk)^{\frac{1}{\alpha}}} H(np_i) \leq \left(\frac{3}{2}nk\right)^{\frac{1}{\alpha}}\frac{1}{e}.$$

3. $np_i \geq 2 \Leftrightarrow i \leq (2nk)^{\frac{1}{\alpha}}$. Then $\frac{nk}{i^\alpha} - \frac{nk}{(i+1)^\alpha} \geq \frac{\alpha nk}{i^{\alpha+1}} \geq \frac{2\alpha}{i} \geq \frac{2\alpha}{(2nk)^{\frac{1}{\alpha}}}$. From Lemma 3 it follows $H(np_i) \leq 2 \cdot 2^{-np_i}$. This implies

$$\sum_{i=1}^{(2nk)^{\frac{1}{\alpha}}} H(np_i) \leq \sum_{i=1}^{(2nk)^{\frac{1}{\alpha}}} 2 \cdot 2^{-np_i} \leq \sum_{i=1}^{\infty} 2 \cdot 2^{-\frac{2\alpha i}{(2nk)^{\frac{1}{\alpha}}}} \leq \frac{2}{1 - 2^{-\frac{2\alpha}{(2nk)^{\frac{1}{\alpha}}}}} \leq \frac{(2nk)^{\frac{1}{\alpha}}}{\alpha}$$

If $m \geq n$, then in the second and third case the sums change only by a constant factor which implies the claim.

We now consider the case $m \leq n$ differentiate as following:

1. $np_i = \frac{nk}{i^\alpha} \leq \frac{1}{2} \Leftrightarrow i \geq (2nk)^{\frac{1}{\alpha}}$
   Analogous to the above, we get an upper bound for the expectation of

$$\frac{1}{2}mp_i k \leq \frac{1}{2}\frac{m}{n}np_i k \leq \frac{1}{2}\frac{m}{n}k \ .$$

2. $\frac{1}{2} \leq np_i \leq 2 \Leftrightarrow \left(\frac{nk}{2}\right)^{\frac{1}{\alpha}} \leq i \leq (2nk)^{\frac{1}{\alpha}}$
   For the expectation this implies an upper bound of

$$\left(\frac{3}{2}nk\right)^{\frac{1}{\alpha}}\frac{1}{e}mp_i \leq \frac{m}{n}\left(\frac{3}{2}nk\right)^{\frac{1}{\alpha}}\frac{2}{e}$$

3. $np_i \geq 2 \Leftrightarrow i \leq (2nk)^{\frac{1}{\alpha}}$
   This is the only non-trivial case. Then $\frac{nk}{i^\alpha} - \frac{nk}{(i+1)^\alpha} \geq \frac{\alpha nk}{i^{\alpha+1}} \geq \frac{2\alpha}{i} \geq \frac{2\alpha}{(2nk)^{\frac{1}{\alpha}}}$

$$\sum_{i=1}^{(2nk)^{\frac{1}{\alpha}}} H(np_i)\min\{1, mp_i\} \leq \underbrace{\sum_{i=1}^{(2mk)^{\frac{1}{\alpha}}} H(np_i)mp_i}_{A} + \underbrace{\sum_{i=(2mk)^{\frac{1}{\alpha}}}^{(2nk)^{\frac{1}{\alpha}}} H(np_i)}_{B}$$

From Lemma 3 follows $H(np_i) \leq 2 \cdot 2^{-np_i}$. This implies for the sum A:

$$\sum_{i=1}^{\infty} H(np_i)mp_i \leq \frac{m}{n}\sum_{i=1}^{\infty} H(np_i)np_i \leq \frac{m}{n}\frac{(2nk)^{\frac{1}{\alpha}}}{\alpha}$$

For the sum term B we get the following upper bound

$$\sum_{i=(2mk)^{\frac{1}{\alpha}}}^{(2nk)^{\frac{1}{\alpha}}} H(np_i) \leq \sum_{i=(2mk)^{\frac{1}{\alpha}}}^{(2nk)^{\frac{1}{\alpha}}} 2 \cdot 2^{-np_i} \leq 2 \cdot 2^{-\frac{n}{2m}}\cdot\sum_{i=0}^{\infty} 2 \cdot 2^{-np_i}$$

$$\leq 4\frac{m}{n}\sum_{i=0}^{\infty} 2 \cdot 2^{-np_i} \leq 4\frac{m}{n}\frac{(2nk)^{\frac{1}{\alpha}}}{\alpha}$$

For this estimation we used $x \leq e^x$ and $2^{-\frac{n}{2m}} = \frac{1}{e^{\frac{\ln 2}{2}\cdot\frac{n}{m}}} \leq \frac{m}{n}\cdot\frac{2}{\ln 2}$.

$\square$

Summary. All distributions perform best when the number of receivers is at least the number of senders. The uniform distribution is the best choice if $n \approx C$, yet such a ratio cannot be taken for granted. The geometric distribution cannot be recommended, while the factorised geometric distribution achieves the best results with few senders and many channels. The success of the Pareto distributions depends on the factors C, α

Table 1. Performance for Receivers > Senders

Distribution	Performance	Use in Case
Uniform	$\Theta(n)$	sender $\approx$ channels
Pareto	$\Theta\left((n \cdot k_{\alpha,n})^{\frac{1}{\alpha}}\right)$	senders > channels
Factorised	$\Theta(s)$	senders $\ll$ channels
Geometric	$\Theta(1)$	one sender

and the resulting k. So if s is larger than $O\left((n \cdot k_{\alpha,C})^{\frac{1}{\alpha}}\right)$, then the factorised geometric distribution should be used. For an overview see Table 1.

From the analytical insights of the proofs there is high evidence that in fact the factorised geometric distribution is even optimal for an unlimited number of channels. Very recently, we have verified the following conjecture which is proven in a subsequent paper currently under submission.

Conjecture 1. *The factorised geometric probability distribution is asymptotically optimal if the number of channels is unlimited, i.e. there exists $c > 0$ for all $n_0 \leq N_0 \leq m_0 \leq M_0$ such that for $s = \lfloor \frac{n_0}{2} \rfloor$*

$$\min_{n \in [n_0, N_0]} \min_{m \in [m_0, M_0]} \mathbf{E}\left[M_{p^{s\text{-}geo}, p^{s\text{-}geo}}^{n,m}\right] \geq c \cdot \sup_{p,p'} \min_{n \in [n_0, N_0]} \min_{m \in [m_0, M_0]} \mathbf{E}\left[M_{p_i, p_i'}^{n,m}\right].$$

We have not explained yet why senders and receivers should use the same probability distribution. If the senders choose the uniform distribution, each channel has the same probability. So, there exists no frequency on which receivers meet a sender with a higher probability than on other frequencies. Such channels are important if there are only few senders. However, situations with many receivers may result in duplicates. Additionally, if the receivers use the factorised geometric and the senders the uniform distribution, with a increasing number of senders more messages will be unheard. That is why the receivers should also use the uniform distribution.

If senders use the factorised geometric distribution, the first s frequencies have the highest probability. So, in situations with few senders and few receivers the communication participants meet with a higher probability than if both use the uniform distribution. However, in situations with many senders interference occurs on these channels with the highest probability. If additionally only few receivers are available, the factorised geometric distribution should be factorised with a higher factor, e.g. factor $3s$ for receivers and s for senders, and of course, the same channel numbering. If there are many receivers a uniform distribution achieves best results because receivers have the same probability for each channel. So receivers also choose channels on which interference is not very probably.

As described above it depends on the ratio between senders and receivers whether or not nodes should use the same distribution for sending and receiving. Due to the fact that this ratio is unknown and can vary widely, all nodes should use the same probability distribution.

4.3 Empirical Results

In this section we compare the results of the uniform, the factorised geometric and Pareto one distribution in different situations. With the aid of a defined ratio (success ratio) of senders and expected values as well as of receivers and channels the performance is compared in different situations. The average success ratio was calculated for 96, 120 or 196 channels[1] and from 10 to 60 senders and 5-35 receivers. We have only studied situations in which the number of senders never rises above the number of channels.

The factorised geometric distribution outperforms the uniform distribution. As mentioned before, the factor s in the factorised geometric should lay close-by 12 and 14.

Example 1. *In figure 2 the success ratio of the factorised geometric distribution for 96 channels is shown on the vertical axis. On the horizontal axis the stretch factor proceeds. The Pareto one distribution achieves an average success of 0.58097 where $k = 0.1943$. An average success ratio of 0.6468 has the uniform distribution. For dilation factor 12 maximises the factorised geometric distribution with an success ratio of 0.7608.*

These results argue for the use of the factorised geometric distribution. However, if there are many communication participants, the uniform distribution seems to be the best choice.

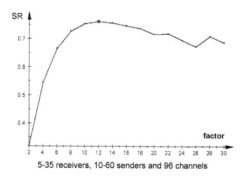

5-35 receivers, 10-60 senders and 96 channels

Fig. 2. Success Ratio of the Factorised Geometric Distribution

Graph plots a) and b) in Figure 3 show the expectations on the vertical axis and the number of senders on the horizontal axis. They compare the uniform distribution, the factorised geometric distribution, and the Pareto distribution with $\alpha = 1$.

The figure shows that the factorised geometric distribution achieves worse results when the number of senders increase. For this reason a combination of the uniform and the factorised distribution should be used in a routing protocol. The Pareto one distribution outperfoms the factorised if there exist many senders. However, in those situations the uniform distribution performs at its best.

---

[1] These values are estimations for a real-world supermarket scenario.

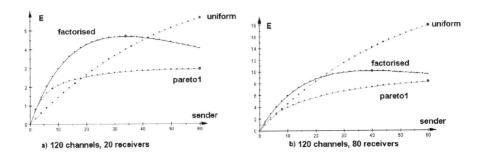

Fig. 3. Expectations with few (a) and many (b) Receivers

4.4 Distributions and Synchronisation

For synchronised nodes the usage of the factorised geometric distribution offers many advantages: In a synchronised network all senders send at the same time. So all nodes in sending range are potential senders. The throughput of messages can be increased if a dynamic measure for the hop distance to the server is used. Furthermore, it should realises that nearly the same number of nodes are in each distance.

To reduce the probability of an interference caused by senders in different hop distances, the factorised geometric distribution is a very good instrument: The permutation of the channel numbering can depend on the distance. For example, the sending range includes only nodes of two other distances: We take a look at a node of distance 5, it can communicate only with nodes of distance 4 and 6 (see figure 4). To reduce the probability that an interference occurs due to two senders of different distances, the channel numbering permutes. The first s channels have the highest probability to be chosen for sending if the node has an even distance. Otherwise the last s channels have the highest probability. If nodes try to receive a message, the reverse order of channel numbering has to be used to reach a good throughput. With this modification interference caused by senders of different distances are significant less probably than those caused by senders of same distance.

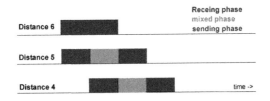

Fig. 4. Modified global synchronisation

A modification of the (global) synchronisation can realise that only nodes of two different distances are sending at the same time (see figure 4). For this feature, the wake up time depends on the distance and if a message is in the buffer. The first rounds are only for receiving data. If a node has a not delivered messages in sleep mode, it can continue sleeping in these receiving phases. After some receiving rounds all nodes which have a message try to send while other nodes still listen for a message. Therewith a **local** synchronisation of the wake up time is realised.

5 Conclusion

We have seen that the throughput of messages can be increased by a combination of the uniform and the factorised geometric distribution. In doing so, a high number of channels is desirable. Currently, only few frequencies are supported by radio transceivers. This restriction, however, is merely caused by the current paradigms in wireless networking and national laws on radio frequency utilisation, which should not hinder research from considering wider frequency spectra.

The developed channel selection method can be used in every MAC Layer. We have only assumed that nodes are synchronised with a standard synchronisation scheme. If collisions are not explicitly avoided and messages are small, sending and receiving actions take a short time. This is advantageous regarding latency and energy consumption. Because of the short transmission, the energy costs of internal rounds are negligible. The shorter the messages the more channels can be simulated with the same energy consumption.

The uniform distribution for the channel selection does not always achieve the best expectations. The factorised geometric distribution has advantages in situations with only few communication participants and supports multiple round protocols by the permutation of channel ordering: Nodes with different distances to the central station use different probabilities for the same channels. In that way the probability of interference caused by senders of different distances is reduced. The throughput rises with a more probabilistic successful transmission. Messages need less sending attempts for successful transmitting and so energy is saved.

Here, we have achieved one of the important aims by developing algorithms for sensor networks: Messages need less sending attempts for successful transmitting and so energy is saved. A recent, yet unpublished, paper of the authors proves the conjecture stated in section 4.2 to be true: The factorised geometric distribution is asymptotically optimal for one sending attempt if the number of channels are unlimited. In the same paper, the choice of probability distribution is improved for a multiple round model.

References

1. Shih, E., Cho, S., Ickes, N., Min, R., Sinha, A., Wang, A., Chandrakasan, A.: Physical Layer Driven Protocol and Algorithm Design for EnergyEfficient Wireless Sensor Networks. International Conference on Mobile Computing and Networking, Proceedings of the 7th annual international conference on Mobile computing and networking, 272-287 (2001)
2. Warneke, B., Last, M., Liebowitz, B., Pister, K.S.J.: Smart dust: Communicating with a cubic-millimeter computer. Computer **34** (2001) 44–51
3. Hill, J., Szewczyk, R., Woo, A., Hollar, S., Culler, D., Pister, K.: System architecture directions for networked sensors. In: ASPLOS-IX: Proceedings of the ninth international conference on Architectural support for programming languages and operating systems, New York, NY, USA, ACM Press (2000) 93–104

4. Jeong, J., Ee, C.T.: Forward Error Correction in Sensor Networks. UCB Technical Report (2003)

5. Langendoen, K., Halkes, G. In: Embedded System Handbook, R. Zurawski (editor). CRC Press (2005) Chapter: Energy-efficient Medium Access Control.

6. Han, Q., Lazaridis, I., Mehrotra, S., Venkatasubramanian, N.: Sensor data collection with expected reliability guarantees. In: PERCOMW '05: Proceedings of the Third IEEE International Conference on Pervasive Computing and Communications Workshops (PER-COMW'05), Washington, DC, USA, IEEE Computer Society (2005) 374–378

7. Heinzelman, W.R., Chandrakasan, A., Balakrishnan, H.: Energy-efficient communication protocol for wireless microsensor networks. In: HICSS '00: Proceedings of the 33rd Hawaii International Conference on System Sciences-Volume 8, Washington, DC, USA, IEEE Computer Society (2000) 8020

8. Handy, M., Grassert, F., Timmermann, D.: Dcp: A new data collection protocol for bluetooth-based sensor networks. In: DSD '04: Proceedings of the Digital System Design, EUROMICRO Systems on (DSD'04), Washington, DC, USA, IEEE Computer Society (2004) 566–573

9. Awerbuch, B., Holmer, D., Rubens, H.: The Pulse Protocol: Energy Efficient Infrastructure Access. (IEEE Infocom)

10. Awerbuch, B., Holmer, D., Rubens, H., Wang, I.J., Chang, K.: The Pulse Protocol: Sensor Network Routing and Power Saving. MILCOM (2004)

11. Intanagonwiwat, C., Govindan, R., Estrin, D.: Directed diffusion: a scalable and robust communication paradigm for sensor networks. In: MobiCom '00: Proceedings of the 6th annual international conference on Mobile computing and networking, New York, NY, USA, ACM Press (2000) 56–67

12. Singh, S., Raghavendra, C.: PAMAS - Power Aware Multi-Access protocol with Signalling for Ad Hoc Networks. SIGCOMM Comput. Commun. Rev. **28** (1998) 5–26

13. Karn, P.: MACA- a New Channel Access Method for Packet Radio, S.134-140. ARRL/CRRL Amateur Radio 9th Computer Networking Conference (1990)

14. Schiller, J.: Mobile Communications, Pages 37-42 and 61-81. (2000)

15. Schindelhauer, C., Liu, M.J., Ruehrup, S., Volbert, K., Dierkes, M., Bellgardt, A., Ibers, R., Hilleringmann, U.: Sensor Networks with more Features using less hardware. GOR International Conference Operations Research, Page 63 (2004)

16. Roberts, L.G.: Aloha packet system with and without slots and capture. SIGCOMM Comput. Commun. Rev. **5** (1975) 28–42

17. Vanderplas, C., Linnartz, J.P.M.: Stability of mobile slotted ALOHA network with Rayleigh fading, shadowing and near-far effect. IEEE Trans. Vehic. Technol. 39, 359-366 (1990)

18. Namislo, C.: Analysis of mobile radio slotted ALOHA networks. IEEE J. Selected Areas Commun. 2, 199-204 (1984)

19. Borgonovo, F., Zorzi, M.: Slotted aloha and cdpa: a comparison of channel access performance in cellular systems. Wirel. Netw. **3** (1997) 43–51

20. Zheng, T., Radhakrishnan, S., Sarangan, V.: Pmac: An adaptive energy-efficient mac protocol for wireless sensor networks. In: IPDPS '05: Proceedings of the 19th IEEE International Parallel and Distributed Processing Symposium (IPDPS'05) - Workshop 12, Washington, DC, USA, IEEE Computer Society (2005) 237.1

21. Group, B.S.I.: Specifications of the Bluetooth System Vol. 1, v. 1.0B 'Core' and Vol.2, v. 1.0B 'Profiles'. (1999)

22. Golmie, N.: Bluetooth dynamic scheduling and interference mitigation. Mob. Netw. Appl. **9** (2004) 21–31

23. Siegemund, F., Rohs, M.: Rendezvous layer protocols for bluetooth-enabled smart devices. Personal Ubiquitous Comput. **7** (2003) 91–101

Enhanced Power-Aware Routing for Mobile Ad Hoc Networks*

Il-Hee Shin and Chae-Woo Lee

School of Electrical and Computer Engineering, Ajou University,
San 5 Wonchon-dong Yeoungtong-gu, Suwon, Korea
ilshin@ajou.ac.kr, cwlee@ajou.ac.kr

Abstract. The Route re-establishment methods that intend to extend the lifetime of the network attempt to find a new route periodically not to overly consume the energy of certain nodes. They outperform other algorithms in the network lifetime aspect, however, they require heavy signaling overheads because new routes are found based on the flooding method and route re-establishments occur frequently as a result. Because of the overhead they often can not extend the lifetime of the network as much as they want. In the paper, we propose a new maintenance algorithm which considers the costs associated with packet transmission and route re-establishment at the same time. Since the proposed algorithm considers packet transmission and future route re-establishment costs at the same time when it initially finds the route, it spends less energy to transmit the packets while evenly consuming the energy of the node as much as possible. Simulation results show that the proposed algorithm outperforms other route re-establishment methods.

1 Introduction

In ad hoc networks, studies on the routing protocols that consume less energy are very active because a lot of energy is consumed to discover, maintain, and re-establish the route, which results in shorter network lifetime. For example,if a mobile node with low energy forwards a lot of packets, in short amount of time it will not be able to function as an intermediate node because of its energy depletion. As the number of such nodes increases, the network will be more likely to be partitioned in that there may exist a source which cannot transmit the packet to the destination because no routes are available to the destination. To devise energy efficient routing algorithms, there have been numerous researches such as the scheme using the energy level as a route selecting condition [3], a power control method which tunes the transmission range to reduce energy consumption [9], a route discovery scheme reducing the overhead by using GPS

---

* This work was supported by the grant No.R01-2003-000-10724-0 from Korea Science & Engineering Foundation.

V.R. Sirotiuk and E. Chávez (Eds.): ADHOC-NOW 2005, LNCS 3738, pp. 285–296, 2005.

information [5], and the routing method distributing the routing load by route re-establishment [1]. .5

Among them the studies on the routing protocols that consume less energy are very active because a lot of energy is consumed to discover, maintain, and re-establish the route, which results in shorter network lifetime. In ad hoc networks, if a mobile node which has a low energy forwards a lot of packets, then in short amount of time it will not be able to function as an intermediate node because its energy is depleted. As the number of such nodes increases, the network will be more likely to be partitioned in that there may exist a source which cannot transmit the packet to the destination because no routes are available to the destination. If the energy is unevenly consumed, it will result in asymmetric link characteristics in that the energy level of the route is different as the direction of the route changes or the communication is possible in one way only. Thus, the routes that consider the energy level of each node are more stable and can transmit more data. To devise energy efficient routing algorithms, there have been several researches such as the scheme using the energy level by route selecting parameter [3], a power control method which tunes the transmission range to reduce energy consumption [9], a route discovery scheme reducing the overhead by using GPS information [5], and the routing method distributing the routing load by route re-establishment [1]. .5

Among the energy efficient routing algorithms, the route re-establishment method is very effective in that it changes the route periodically to distribute the routing load and thus to extend lifetime of the network by considering into account the energy information of each immediate node at the route maintenance phase as well as the initial route discovery phase [1]. In the method, after working as an intermediate router for a while or consuming a fixed amount of energy while functioning as an intermediate router, to save its energy the node informs the source to find a new route. Then the source enters the route re-establishment phase to select a new energy-efficient route and use other nodes for intermediate routers. Because the possibility of a network partition becomes higher if the same nodes are continuously used to transmit packets, the algorithm proposes to periodically re-establishes a new route. Though this scheme distributes the energy consumption to many other nodes instead of concentrating it to a limited number of nodes, it has large signaling overheads by the repeated route re-establishment. .4

In this paper, we propose a new routing algorithm called Enhanced Route Selection Algorithm (ERSA) in which we aim not only to protect the node from overly consuming the energy compared to the other nodes in the network, but also to reduce the energy required to transmit the packets associated with the route re-establishment. The rest of this paper is organized as follows. Section 2 briefly summarizes the existing route selection algorithms and the route maintenance schemes. Section 3 describes the proposed route selection algorithm and section 4 presents a performance analysis of the algorithm using simulations. Section 5 summarizes the paper.

2   Related Work

Since the routing algorithm consists of route selection and maintenance parts, we briefly explain the existing route selection schemes [3][6][8] and the route maintenance schemes in ad hoc networks before describing the proposed algorithm.

2.1   Route Selection Algorithms

Minimum hop routing protocol [6]. Minimum Hop Routing Protocol (MHRP) is the basic scheme which selects the minimum hop route among candidate routes between the source node and the destination node [6]. As it uses the route which has the minimum number of hops to transmit the packet to the destination node, the transmission delay of MHRP is smaller than that of other algorithms. In this algorithm, the best route remains the same as long as the network topology does not change. Network topology changes when a link fails by the mobility of the intermediate node or by the failure of node due to power shortage. Thus this algorithm keeps relatively stable route from the source to the destination. In this algorithm, if all candidate routes have the same number of hops from the source to the destination, there is no priority among them. Since MHRP forwards the packet using the chosen minimum hop route until it is unavailable, all the routing burden is given to the selected intermediate nodes. This speeds up the network partitioning because the energy of the selected node will deplete soon. Figure 1 shows a simple example of an ad hoc network to compare the routing algorithms explained in the paper [8]. In the figure E and C respectively represent the energy of the node and the packet transmission cost. In the figure the route selected by MHRP is *"Source − Node D − Destination"*.

Maximum energy routing protocol [2]. Maximum Energy Routing Protocol (MERP) forwards the packets using the path in which the sum of the energy of each intermediate node is the largest among all candidate routes. Since it selects the route that has the maximum energy level, it relieves routing burden from the nodes with low energy and attempts to evenly consume the energy of each node. In figure 1, the route selected by MERP is denoted as *"Source − Node B − Node C − Destination"*. In MERP, though theroute selected to transmit the packets has the maximum energy level among the candidates, it one or more nodes along the path involved in other candidate routes, then the route is not selected because if selected such nodes may consume more energy. In MERP, it is necessary to consider an additional factor so that it can limit the number of hops of the routes; the routes that traverse more hops are likely to have more energy. In figure 1, the route of *"Source − Node A − Node B − Node C − Destination"* has the maximum summation of the remaining energy of each intermediate node among the candidate routes from Source to Destination. However, it cannot be selected by the optimum route because it includes another route of *"Source − Node B − Node C − Destination"*. Instead *"Source − Node B − Node C − Destination"* is selected by MERP.

Min-Max energy routing protocol [2]. Min-Max Energy Routing Protocol (MMERP) selects the route by using the remaining energy level of each node as a route selection criterion like MERP. It selects the route in which the node with minimum energy level in each candidate route is the maximum among them. MMERP can prevent using the nodes with low energy level node as intermediate nodes because a selected route attempts not to include such nodes when possible. This scheme prevents early network partition. In figure 1, the route *"Source – Node D – Destination"* is selected as the proper route by MMERP.

Minimum transmission cost protocol [7]. Minimum Transmission Cost Protocol (MTCP) regards the energy parameters as the link cost and depending on the situation, it selects the route based on the energy cost such as transmission power, receiving power, etc. This algorithm selects the route consuming the minimum energy to transmit a packet from a source node to a destination node. Since the route is selected based on the transmission cost, it consumes the smallest amount of energy per packet than the above-mentioned algorithms. In figure 1, when MTCP is used, the selected route is *"Source – Node E – Node F – Destination"*. Since this algorithm considers the energy level of the node as the sole criterion in selecting a route, it does not reflect other factor which may affect the performance of the ad network such as network lifetime. There have been several studies to solve the problems of MTCP, e.g., those that consider the energy consumption rate, the transmission delay, traffic volume, the queue length, and etc.

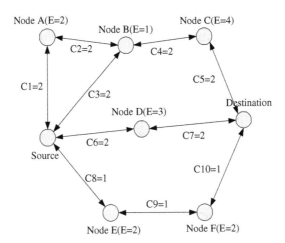

Fig. 1. The network topology to compare the route selection algorithms

2.2   The Route Maintenance Method

In ad hoc networks, after the route is initially selected, the route is maintained by the two methods: 1) Continuing Route Maintenance (CRM) method which continuously uses the route selected until the session is terminated, and 2) Re-Establish Route Maintenance (RERM) method which searches for a new route when the packet transmission cost increases by the fixed amount or a certain amount of time has passed[1]. In CRM, the selected route is used until any of the intermediate nodes cannot forward the packet because of a link failure either from the mobility or energy depletion of the node. CRM is relatively simple to implement and the delay variation for the packet is not large. This method, however, can result in earlier network partition because the energy consumption is concentrated on the initially selected intermediate nodes.

This problem can be solved by RERM method, which searches for a new route when the selected routed is used for a certain amount of time or the cost along the route is increase by a certain limit. RERM method distributes the routing burden to other nodes through the route re-establishment. In RERM, there are following two schemes to re-establish a new route [1].

- Semi-Global Approach (SGA): SGA is a route maintenance method in which the source node requests the intermediate nodes to acknowledge their status and renews its routing cache by route re-establishment when the packet transmission cost is increased by a certain amount. This method has large overhead in that the signaling packets must be exchanged between the source and a number of intermediate nodes.
- Local Approach (LA): In LA, all intermediate nodes of the route monitor continuously their remained energy or packet transmission cost. If any of the intermediate node cannot performs its duty as a router, it transmits the packet which notifies route error to the source node. Upon receiving the error notifying message, the source node begins to search a new route. The error notifying message is sent when the following condition is met.

$$Cost_i(t) - Cost_i(t_0) \geq \delta \tag{1}$$

where the parameter δ denotes the maximum allowed cost increase between time t_o to t at the intermediate node i.

LA method allows the source node to properly use the current status of the intermediate nodes in maintaining the route. For example, if the energy of the intermediate node decreases, the link cost increases also. If the cost increase is larger than the parameter δ since the last route re-establishment, the intermediate node transmits the error notifying packet to re-establish a new route toward a source node. If the selected route includes the intermediate node of which cost increases rapidly, the route re-establishment may occur immediately. In this case, the route re-establishment overhead becomes larger.

To utilize the benefits of LA while reducing the signaling overhead, in section 3 we propose Enhanced Route Selection Algorithm which considers the packet transmission and the route re-establishment costs when establishing a route.

3   The Proposed Algorithm

The proposed Enhanced Route Selection Algorithm, based on Dynamic Source Routing (DSR) algorithm, considers the packet transmission and the signaling overhead cost at the same time when establishing a route. The proposed algorithm consists of three phases, i.e., 1) the route discovery phase in which the source node searches for the route from the source node to the destination node for the first time, 2) the route selection phase in which the proper route is selected among several candidate routes, and 3) the route re-establishment phase which is activated by the increase in cost of the intermediate nodes.

3.1   The Route Discovery Phase

In ERSA, the route discovery phase in which the source is searching an energy-efficient route is similar to that of DSR. A source node that has the traffic destined to a destination node broadcasts Route REQuest (RREQ) message to the neighborhood nodes to find out a proper route from the source to the destination. RREQ includes the addresses of the source and the destination nodes, and $RREQ\_ID$ (IDentification) which prevents the intermediate nodes from retransmitting the duplicated RREQ. RREQ is flooded and the neighborhood node receiving the RREQ broadcasts it again by adding its address and cost. To broadcast the received RREQ, each intermediate node checks $RREQ\_ID$ of each RREQ packet and gathers the RREQs with the same ID for the predefined amount of time. Then after comparing the received RREQs, the intermediate node decides the best route from the source to itself and broadcasts a new RREQ message with the route. By using $RREQ\_ID$ we can keep the number of candidate routes from growing too much and reduce the route discovery signaling overhead. This is because the intermediate node does not broadcast RREQ as it receives but it accumulates the incoming RREQ for the predefined amount of time and broadcasts it once.

The cost associated with the route can be divided into the Packet Transmission Cost (PTC) and the Route Re-Establishment Cost (RREC).

Packet Transmission Cost. The Packet Transmission Cost (PTC) means the cost which is necessary to transmit a packet from a node to the other. PTC at an intermediate node i is calculated as follows.

$$PTC_i = C_1(E_{RX} + E_{TX}) + C_2/E_r + C_3E_c + C_4N_{link} + C_5T_{RX}, \qquad (2)$$

where E_{RX}, E_{TX}, E_r, E_c, N_{link}, T_{RX} denote respectively, at node i, the amount of energy consumed in receiving and transmitting a packet, the energy remained, the current energy consumption rate, the number of links, and the current data rate. C_i's (i=1, 2, 3, 4, 5) are weighting factors and dependent on applications or what to achieve in the ad hoc network, e.g., maximum network lifetime, or minimum delay, etc. By properly choosing the weighting factors, we can find better routes than the algorithms that simply finds the route with the largest remaining energy.

Route Re-Establishment Cost. Generally, the existing algorithms use only the packet transmission cost to find a energy-efficient route. However, the route selected by the PTC only can not be best one because the existing algorithms can not predict how long they can use the route. If they can not use the route for a long time because the cost associated with the nodes along the route increases fast, then they will have to find a new route soon, which incur route discovery overhead. As route re-establishment occurs more frequently, it is likely to shorten the network lifetime. Thus when LA approach is used, the route selection algorithm must consider how long it can use the selected route.

In this paper, we introduce the notion of Route LifeTime (RLT) and Route Re-Establishment Cost (RREC) to consider the expected route usage time when the route is selected. RLT of each node is defined as the maximum time that it can support the route before its cost increases to the predefined amount. Then the route can exist for the minimum of all RLTs of the node that constitute the route. RREC is the cost associated with the route maintenance. Generally RLT and RREC are inversely proportional. In ERSA, each intermediate node estimates its own RREC and adds it to PTC when it calculates its cost to support the route. As explained earlier, the node of which cost increased by the predefined amount (δ) notifies this to the source and requests route re-establishment.

If we interpret RLT of node i differently, it is the estimated time until the PTC increases by the parameter δ after it serves as a router for the route. Therefore, RLT is simply calculated as follows.

$$RLT_i = \delta \cdot \left(\frac{d(PTC_i)}{dt}\right)^{-1} \tag{3}$$

RREC is defined as how often the route re-establishment occurs at the node i. We can simply calculate RREC as the inverse of RLT. Mathematically it is written as follows.

$$RREC_i = \frac{1}{RLT_i} = \frac{1}{\delta} \cdot \frac{d(PTC_i)}{dt} \tag{4}$$

The proposed cost metric. Each intermediate node calculates its cost to be inserted to RREQ message by combining PTC and RREC. PTC and RREC represent respectively the costs associated with transmitting a packet and re-establishing a route. The total cost of node i (TC_i) used by the proposed algorithm is calculated as

$$TC_i = C_6 \cdot PTC_i + (1 - C_6) \cdot RREC_i, \qquad (0 \le C_6 \le 1), \tag{5}$$

where we can adjust relative weight for PTC and RREC by choosing C_6 properly. The overhead of the route re-establishment becomes larger as the average duration of a session becomes longer and the number of mobile nodes increases.

3.2   The Route Selection Phase

After receiving the first RREQ, intermediate nodes and the destination node collect, for the predefined amount of time, RREQs having the same ID which carry the information on other candidate routes. After collecting the candidate routes, the node selects the route which has the minimum cost among all candidate routes. Among m candidate routes, the optimal route (R_{opt}) is selected as follows.

$$R_{opt} = \arg\min_{j}\{\sum_i TC_i^1, \ldots, \sum_i TC_i^j, \ldots, \sum_i TC_i^m\}, \qquad (6)$$

where the index i denotes each intermediate node over the route and $\sum_i TC_i^j$ means the summation of the TC_i's of all nodes over the route j.

3.3   The Route Maintenance Phase

The proposed ERSA uses LA as the route maintenance method. As the traffic is forwarded from the source node to the destination node using the selected route, PTC of each intermediate node increases gradually. When the cost of node i increase more than the threshold δ, the node transmits the route error reporting packet toward the source node. The other intermediate node receiving the packet looks up its routing caches and if it finds a redundant route from it to the destination node, the path is used as the part of the new route. Otherwise, it forwards the error reporting packet backward to the source. The redundant route saved at the cache was created from the information received recently. If the network is very dynamic, e.g., the velocity of nodes is fast, then the caching time for the redundant routes must be small. If the source and the intermediate nodes between the error reporting and the source nodes do not have a redundant route in their cache, then the source node begins the re-establishment phase and searches for a proper route again.

4   Simulation Results

In this section, we analyze the performance of the proposed algorithm and compare it with those of other algorithms by simulation. As performance metrices we use the network energy which denotes the total summation of each node after simulation, and the energy consumed for exchanging signaling packets such as RREQ and error reporting packets.

4.1   Simulation Scenario

We use a simple network topology as shown in figure 2 which has $N \times N$ nodes. In the figure, the arrow represents packet transmission direction. Though real networks may not have this kind of simple topology, it can help us to understand with ease the difference between the proposed and the other algorithms. We assume that each node has the same initial energy when the network is constructed. For simplicity we assume that is only a pair of a source node and a destination node and the data rate is set to 5 packets/s with constant bit rate

(CBR). Each node is assumed to broadcast the RREQ packet only once. We apply LA maintenance method to all algorithms used in our simulation, i.e., ERSA, MERP and MMERP. To clealy show the performance differences between the algorithms, we set the cache usage time as 0 for all nodes. The initial distribution of RLT is set to the exponential with the average 115s, and its minimum value to 5s. We also assume that the cost for transmitting and receiving the packet is the same for all nodes and set to 1. Table 1 summarizes several important simulation parameters.

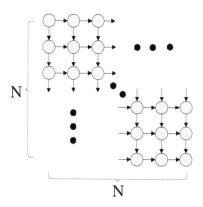

Fig. 2. Network topology for the simulation

Table 1. The parameters used in the simulation

Parameters	Values
Caching time(s)	0
Data rate(packet/s)	CBR, 5(packet/s)
Network size	$5 \times 5 \sim 50 \times 50$
Route maintenance-minimum time(s)	5
Route maintenance-average initial time(s)	115
PTC related parameters	$C_2 = 1$, others=0
Cost related parameter	$C_6 = 0.5$
Initial energy of each node	1,000,000
Energy consumed per packet transmission	1

4.2   Simulation Results

Energy remained in the network after simulation. We analyze the network energy remained at the network after the source node connects 500 sessions to the destination node one by one. The average length of a session is set to 300s and follows the exponential distribution. Figure 3 shows the network energy remained after simulation as the network size varies. The network energy ratio is defined as follows.

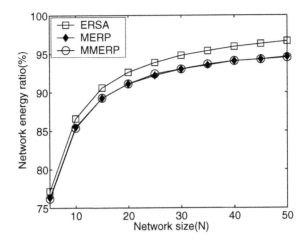

Fig. 3. Comparison of the remaining network energy

$$\text{Network Energy Ratio} = \frac{\text{Total Current Energy}}{\text{Total Initial Energy}} \tag{7}$$

In figure 3, we can see that ERSA outperforms the other algorithms in the network energy aspect. Since the re-establishment overhead resulting from RREQ flooding increases as the network size grows, the performance difference becomes larger as the network size increases. If the average length of session becomes longer or the number of nodes becomes larger, we will be able to see similar performance difference.

Signaling overhead. Figure 4 shows the percentage of the energy consumed for the route re-establishment phase. From the figure we can see that the proposed algorithm consumed the least amount of energy for exchanging route discovery and other two consumed almost the same amount regardless of the network size. As expected, in this figure, we can also see that signaling overhead increases as the network size becomes larger.

The proposed algorithm that considers both PTC and RLT outperforms the others in the network energy and the signaling overhead aspects in the simulation. ERSA has advantages in that it prevent the energy consumption concentrated to a few nodes by using LA maintenance method and it reduces the signaling overhead by considering the re-establishment cost.

In the simulation, we used the topology of $N \times N$ matrix form. The real ad hoc networks, however, have more dynamic topology and the mobility of the node can be large. In the simulation the density of the node was constant, however, in real ad hoc networks it may not be constant. However, we expect that the result will be similar in real ad hoc networks because in the simulation we assumed that the network is static and as the mobility of the nodes increases we may observe more frequent route re-establishment, which implies greater

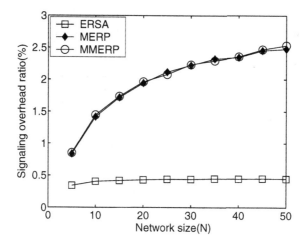

Fig. 4. Comparison of the signaling overheads

performance difference. Even if the node densities become irregular, it is more likely that the node which have low connectivity may not a candidate route in its cache, which also implies more frequent route re-establishment.

5   Conclusion

In this paper, we proposed an enhanced route selection algorithm which considered not only the packet transmission cost but also the route re-establishment cost. The proposed ERSA prevents the early network partition because it distributes routing overhead and extends the network lifetime because it reduces the energy needed to exchange route discovery packets, compared to other route selection algorithms such as Maximum Energy Routing Protocol and Min-Max Energy Routing Protocol. Simulation results showed that the proposed algorithm outperformed the other algorithms, especially as the network size or the average session length increases the performance difference became greater. In this paper, we concentrated on describing the basic idea of the algorithm and briefly showed the simulation results. Since the selection of the parameter δ affects much on the performance of the algorithm, a good selection of δ will give better simulation results. Studies on determining the weighting factors may be also necessary in the future.

References

1. Maleki,M.,Dantu,K.,Pedram, M.: Power-aware Source Routing Protocol for Mobile Ad Hoc Networks. Proc. ISLPED, (2002) 72–75
2. Kim, D.K., Garcia-Luna-Aceves, J.J.,Obraczka, K.: Performance Analysis of Power-Aware Route Selection Protocols in Mobile Ad Hoc Networks. Proc. IEEE Networks, Auguest (2002)

3. Youssef, M.A., Younis, M.F., Arisha, K.A.: Performance Evaluation of an Energy-Aware Routing Protocol for Sensor Networks. Proc. ICWN 2002, June (2002)
4. Hong, x.,Xu, K.,Gerla, M.: Scalable Routing Protocols for Mobile Ad Hoc Networks. Network, IEEE, vol:16, July-Auguest (2002) 11–21
5. Mauve, M.,Widmer, J.: A Survey on Position-Based Routing in Mobile Ad Hoc Networks. Network, IEEE, vol:15, November-December (2001) 30–39
6. Royer, E.M., Toh, C.K.: A Review of Current Routing Protocols for Ad Hoc Mobile Wireless Networks. Personal Communications, IEEE, vol:6, April (1999) 46–55
7. Toh, C.K.: Maximum Battery Life Routing to Support Ubiquitous Mobile Computing in Wireless Ad Hoc Networks. Communications Magazine, IEEE, vol:39, June (2001) 138–147
8. Akyildiz, I.F., Su, W., Sankarasubramaniam, Y., Cayirci, E.: A Survey on Sensor Networks. Communications Magazine, IEEE, August (2002) 102–114
9. Tsudaka, K., Kawahara, M.,Matsumoto, A., Okada, H.: Power Control Routing for Multi Hop Wireless Ad-hoc Network. GLOBECOM01, vol:5, November (2001) 2819–2924

Message Stability and Reliable Broadcasts in Mobile Ad-Hoc Networks

Kulpreet Singh, Andronikos Nedos, Gregor Gärtner, and Siobhán Clarke

Distributed Systems Group, Trinity College, Dublin, Ireland

Abstract. Many to many reliable broadcast is useful while building distributed services like group membership and agreement in a MANET. Efforts in implementing reliable broadcast optimised for MANETs have resulted in new protocols that reduce the number of transmissions required to achieve reliable broadcast. A practical implementation of reliable broadcasts requires the ability to detect message stability, and there is still a need to develop protocols that efficiently support message stability determination in a MANET. In this paper we describe such a protocol that is independent of the broadcast optimisation being used, and focuses on providing efficient message stability. As the main idea of the protocol, we define a message dependency relationship and use this relationship to implement reliable broadcast as well as message stability detection. Simulations for mobile and static scenarios show our protocol has only a minimal performance degradation with node mobility.

Keywords: message stability, reliable broadcast, MANET

1 Introduction

In mobile ad-hoc networks (MANETs) frequent topology changes and changing membership pose new challenges for application developers. Distributed services like replication, agreement and group membership help developers build applications for MANETs. Many to many reliable broadcast is useful while building these distributed services.

A practical implementation of many to many reliable broadcast service requires addressing the problem of determining when a message has become stable at the participating nodes. A message is said to be stable when all nodes participating in the reliable broadcast have received the message; guaranteeing that no node will ask for the message again. This allows nodes to delete the stable message from their buffers. In this paper we describe a many to many reliable broadcast protocol that supports efficient determination of message stability in a MANET, and show how it scales with node mobility and node density.

The main idea behind the protocol is the use of dependency relationship between messages broadcast over multiple hops and using these relationships to implement reliability and determine message stability. In this paper we describe this dependency relationship in detail and show how this relationship is used to implement reliable broadcast and further how it allows each participating node to

V.R. Sirotiuk and E. Chávez (Eds.): ADHOC-NOW 2005, LNCS 3738, pp. 297–310, 2005.

determine independently whether a message has been received and delivered at all other nodes. Finally, we show how this ability is easily extended to determine message stability.

In the course of developing reliable broadcast protocols for MANETs, numerous broadcast optimisations have been developed. These optimisations are required to handle the broadcast storm problem as first presented in [1]. The same paper also presented some initial solutions to the problem. The protocol described in this paper is architectured so that it allows any of the optimisations to be deployed below our broadcast and the stability determination algorithm. This allows for flexible use of our scheme, depending on the optimisation available in the MANET.

Given the exorbitant energy costs of message transmissions [2] in a MANET, we trade reduced message transmissions for increased computation for every message stabilised. We show how we achieve the same by using implicit acknowledgements through message dependencies. This requires working with large complex data structures, but saves on the number of message transmissions.

The protocol we present can be easily extended to provide an agreement protocol and further a virtually synchronous membership service, but we do not describe these in this paper. The theme of this paper remains achieving message stability in a MANET. Simulation results show that our protocol is not adversely affected by node mobility; and the number of message transmissions required to determine message stability is dependent only on the size of the network.

2   Related Work

Given the importance of reliable broadcast for building applications for MANETs, and recognising the challenges presented by the shared medium in a MANET, a lot of effort has been focused towards reducing redundant broadcasts and improving the reliability of broadcasts in a MANET.

A redundant broadcast is called redundant as it does not communicate a broadcast message to any additional nodes. Numerous techniques to reduce redundant broadcasts and improve reliability have been developed. These techniques vary in the amount of information utilised to recognise a broadcast as redundant or not. The simplest techniques are probabilistic flooding as presented in [3][4] and the counter based technique presented in the seminal paper identifying the broadcast storm problem [1]. Other techniques as [5][6][7][8] involve keeping track of neighbourhood information and possibly the path that a message took to the current node. Other techniques such as [9] utilise the information available from the underlying routing protocol.

Work has also gone into comparing these techniques through different simulations parameters. [10] and [11] present detailed study of how various techniques compare against each other. [11] also presents a clean characterisation of various techniques depending on the knowledge utilised by a node to determine whether it should rebroadcast a received message.

For practical implementations of reliable broadcasts and other distributed services it is imperative to efficiently determine message stability. To our knowledge [12] is the only system implementation presented in literature that addresses the problem of message stability detection in MANETs. The system uses a gossip based approach to detect message stability. The gossipped messages act as explicit acknowledgements for messages, in contrast, our system uses implicit acknowledgements implemented via message dependencies. Baldoni et al. in [13] discuss a session based approach to implement message stability in partitionable MANETs.

3   Reliable Broadcast and Message Stability

In this section we present in detail our approach to implementing a reliable many to many broadcast that allows efficient determination of message stability. We first describe our approach towards using an optimised broadcast to implement a many to many reliable broadcast. Then we show how this reliable broadcast is used to determine when a message has been received and delivered at another node. Finally, we show how this knowledge is collected to determine message stability.

3.1   Implementing Reliable Broadcasts

As mentioned earlier, we allow any broadcast optimisation to be utilised to implement our reliable broadcast and thus the message stability protocol. For the purposes of this paper we describe the use of a counter based scheme [1] to reduce redundant broadcasts; the same scheme is used in our simulation experiments as well.

Reliability is implemented using negative acknowledgements (nacks). A transmitting node assumes that the messages are received at its neighbours. This results in an optimistic approach for sending messages, wherein a node keeps transmitting messages assuming the earlier messages have been received by one or more of its neighbours. When a neighbour of the sending node realises it has missed a reception, it sends a nack for the missing message.

Apart from the information in message headers required to implement the broadcast optimisation we provide additional information in the headers to implement reliability and determine message stability. This extra information includes the message identifier (mid) of the message being sent. A message identifier is composed of the id of the sending node and the *sequence number* of the message; thus p_i is a message sent by p with sequence number i.

Before we describe the message dependency relationship in detail and the working of our protocol, we provide some terminology. We say a message is *sent* by a node if the node originates the message and is the first node to transmit the message. A message is said to be *forwarded* by a node if it receives a message sent by some other node and then retransmits it. A message is said to be *delivered* at a node if it has been received by the node and after fulfilling certain conditions is handed to the application at the node. The conditions required to be satisfied before a message is delivered are described in Section 3.3. The next section now describes message dependencies.

3.2   Message Dependencies

A message p_i is said to be dependent on a message q_j if p receives and delivers q_j before sending p_i or if $q = p \land j = i - 1$. The dependency relationship is transitive and is described in detail next.

If p sends a message p_i, p_{i-1} is a dependency of p_i, we call p_{i-1} the *last sent* dependency of p_i. A reliability scheme based only on the *last sent* dependency will work fine, but since we know a sending node has also received and delivered other messages from other nodes, we compound the message header with another dependency. This allows a receiving node to send a nack for other messages that the sending node delivered before sending p_i.

The other dependency is the message last delivered by the sending node; that is if a node p receives and delivers a message q_k and then the first message it sends is p_i, it includes a dependency for q_k in the header for p_i. This dependency is called the *last delivered* dependency of p_i. Our protocol, as described in Section 3.3, requires that a node that receives p_i delivers both *last sent* and *last delivered* dependencies of p_i before delivering p_i.

The dependency relationship is transitive and thus allows nodes to work towards synchronising their progress on message exchanges. Since all participating nodes act as both senders and receivers, the dependency relationship is used to implicitly gather acknowledgements. In the next section we show how reliability is implemented using the dependency relationship.

3.3   Reliability via Dependencies

If a message m is a dependency of n, we write $m \rightarrow n$. If $m \rightarrow n \land n \rightarrow n_1 \ldots n_l \rightarrow o$ then m is called the *eventual dependency* of o, written as $m \triangleright o$. When we write a single letter like m for a message, it signifies that the sender of the message is not important in the current context.

We now state the two fundamental rules of our protocol —

$\mathcal{R}1$ A message received by a node is not delivered until all its dependencies have been received and delivered at the receiving node.

$\mathcal{R}2$ A message is not forwarded by a node till it has been delivered at that node.

From the definitions and the rules described till now, we state the lemmas —

Lemma 1. *A message m delivered by a node p, is an eventual dependency of any message, p_k, sent by p after delivering m.*

Proof. From the definition of dependencies, the first message p_i sent by p after delivering m includes m as a dependency; by the definition of eventual dependency, all messages p_j sent by p such that $j > i$ will have m as an eventual dependency.

Lemma 2. *A node p delivers a received message n only if all messages m such that $m \triangleright n$ have been delivered by p.*

Proof. Follows from the definition of eventual dependency and the rule $\mathcal{R}1$. □

Corollary 3. *If a node p delivers a message n it has delivered all messages m such that $m \triangleright n$.*

Negative Acknowledgements. Apart from the rules $\mathcal{R}1$ and $\mathcal{R}2$, we now describe how negative acknowledgements are generated when a node receives a message while some of the message's dependencies have not been delivered at the receiving node. Suppose a node p receives a message q_i from q with two dependencies as q_{i-1} and s_j. From rule $\mathcal{R}1$, p will deliver the message only if p has received and delivered both q_{i-1} and s_j.

If p can not deliver q_k, it is because some eventual dependency of q_k has not been received by p. This requires p to nack for some eventual dependency of q_k. To find which eventual dependency to nack for, p checks if it has received either of the two dependencies of q_k. If it has not received any one of two dependencies, p nacks for the missing dependency. If it has received the dependencies, p recursively checks if the dependencies of the already received dependencies have been received. When a dependency is found which has not yet been received at p, p sends a nack for the same. We now state the third rule for our protocol.

$\mathcal{R}3$ For every message received that cannot be immediately delivered, the receiving node sends a nack for some eventual dependency of the received message which has not yet been received.

Our system requires nodes to send messages at regular intervals to allow for nodes to receive messages which help them determine missing dependencies. This regular timeout messages sent by each node makes our system best suited for applications that require all-to-all streams of regular messages, like group membership, replication and consistency management.

Reliability Proofs. In this section we present the theorems for reliability and arguments for their correctness; but first we define the concept of reachability. Two nodes p and q are said to be *reachable* from each other if they can send and receive messages to and from each other, otherwise they are said to be *unreachable*. Nodes can become unreachable from each other due to mobility related link failures or processor failures. As an example, consider nodes p and q being able to send and receive messages through an intermediate node r. If r failstops and if there are no other intermediate nodes that forward their messages, then p and q become unreachable from each other.

Further, we assume for the sake of our correctness arguments that we do not run into the fairness problem [14] encountered with the IEEE 802.11 MAC protocol. The eventual fair broadcast assumption guarantees that every participating node broadcasts and receives messages and is stated as —

Eventual fair broadcast assumption — In an infinite execution, each node broadcasts messages infinitely often, the immediate neighbours of the node receive infinitely many of those messages and, if a message is re-broadcast infinitely often, all neighbours of the sending node eventually receive that message.

To allow a broadcast message to be propagated to the network we introduce the final rule for our protocol.

$\mathcal{R}4$ Every message that is received and delivered by a node is forwarded by the receiving node.

Rules $\mathcal{R}2$ and $\mathcal{R}4$ work together to first make sure a message is not forwarded till it has been delivered, and at the same time a message is guaranteed to be forwarded when it has been delivered. From the assumptions and rules described, we now state the property that guarantees the dissemination of a message.

Eventual Reception Property — For any message, if any node p broadcasts or receives a message, then every node that remains reachable from p, eventually receives it.

To establish this property we prove Theorems 4, 5 and 6. But before we do so, we state one final assumption —

No spurious messages assumption — All messages received by any node have been sent by some participating node.

Theorem 4. *Given two nodes p and q remain neighbours, if q receives a message m forwarded by p, q eventually delivers the message.*

Proof. From Corollary 3, p has received and delivered all eventual dependencies of m, and from rule $\mathcal{R}3$, q nacks for missing dependencies of m. Given p and q remain neighbours and the eventual broadcast assumption, q eventually delivers m. □

Theorem 5. *Every message sent by a node p is eventually received by every node that remains reachable from p.*

Proof. The proof for this theorem is presented in Appendix I.

Theorem 6. *Every message received by a node p is eventually received by every node that remains reachable from p.*

Proof. When p receives the message m, by Theorem 4, p delivers m; by rule $\mathcal{R}4$, p forwards m; from the no spurious messages assumption m has been broadcast by some participating node and finally from Theorem 5 all nodes receive m. □

The dependencies and the many to many reliable broadcast allow us to implement an useful facility called the "Observable Predicate for Deliver", or OPD. The OPD allows a node to determine if another node has yet received and delivered a message sent by any of the participating nodes. The next subsection describes how we implement the OPD and then we describe how the OPD is used to determine message stability.

3.4   Observable Predicate for Delivery

The *Observable Predicate for Delivery*, or OPD, allows a node p to determine if the sender of a certain message q_i has received and delivered another message r_j before sending q_i. We say, $OPD(p, r_j, q_i)$ is true if node p can determine that the node q has delivered the message r_j before q sent q_i. The Trans broadcast system in [15] first employed the property to build a partial relation between messages and to determine message stability.

For the purposes of evaluating the OPD the protocol maintains a directed graph at each node. The directed graph is a graph of message identifiers. We call this directed graph the "Delivered Before Graph", or DBG. A message identifier for message s_k sent by s has two edges to it in the DBG, one from each of s_k's dependencies. Thus there are edges in to s_k from s_{k-1} and the message last delivered at s when s_k was sent.

The main idea for determining the OPD is due to transitivity relationship in message dependencies and the way DBG is constructed. In a DBG there is a path from a message m to n if and only if $m \triangleright n$. Also from rule $\mathcal{R}2$, if $m \triangleright n$, we know that the sender of n is guaranteed to have delivered m before sending n. We later prove these statements as Theorem 7, but first we state the conditions under which $OPD(p, r_j, q_i)$ evaluates to true. The $OPD(p, r_j, q_i)$ is true if —

1. p has delivered a sequence of messages starting from message r_j and ending in message q_i.
2. There is a path from r_j to q_i in p's DBG.

Theorem 7. *If there is a path from message p_i to q_j in a node r's DBG, then q has received and delivered p_i before broadcasting q_j.*

Proof. From the the definition of DBG, $p_i \triangleright q_j$; from rule $\mathcal{R}2$ q has delivered q_j; finally from Corollary 3, q has delivered p_i. □

3.5   Message Stability

In this section we describe how message stability is determined using OPD. We assume the membership of the mobile ad-hoc network under consideration is known to be the set of nodes, $\mathcal{M}$, and remains the same during our system runs. We take the liberty of this assumption for this paper, but our broadcast scheme allows tracking of virtual synchronous membership as elaborated further in [16]. For the simulations, we start measuring message stability only after an initial membership has been installed. Since we are interested only in measuring message stability characteristics, we chose scenarios where there are no partitions and no nodes fail.

Given the OPD can be evaluated at all the participating nodes, determining message stability becomes straight forward. On reception of a message, the receiving node uses the OPD mechanism to evaluate if any of the unstable messages has now become stable.

To determine the same we introduce the "Message Last Delivered" mld array. This array maintains references to the messages last delivered at each node from all other nodes in $\mathcal{M}$. Thus, $mld[p]$ at node q is the message last delivered from p at node q. Given the definitions of mld and OPD we have the following lemma —

Lemma 8. *A message q_i is determined as stable at node p when $\forall n \in \mathcal{M}$ $OPD(p, q_i, mld[n]) = true$.*

Proof. The proof follows from the definition of OPD. When the condition in the lemma is true, p knows that the message q_i has been delivered at all nodes participating in the broadcast, and thus q_i is stable at p.   □

Next we prove the liveness of the protocol for determining message stability. We state the theorem —

Theorem 9. *In an infinite execution of the protocol, when nodes do not become unreachable from each other, a message q_i eventually becomes stable at all participating nodes.*

Proof. From Theorem 6, in an infinite execution the message q_i is eventually received at all participating nodes. Further from the eventual broadcast assumption, all participating nodes will broadcast messages which by Lemma 1 will have q_i as an eventual dependency. Again from Theorem 6, these messages will be received at all other nodes, thus resulting in q_i being determined stable at all participating nodes.   □

Our message stability approach involves checking for paths in the DBG. This is the computationally intensive; one that we consider as a good trade off for reduction in number of message transmissions.

4   Simulations

The goal of the simulation experiments is to verify that the OPD based mechanism for determining message stability scales well with the number of nodes in the network, the number of nodes per unit area (node density) and node speeds.

We use the ns2 simulator (v 2.27) [17] with the two-ray ground model and a transmission range of 250m, using the random way point mobility model. The number of nodes vary from 4 to 36, and simulations are run for both static and mobile scenarios. For the mobile scenarios we compare runs using speeds of 2 m/s and 8 m/s. Each of the scenarios are run 10 times and all values shown are an average of the results generated from the scenarios. We do not run simulations for more than 36 nodes as it runs into scalability issues with ns2 when all-to-all broadcasts are simulated.

To observe the effects of network size and node density we ran simulations with static nodes while varying the distance between the nodes. We use a square grid of $n \times n$ nodes and vary the distance between nodes along the length and the breadth of the grid. We use 40, 100, 140 and 200 meters as the inter-node distance, d. The maximum distance between nodes is thus the distance between diagonally opposite nodes, viz $\sqrt{2}(n-1) * d$. So for the 36 node, 40m scenario, the maximum distance between nodes is 283m. With the transmission radius set to 250m, a single transmission from a node covers almost all nodes in the most dense and largest network. Whereas for the least dense network, with $d = 200m$, the maximum distance between nodes is about 1414m, requiring a number of forwards of a message to propagate the message across the network.

To isolate the effects of node density from mobility scenarios we chose the mobility scenarios such that the ratio of the aggregate area covered by the radio ranges of all nodes to the geographical area is the same. We call this ratio the "coverage ratio". For the mobility scenarios we kept this ratio equal to 10. Thus given the number of nodes and the transmission radius, we can determine the geographical area to run the simulations.

To compare the mobile cases to the static case, we ran simulations for a static network with the coverage ratio same as that chosen for the mobile network. With inter-node distance, $d = 140m$, we get the coverage ratio of 10 in the static scenarios, the same as that for the mobile scenarios. In the figures below we have plotted this static case both in the graph for static and mobile network, this helps us observe the affect of mobility, given networks with same node density.

Timeouts. For the simulations each node sends a message at variable time intervals; all participating nodes act as both senders and receivers. The time interval between message transmissions by a node is made proportional to the number of neighbours of the sending node. Each node starts by using a randomly selected time period between 0.1 and 0.15sec. Given the number of neighbours[1], $|neighbours|$, a sending node then chooses a random timeout interval from between $0.1 \times |neighbours|$ and $0.15 \times |neighbours|$. Such an approach results in timeout intervals of up to 4 sec in a high density network. The timeout approach described above reduces the probability of collisions, but increases end to end latency of communication.

4.1   Results

All figures present two graphs, one for the static network and the second for the mobility scenario. The graph for the mobility scenarios includes one curve from the static scenarios with the same coverage ratio. This way we are also able to compare the performance of the mobile scenario with static scenarios with different node density.

One of the metrics we use is the number of messages transmitted in the entire network while stabilising a message. To measure the same, we count the average number of message transmissions in the entire network after a message is sent, till the message is stabilised at all nodes. To capture the latency aspect

---

[1] A node easily determines the set of its neighbours by looking at the messages received and whether they were forwarded to reach this node or not.

of our system's performance we later present results for time required to deliver messages across participating nodes.

Figure 1 shows the number of messages required to determine message stability. For the static cases we see a linear increase in the number of transmissions with the increase in the number of nodes in the network. We also see that the number of message transmissions required does not show much variation with the density of the network. This comes as a surprise to us, as we expected high density network to require fewer retransmissions. This could be a result of either the counter based scheme not saving too many rebroadcasts or the minimal broadcast optimisations provided by a counter based scheme being counteracted by the increase in number of collisions for dense networks.

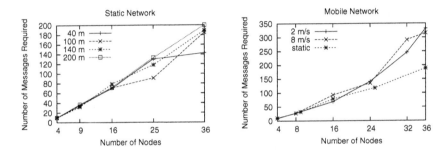

Fig. 1. Average number of messages required to determine stability

The mobile scenarios show an interesting behaviour as well. We see a similar performance up till about 25 nodes for both the mobile cases and the static cases. But at 32 nodes, the number of transmissions required for both the mobile scenario are similar and only about 50% higher than those for the static case. This shows our protocol handles mobility quite well.

Figure 2 shows the average number of times each message is forwarded before it stabilises on all the participating nodes. Again we see a linear increase over the

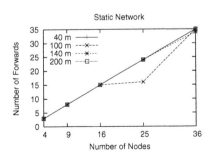
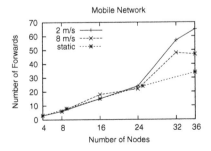

Fig. 2. Average number of times a message is forwarded

static scenarios and no change with the density of the network. The interesting result is again when we compare the mobile and static scenarios with similar node density. The mobile and the static cases show similar performance till about 25 nodes, and the mobile cases then show a 50% to a 100% increase in the number of times a message is forwarded as compared to the static cases. We also see the $8m/s$ scenario showing a better performance than the $2m/s$ scenario, this is probably because of spatial reuse [18] caused by higher mobility.

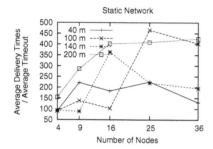 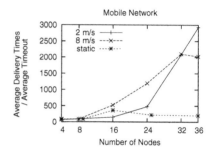

Fig. 3. Average Delivery Times

Figure 3 shows how the message delivery time is affected. To measure the delivery time we measure the average time elapsed between a message send and the times the message is delivered at all nodes. We then take the ratio of this average delivery time of a scenario and the average timeout used by all nodes in that scenario. This way we normalise the effect of the timeouts used for different scenarios. In other words, the graphs show the delivery times for the scenarios if the timeout was 1 *sec*.

The results for the static case show a wide variation over the node density, yet there is a tendency for the delivery times to normalise as the number of nodes increases. The graphs shows that the delivery times are affected by the node density. An interesting observation is that the normalised delivery time for the static scenarios are always between 100 and 450 *msec*. For the mobile cases the delivery times are close to those of the static scenarios till about 25 nodes, but are substantially higher for more than 25 nodes. This shows that for a large network, mobility does affect the delivery times of a message; even while the effort required to stabilise a message is not affected that much.

5   Conclusions

In this paper we presented an approach to providing a reliable broadcast using message dependencies. The approach does not require explicit acknowledgements, instead message dependencies provide implicit acknowledgements. The reliable broadcast allows the use of any existing broadcast redundancy reduction techniques. We further presented how this reliable broadcast which uses message dependencies can be used to determine message stability, an important property for practical implementations of a broadcast protocol.

Our simulation results show that our protocol scales well with node density of the network and node mobility. In fact for networks with up to 25 nodes the results for static and mobile cases are very similar. The simulation results give us confidence that our broadcast scheme should be used to implement higher level services like group membership and agreement. We are implementing our protocol that will allow us to compare it with the one presented in [12].

References

1. Tseng, Y.C., Ni, S.Y., Chen, Y.S., Sheu, J.P.: The broadcast storm problem in a mobile ad hoc network. Wirel. Netw. **8** (2002) 153–167
2. Liang, W.: Constructing minimum-energy broadcast trees in wireless ad hoc networks. In: MobiHoc '02: Proceedings of the 3rd ACM international symposium on Mobile ad hoc networking & computing, New York, NY, USA, ACM Press (2002) 112–122
3. Sasson, Y., Cavin, D., Schiper, A.: Probabilistic broadcast for flooding in wireless mobile ad hoc networks. In: Proceedings of IEEE Wireless Communications and Networking Conference. Volume 2. (2003) 1124 – 1130
4. Eugster, P.T., Guerraoui, R., Handurukande, S.B., Kouznetsov, P., Kermarrec, A.M.: Lightweight probabilistic broadcast. ACM Trans. Comput. Syst. **21** (2003) 341–374
5. Wu, J., Dai, F.: Broadcasting in ad hoc networks based on self-pruning. In: Twenty-Second Annual Joint Conference of the IEEE Computer and Communications Societies. Volume 3., IEEE (2003) 2240 – 2250
6. Lou, W., Wu, J.: On reducing broadcast redundancy in ad hoc wireless networks. IEEE Transactions on Mobile Computing **1** (2002) 111– 122
7. Wei, P., Xicheng, L.: Ahbp: An efficient broadcast protocol for mobile ad hoc networks. Journal of Computer Science and Technology (2001)
8. Sun, M., Feng, W., Lai, T.H.: Location aided broadcast in wireless ad hoc networks. In: Proceedings of IEEE Global Telecommunications Conference. Volume 5. (2001) 2842 – 2846
9. Jo, J., Eugster, P.T., Hubaux, J.: Route driven gossip: Probabilistic reliable multicast in ad hoc networks. In: Twenty-Second Annual Joint Conference of the IEEE Computer and Communications Societies. Volume 3. (2003) 2229 – 2239
10. Kunz, T.: Multicasting in mobile ad-hoc networks: achieving high packet delivery ratios. In: Proceedings of the 2003 conference of the Centre for Advanced Studies on Collaborative research, IBM Press (2003) 156–170
11. Williams, B., Camp, T.: Comparison of broadcasting techniques for mobile ad hoc networks. In: MobiHoc '02: Proceedings of the 3rd ACM international symposium on Mobile ad hoc networking & computing, New York, NY, USA, ACM Press (2002) 194–205
12. Friedmann, R., Tcharny, G.: Stability detection in mobile ad-hoc networks. Technical report, Israel Institute of Technology (2003)
13. R.Baldoni, Ciuffoletti, A., Marchetti, C.: A message stability tracking protocol for mobile ad-hoc networks. In: 5th Workshop on Distributed Systems: Algorithms, Architectures and Languages (WSDAAL 2000), Ischia (Naples), Italy (2000) 18–20

14. Barrett, C.L., Marathe, M.V., Engelhart, D.C., Sivasubramaniam, A.: Analyzing the short-term fairness of ieee 802.11 in wireless multi-hop radio networks. 10th IEEE International Symposium on Modeling, Analysis, and Simulation of Computer and Telecommunications Systems (MASCOTS'02) (2002) 137 – 144
15. Moser, L.E., Melliar-Smith, P., Agarwal, V.: Processor membership in asynchronous distributed systems. IEEE Transaction on Parallel and Distributed Systems **5** (1994) 459–473
16. Singh, K.: Towards virtual synchrony in manets. Fifth European Dependable Computing Conference - Student Forum (2005)
17. Breslau, L., Estrin, D., Fall, K., Floyd, S., Heidemann, J., Helmy, A., Huang, P., McCanne, S., Varadhan, K., Xu, Y., Yu, H.: Advances in network simulation. IEEE Computer **33** (2000) 59–67
18. Guo, X., Roy, S., Conner, W.: Spatial reuse in wireless ad-hoc networks. In: IEEE 58th Vehicular Technology Conference. Volume 3. (2003) 1437–1442

Appendix I

Here we present the proof for Theorem 5. To prove the theorem we consider the four cases as presented in Figure 4. The cases (i) and (ii) assume static nodes and we use induction to prove the theorem under the static nodes assumption. Cases (iii) and (iv) allow for mobile nodes and we again use an induction to prove the theorem.

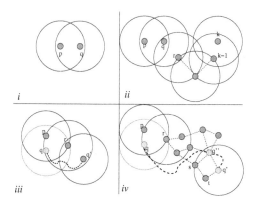

Fig. 4. Node p sends a message, q eventually receives the message

If a node p sends a message p_i, nodes that stay in the immediate neighbourhood of p eventually receive p_i. This is the case in Figure 4 (i).

case (i) From the eventual fair broadcast assumption, q eventually receives some message p_k from p, such that $k > i$. From Lemma 1, $p_i \triangleright p_k$, thus from rule $\mathcal{R}3$, q nacks for some dependency of p_k. By the definition of dependencies, q eventually nacks for p_i and then again from the eventual fair broadcast assumption, receives the same.

Next we prove under the assumption of static nodes, that if a node p sends a message, all nodes that remain reachable from p receive the message. The proof is an induction on the series of neighbouring nodes that a message goes through.

case (ii) Figure 4 (ii) presents this case. If the node p sends a message p_i, from case (i), q eventually receives the message, this is the base case for the induction. As the inductive step, we assume $(k-1)^{th}$ neighbour receives the message and prove that whenever $(k-1)^{th}$ neighbour receives p_i, $(k)^{th}$ neighbour receives p_i. Assume that the $(k-1)^{th}$ and k^{th} neighbours have identities $k-1$ and k, as shown in the figure. Then from Theorem 4 and rule $\mathcal{R}4$, $k-1$ delivers and forwards p_i. By Lemma 1 and the argument for case (i), k eventually receives p_i via k.

Next we consider mobility and show that if a node broadcasts a message all nodes that remain reachable from p receive the message, even if they are mobile and do not stay in the immediate neighbourhood of the sending node. First we prove that if a node q moves away to q' shown in Figure 4 (iii), q eventually receives the message sent by p.

case (iii) By case (i), r receives p_i and by Theorem 4, r delivers p_i. Then by $\mathcal{R}4$, r forwards p_i. This case then reduces to case (i).

Finally, we prove that if q moves such that there is a sequence of nodes that have to forward p_i, as in Figure 4 (iv), q still eventually receives the message. The proof for this case is an induction on the series of nodes the message is forwarded through to reach q.

case (iv) Case (iii) is the base case for the induction; the length of the series of intermediate nodes is 1. For the inductive step, we assume that if q moves to location shown as q'' as shown in Figure 4 (iv), and r is replaced by a series of nodes of length $|l-1|$, such that the last node in the series is s, q still receives p_i. The proof for the inductive step when the length of the series is l, and q moves to q' such that the last node in the series is t is shown now. Given the assumption for the series with length $l-1$, s has received p_i. From Theorem 4 and rule $\mathcal{R}4$, s delivers and forwards p_i. Again from the argument in case (i), t receives and delivers p_i via s and then q' receives p_i via t. □

A Detection Scheme of Aggregation Point for Directed Diffusion in Wireless Sensor Networks

Jeongho Son, Jinsuk Pak, and Kijun Han[*]

Department of Computer Engineering,
Kyungpook National University, Korea
{jhson, jspak}@netopia.knu.ac.kr, kjhan@knu.ac.kr

Abstract. In this paper, we propose an aggregation scheme, which is called Combine Early and Go Together (CEGT), to reduce the energy cost to transmit sensor data from sources to the sink. In the CEGT, the aggregation point is chosen in such a way that the sum of distances from source nodes to the sink via the aggregation point measured in the number of hops is minimized. Unlike most aggregation schemes, CEGT does not need to construct a tree from each source node to the sink. Simulation results show that our scheme can prolong the network life time and improve the path durability than the conventional schemes such as the gradient path and the shortest path schemes.

1 Introduction

Directed diffusion is a data-centric protocol used in wireless sensor networks. It does not use the traditional address. A sink node requests data by sending interest for named data. Data matching the interest is then "drawn" down toward the sink. Intermediate nodes can cache, or transform data and may direct interests based on previously cached data that is named gradient. So, directed diffusion does not use routing tables for routing, instead it finds a path by flooding which is very expensive in sensor networks[1].

Aggregation is a way to save the energy by preventing redundant transmission of the data from sources to sink for direct diffusion in sensor networks. The idea is to combine the data coming from different sources eliminating redundancy, minimizing the number of transmissions and thus saving energy[2]. Many solutions for determining where to aggregate data to reduce the energy cost have been proposed.

In the Shortest Paths Tree (SPT), each source sends its information to the sink along the shortest path between the two, and overlapped paths are combined to form the aggregation tree. It is possible when a sink node knows positions of all nodes and each node has a global address. It needs the state information of all nodes to route data.

In the Greedy Incremental Tree (GIT) scheme, the aggregation tree initially consists of only the shortest path between the sink and the nearest source. And, the tree is expanded in such a way that the next source closest to the current tree is connected to the tree. This scheme can aggregate data, but its aggregating strategy is not optimal because it mainly puts a focus on finding the shortest path for routing without considering aggregation[10].

---

[*] Correspondent author.

V.R. Sirotiuk and E. Chávez (Eds.): ADHOC-NOW 2005, LNCS 3738, pp. 311–319, 2005.
© Springer-Verlag Berlin Heidelberg 2005

In the Gradient Path scheme, a gradient which is information where to delivery information for data is constructed by means of flooding in directed diffusion. If data sent by two more sources pass on one node, the node combines the data. This scheme does not have global addresses. It makes an aggregation point, but it is not optimal in terms of power saving since it should pay flooding overheads for finding paths[9].

In this paper, an aggregation scheme, called Combine Early and Go Together (CEGT), is proposed to reduce the energy cost for the directed diffusion protocol. In our scheme, source nodes select an aggregation point in such a way that the total hop count from the sources to the sink is minimized.

In Section 2, we explain our proposed scheme (CEGT) in detail. In Section 3, we evaluate the performance of CEGT through simulation. The last section provides some concluding remarks.

2 Combine Early and Go Together (CEGT)

Directed diffusion uses gradient information to select an aggregation point as shown in Fig. 1. The sink sends an interest message that describes its requests. All nodes receiving the message establish gradient that directs backward node. The sink receives data from sources and then selects one or more path, called reinforcement. After this process, the sources send data through the reinforced path.

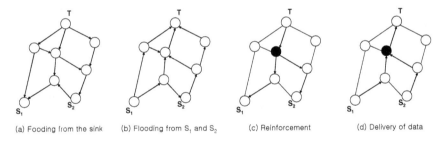

(a) Fooding from the sink   (b) Flooding from S₁ and S₂   (c) Reinforcement   (d) Delivery of data

Fig. 1. Selection of an aggregation node on directed diffusion

In this paper, a scheme to optimally select aggregation points for directed diffusion is presented. In our scheme, aggregation points are selected in such a way that the total number of hops from sources to the sink can be minimized. Our scheme can be formally described as follows.

Select a node as the aggregation point such that

$$\left(\sum d(S_i\,,\,N_a) + d(N_a\,,T) \right)$$

can be minimized, where $d(S,N)$ means the distance between S and N, S_i is the i-th source node, N_a is an aggregation node, and T is the sink.

In Fig. 2, for example, the number of hops from S_1 and S_2 to the sink T can be minimized if the sources send their data to the sink aggregating at N_a. So, N_a is chosen as aggregation point by our scheme.

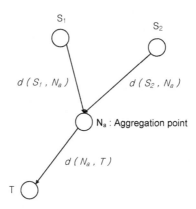

Fig. 2. Basic concept of the proposed aggregation scheme

Table 1. Control messages for CEGT

Message Type	Parameters	Delivery Method
Search Message	Source ID, Radius of Search Area, Destination ID	Limited flooding
Advertisement Message	Source ID, Destination ID	Gradient path

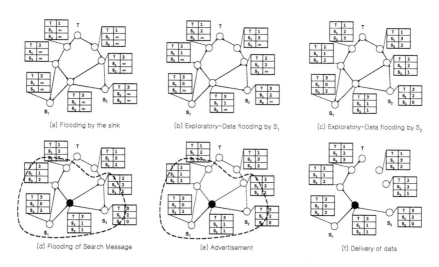

Fig. 3. The procedure of CEGT

In our scheme, searching the aggregation point is initiated by one of source nodes which is selected by the sink. The sink knows already its interesting areas. The source node finds the aggregation point by flooding a control message, called 'Search Message'. The Search Message is flooded only within a limited area around the source node (for example, 2 or 3 hops apart from the source). The number of hops as limited area is determined by exploratory data from each node. The node receiving the Search Message returns the hop counts between itself and the source nodes. After collecting these information, the source node selects an aggregation point node based on the distance (measured in the number of hops) between two sources and the sink. A node with minimum hop count is selected as aggregation point node. Finally, the source node advertises the information on the aggregation point using a Notification Message to allow the other source nodes to update their gradient information. Control messages used in the CEGT are listed in Table 1.

3 Simulation Results

The proposed scheme is evaluated via simulation. We investigate the number of transmissions from two source nodes to the sink and the path availability. We assume that the sink always receives data from two source nodes. These two source nodes send data to the sink until dieing. After two source nodes are dead, two other nodes are selected randomly to send a sensing data until all nodes cannot send to the sink. We do not consider the energy cost for path recovery because the directed diffusion does not find the shortest path.

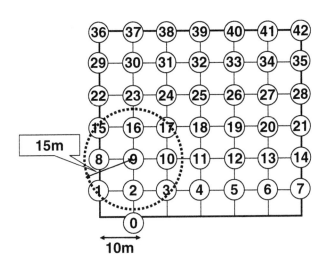

Fig. 4. The network topology for simulation

Simulation is carried out using a grid network configuration in which 42 nodes excluding one sink are deployed with 8 neighbors as shown in Fig. 4. The sink is located at the most left and bottom (the node numbered by 0) and has 3 neighbors, and is assumed to have unlimited power. The network size is 60*m* * 60*m*. We also use a random topology to show the consistency of results. The distance between two nodes is 10*m*, and the transmission range has a radius of 15*m*. We assume that data is aggregated from only two source nodes. We do not consider aggregation of data from 3 or more source nodes. Table 2 shows the network and radio parameters for simulation[4].

Table 2. Network and radio parameters

Parameter	Value
Topology	Grid, Random
Network Size	60*m* * 60*m* in grid topology 100*m* * 100*m* in random topology
The number of Sensor Nodes	43 in grid topology 75, 100, 125, and 150 in random topology
Initial Energy	0.25 *Joule*
Tx Amp Energy	100 *pJ/bit/m²*
Tx / Rx Circuitry Energy	50 *nJ/bit*
Data Size	2000 *bits*

Fig. 5 shows the network lifetime measured as the number of rounds until all nodes fail to communicate with the sink. All results shown here are obtained by averaging over results from 1000 simulation runs. We can see that the network lifetime with CEGT is longer than those with the shortest path or the gradient path schemes.

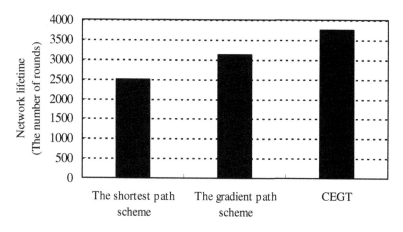

Fig. 5. The network lifetime measured as the number of rounds until all nodes fail to communicate with the sink in a grid network with 43 nodes

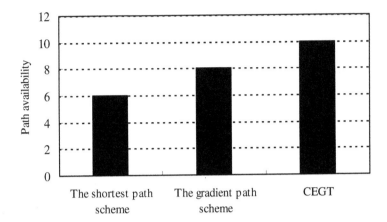

Fig. 6. The number of available paths from the source to the sink in grid network with 43 nodes

Fig. 6 depicts path availability which is defined as the number of paths available from the source to the sink. If a path fails due to power exhaustion at any intermediate nodes from the source to the sink, another path will be established if there is an alternate path through a path recovery process. From this figure, we can see that our scheme offers the highest path availability, which implies that each node consumes its energy very efficiently in our scheme.

Fig. 7 shows the effects of the number of nodes in the network on the network lifetime. Simulation is carried out with a randomly deployed network where all nodes excluding the sink are located randomly. The network size is $100m * 100m$ and other parameters are given the same values listed in Table 2. Simulation is done as we carry the number of nodes from 75 to 150. In Fig. 7, we can see that CEGT can prolong the network lifetime than the shortest path and the gradient path.

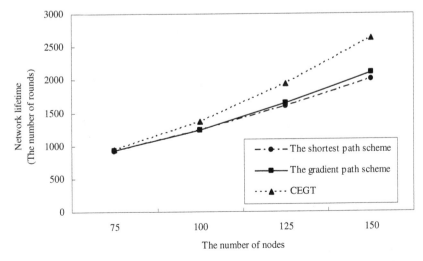

Fig. 7. The network lifetime measured as the number of rounds in randomly deployed network

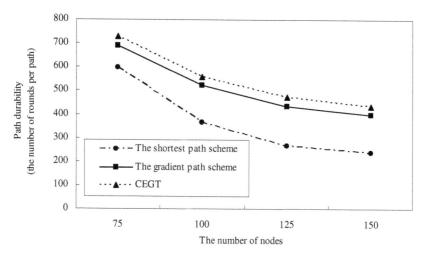

Fig. 8. Path durability (the number of rounds per path)

Fig. 8 shows the effects of the number of nodes in the network on the path durability. The path durability is defined as the duration (counted by the number of rounds) of the path from the time when it is established to the time when it is broken due to power exhaustion at any node over the path. This figure indicates that our scheme offers the longest path duration of three. So, our scheme can contribute to prolonging the network lifetime since it does not require frequent path re-establishment.

We examine the effect of distance between two source nodes on the network lifetime and path availability. In Fig. 9 and 10, we can see that the distance between two

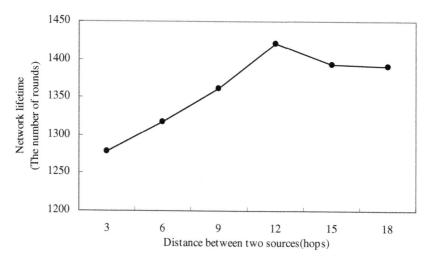

Fig. 9. The network lifetime measured as the number of rounds versus the distance between two sources

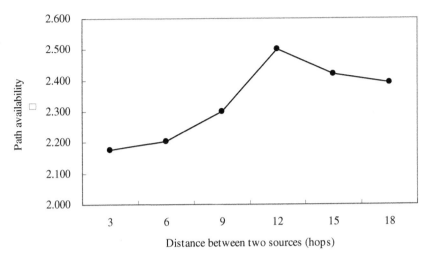

Fig. 10. The number of available paths versus the distance between two sources

nodes strongly influences the network lifetime. As the distance two source nodes becomes larger, the network lifetime becomes longer almost linearly up to any point (say 12). But, after this point, the network lifetime becomes shorter since data from two sources should go over long separate paths to be joined at the aggregation point. Similarly, we can also see that the distance between two source nodes affect the path availability as illustrated in Fig. 10. These two figures indicate that data aggregation does not contribute much to the network performance such as the network lifetime and path availability when two sources are apart too far. So, we need to consider the distance between two source nodes to maximize the effect of aggregation and thus to maximize the network lifetime.

4   Conclusion

We proposed an aggregation scheme, called CEGT, for directed diffusion to save energy in which the aggregation point is selected in such a way that the total number of hops from sources to the sink can be minimized. Our scheme does not require any overhead for constructing a tree to maintain or to find the path from source nodes to the sink. Simulation results showed that that CEGT provides a much better performance than the shortest path and gradient path schemes in terms of the network lifetime and path durability. Our scheme should be appropriately tuned based on the network size and the number of nodes to maximize the effect of aggregation and more analytical results about this will be presented in the near future.

Acknowledgement

This research is supported by Program for the Training of Graduate Students for Regional Innovation.

References

[1] Chalermek Intanagonwiwat, Ramesh Govindan, Deborah Estrin, and John Heidemann, "Directed Diffusion for Wireless Sensor Networking," *IEEE/ACM Transactions on networking*. vol. 11, Feb. 2003.

[2] Bhaskar Krishnamachari, Deborah Estrin, Steephen Wicker, "The Impact of Data Aggregation in Wireless Sensor Networks," *in proc ICDCSW'02*, 2002.

[3] Jin Zhu and Symeon Papavassiliou, "On the Energy-Efficient Organization and the Lifetime of Multi-Hop Sensor Networks," *IEEE Communications letters,* vol. 7, Nov. 2003.

[4] Stephanie Lindsey, Cauligi Raghavendra, and Krishna M. Sivalingam, "Data Gathering Algorithms in Sensor Networks Using Energy Metrics," *IEEE Transactions on parallel and distributed systems,* vol. 13, Sep. 2002.

[5] Dragan Petrovic, Rahul C. Shah, Kannan Ramchandran, and Jan Rabaey, "Data Funneling: Routing with Aggregation and Compression for Wireless Sensor Networks," *IEEE International Workshop,* May 2003, pp. 156-162.

[6] W. Heinzelman, A. Chandrakasan, and H. Balakrishnan, "Energy-Efficient Communication Protocols for Wireless Microsensor Networks," *In proceedings of HICSS,* Jan. 2000.

[7] A. Manjeshwar and D.P. Agrawal. "TEEN:a routing protocol for enhanced efficiency in wireless sensor networks," *In Parallel and Distributed Processing Symposium, Proceedings 15th International,* 2001, pp. 2009-2015.

[8] Deborah Estrin, Ramesh Govindan, John Heidemann, and statish Kumar. "Next century challenges: Scalable coordination in sensor networks," *in proc Mobicom'99,* Aug. 1999.

[9] Utz Roedig, Andre Barroso and Cormac J. Sreenan, "Determination of Aggregation Points in Wireless Sensor Networks," *30th EUROMICRO'04,* 2004, pp. 503-510.

[10] Chalermek Intanagonwiwat, Deborah estrin, Ramesh Govindan, and John Heidemann, "Impact of Network Density on Data Aggregation in Wireless Sensor Networks," *Distributed Computer Systems*, July 2005, pp. 457-458.

Stressing is Better Than Relaxing for Negative Cost Cycle Detection in Networks

K. Subramani*

LDCSEE,
West Virginia University,
Morgantown, WV
{ksmani@csee.wvu.edu}

Abstract. This paper is concerned with the problem of checking whether a network with positive and negative costs on its arcs contains a negative cost cycle. We introduce a fundamentally new approach for negative cost cycle detection; our approach, which we term as the *Stressing Algorithm*, is based on exploiting the connections between the Negative Cost Cyle Detection (NCCD) problem and the problem of checking whether a system of difference constraints is feasible. The Stressing Algorithm is an incremental, comparison-based procedure which is asymptotically optimal, modulo the fastest comparison-based algorithm for this problem. In particular, on a network with n vertices and m edges, the Stressing Algorithm takes $O(m \cdot n)$ time to detect the presence of a negative cost cycle or to report that none exist. *A very important feature of the Stressing Algorithm is that it uses zero extra space; this is in marked contrast to all known algorithms that require $\Omega(n)$ extra space.*

1   Introduction

This paper introduces a new, strongly polynomial time algorithm for the problem of checking whether a network (directed graph) with positive and negative costs on its arcs has a negative cost cycle. Briefly, let $\mathbf{G} = <\mathbf{V}, \mathbf{E}, \vec{\mathbf{c}}>$ denote a network with vertex set $\mathbf{V} = \{v_1, v_2, \ldots, v_n\}$, edge set $\mathbf{E} = \{e_1, e_2, \ldots, e_m\}$ and cost function $\vec{\mathbf{c}} : \mathbf{E} \rightarrow \mathbf{Z}$ that assigns an integer to the edges. The goal of the Negative Cost Cycle Detection (NCCD) problem is to check whether there exists a simple cycle in this network whose edge weights sum up to a negative number. The NCCD problem is one of the more fundamental problems in Network Design, spanning the areas of Operations Research [1], Theoretical Computer Science [3] and Artificial Intelligence. This problem also finds applications in areas such as Image Segmentation and Real-Time scheduling. Note that the NCCD problem is equivalent to checking whether a system of difference constraints is feasible [3].

---

* This research of this author was conducted in part at the Discrete Algorithms and Mathematics Department of Sandia National Laboratories. Sandia is a multiprogram laboratory operated by Sandia Corporation, a Lockheed-Martin Company, for the U.S. DOE under contract number DE-AC-94AL85000.

V.R. Sirotiuk and E. Chávez (Eds.): ADHOC-NOW 2005, LNCS 3738, pp. 320–333, 2005.

Approaches to NCCD can be broadly categorized as relaxation-based [3] or contraction-based [7]. Depending on the heuristic used to select edges to be relaxed, a wide variety of relaxation-based approaches are possible. Each of these approaches works well on selected classes of networks while performing poorly on other classes. This paper discusses a new approach for NCCD called the Stressing Algorithm; this algorithm exploits the connections between networks and systems of difference constraints to achieve the goal of negative cycle detection. A curious feature of the Stressing Algorithm is that it can be implemented in *zero* extra space, assuming that the space in which the input is stored is writable. All algorithms for NCCD, that we know of use at least $\Omega(n)$ extra space on an n-vertex network, to maintain distance labels from the source. At this juncture, it is not clear to us whether the Stressing Algorithm can be modified to solve the Single-Source Shortest Paths (SSSP) problem.

2   Motivation and Related Work

Inasmuch as difference constraints occur in a wide variety of domains, it is hardly necessary to motivate the study of NCCD. Some of the important application areas of NCCD include Image Segmentation, Temporal Constraint Solving, Scheduling and System Verification. Typically, difference constraint systems are solved by finding the single-source shortest paths or detecting negative cost cycles in the associated constraint network. We also note that negative cost cycle detection is an integral subroutine in other network optimization problems such as the min-cost flow problem.

The literature widely contends that NCCD is in some sense, equivalent to the SSSP problem in that an algorithm for one of them can be modified into an algorithm for the other, without additional resources. In this context, all algorithms for both problems require $\Omega(n)$ space and $\Omega(m \cdot n)$ time. We have conjectured that NCCD should be "easier" than SSSP, since the former is a decision problem, whereas the latter is a search problem [7]. While the work in this paper has not been able to reduce the time bounds for NCCD, we have successfully and substantially reduced the space required for NCCD, with respect to all known algorithms for either NCCD or SSSP.

A secondary concern with respect to Relaxation-based methods is that they impose the criterion that *every vertex in the network must be reachable from the source*. For instance, if none of the vertices of a negative cost cycle in a network, is reachable from the source, then a label-correcting approach will not be able to detect the same, since after $O(m \cdot n)$ iterations, the labels corresponding to these vertices will all be set to ∞ and the condition checking for the presence of negative cost cycles fails. If a dummy source is introduced, with arcs of cost 0 to every vertex in the network, then label-correcting approaches will work correctly; however, in this case, the labels associated with the vertices of the network, *do not correspond to the Single-Source shortest paths from the actual source*, but to the shortest paths from the dummy source. So in essence, a label-correcting approach for SSSP must be run twice; the first time to detect negative cost cycles and the second time to compute the shortest paths assuming that negative cost cycles do not exist. It is of course, well known that computing shortest paths in the presence of negative cost cycles is a strongly NP-Hard problem [4]. We remark that the Stressing Algorithm does not require that the input network be connected in any manner; indeed, the notion of a source does not exist, insofar as this algorithm is concerned.

From the practical perspective, the Stressing Algorithm can be easily distributed making it attractive to applications in Adhoc Networks. As we shall see later, a node has to merely monitor its edge to detect the presence of inconsistent constraints.

Approaches to NCCD can be broadly classified into the following two categories:

(a) Relaxation-based methods - Relaxation-based methods (also called Label-correcting methods) maintain an upper-bound of the distance from the source for each vertex; this bound is improved through successive relaxation till the true distance from the source is reached. The relaxation approach of the Bellman-Ford algorithm (BF) is one of the earliest and to date, asymptotically fastest algorithms for NCCD, if the cost function $\vec{c}$ is completely arbitrary. [1,5] present several heuristics which can reduce the number of steps that are executed by the naive BF algorithm; these heuristics include using a FIFO queue (BFFI), a predecessor array (BFPR), and both a FIFO queue and a predecessor array (BFFP).
(b) Contraction-based methods - The Contraction-based methods are predicated on the principle that contracting a vertex in a network preserves shortest paths and hence negative cost cycles, if any. For a detailed proof of the correctness of this technique and its performance profile, see [7].

In this paper, we focus on a new approach that is neither relaxation-based nor contraction-based; the focus of our algorithm is merely the detection of a negative cost cycle; we currently are not aware of a technique that permits the extraction of such a cycle, if it exists. We remark that the Stressing Algorithm and the Contraction-based algorithms apparently contradict the assertion in [2], that every negative cost cycle detection algorithm is a combination of a shortest path algorithm and a cycle detection strategy. Finally, note that the Stressing Algorithm is unique, in that it can be implemented using zero extra space, i.e., no temporary variables in the form of distance labels are required.

3   The Stressing Algorithm

Algorithm 2.1 describes the details of the Stressing Algorithm; at the heart of this algorithm, is the STRESS() procedure (Algorithm 2.2), which is applied to every vertex in each iteration of the outermost for loop. For notational convenience, we shall refer to this loop as $\mathbf{L}$; further the ith iteration of $\mathbf{L}$ will be denoted by $\mathbf{L_i}$.

For reasons that will become apparent later, we choose the convention that the edge from v_j to v_i is referred to as e_{ij} and its cost is referred to as c_{ij}. Thus,

---

Function DETECT-NEGATIVE-CYCLE ($\mathbf{G} =< \mathbf{V}, \mathbf{E}, \vec{c} >$)

1: **for** ($r = 1$ **to** $n - 1$) **do**
2:    **for** (each vertex $v_i \in \mathbf{V}$) **do**
3:        Let c_{ij} denote cost of the lightest (in terms of cost) edge entering v_i.
4:        {If v_i has no incoming edge, $c_{ij} = +\infty$.}
5:        **if** ($c_{ij} < 0$) **then**
6:            STRESS(v_i, c_{ij})
7:        **end if**
8:    **end for**
9: **end for**
10: **if** ($\exists\, e_{ij} \in \mathbf{E} : c_{ji} < 0$) **then**
11:    $\mathbf{G}$ contains a negative cost cyle.
12: **else**
13:    $\mathbf{G}$ does not contain a negative cost cyle.
14: **end if**

Algorithm 2.1. Negative Cost Cycle Detection through Stressing

---

Function STRESS (v_i, c_{ij})

1: Subtract c_{ij} from the cost of each edge entering v_i.
2: Add c_{ij} to the cost of each edge leaving v_i.

Algorithm 2.2. The Stressing Algorithm

to stress a vertex v_i, we first determine the cost of the least cost edge (say e_{ij}) into v_i, We then subtract c_{ij} from the cost of each edge into v_i and add c_{ij} to the cost of each edge leaving v_i.

Some of the salient features of Algorithm 2.1 that bear mention are as follows:

(a) A vertex v_i is stressed by Algorithm 2.1, only if it has an incoming edge of negative cost.

(b) If a vertex v_i has no outgoing edge, then we assume that it has an outgoing edge e_{zi}, such that $c_{zi} = \infty$, where v_z is a specialized "sink" vertex, such that each vertex of $\mathbf{G}$ has an edge into it. Accordingly, if v_i is stressed, the weight of its outgoing edge is not altered. Note that v_z is a virtual vertex, in that it is never part of the actual network and so is the edge e_{zi}.

(c) A given vertex v_i, which is not stressed in the current iteration of $\mathbf{L}$, could get stressed during future iterations.

(d) In a single iteration of $\mathbf{L}$, each vertex is touched exactly once and each edge is touched at most three times (once as an *in-edge*, once as an *out-edge* and once during cost modification.) Thus this iteration can be implemented in $O(m)$ time, from which it follows that the Stressing Algorithm runs in $O(m \cdot n)$.

(e) The Stressing Algorithm does not alter the topology of the network $\mathbf{G}$; however, for all intents and purposes the cost structure of the input network is irretrievably destroyed. We use $\mathbf{G_i}$ to denote the network after the execution of $\mathbf{L_i}$. When the state of the network is not paramount, we use $\mathbf{G}$ to denote the network.

3.1 Space Requirements

Observe that there are no storage variables in either Algorithm 2.1 or Algorithm 2.2. In other words, the Stressing Algorithm takes zero extra space. This is in marked contrast to the label-correcting algorithms, described in [2], which use $\Omega(n)$ extra space and the contraction-based algorithm, described in [7], which could take $\Omega(n^2)$ extra space, in the worst case. Technically, we need two registers: (a) to track the **for** loop index, and (b) for finding the edge with the least cost. However, any reasonable architecture should provide this space, obviating the need for extra RAM space.

The zero extra-space feature of the Stressing Algorithm finds applications in a number of domains such as scheduling web requests and checking consistency in wireless protocols.

4 Correctness

We now establish the correctness of Algorithm 2.1; the arguments used in the proof, require a clear understanding of the connections between networks and systems of difference constraints. Recall that a difference constraint is a relationship of the form $x_i - x_j \leq c_{ij}$ and a system of difference constraints is a conjunction of such relationships.

Lemma 1. *Let* $S' = \{\mathbf{A} \cdot \vec{x} \leq \vec{c}\}$ *denote a system of difference constraints. Then* $S' \neq \phi$ *if and only if the constraint system* $S = \{\mathbf{A} \cdot \vec{x} \leq \vec{c}, \ \vec{x} \leq \vec{0}\} \neq \phi$.

Proof: Given any polyhedron $\mathbf{P}$, defined purely by difference constraints, if $\vec{x} \in \mathbf{P}$, then so is $\vec{x} - \vec{d}$, where d is a positive integer and $\vec{d} = [d, d, \ldots, d]^T$. $\square$

Algorithm 4.1 is an incremental approach for determining the feasible solution of a constraint system $S = \{\mathbf{A} \cdot \vec{x} \leq \vec{c}, \ \vec{x} \leq \vec{0}\}$, constituted of difference constraints only.

Observe that, as specified, Algorithm 4.1 is a non-terminating procedure, since if the initial constraint system $S = \phi$, then it will recurse forever. We proceed to show that if $S \neq \phi$, then Algorithm 4.1 definitely terminates. $\vec{o}$ represents the origin of the current affine space; when the algorithm is called for the first time, $\vec{o}$ is initialized to $\vec{0}$.

Definition 1. *Given a non-empty set of vectors $\mathcal{S}$, a vector $\vec{y} \in \mathcal{S}$, is said to be a maximal element, if $\vec{x} \in S \Rightarrow \vec{y} \geq \vec{x}$, where the $\geq$ relation holds componentwise.*

It is not hard to see that if a set contains a maximal element, then this element is unique. Two elements $\vec{u}$ and $\vec{v}$ in S are incomparable, if neither $\vec{u} \geq \vec{v}$ nor $\vec{v} \geq \vec{u}$.

Function INCREM-DIFF $(\mathbf{A}, \vec{c}, \vec{o})$

1: {Note that the constraint system that we are trying to solve is $\mathbf{A} \cdot \vec{x} \le \vec{c}, \vec{x} \le \vec{0}$.
   Further, the origin of the affine space is $\vec{o} = [o_1, o_2, \ldots, o_n]^T$}. Initially, $\vec{o} = \vec{0}$.
2: **if** $(\vec{c} \ge \vec{0})$ **then**
3:   Set $\vec{x} = \vec{o}$.
4:   **return**($\vec{x}$)
5: **end if**
6: Find a constraint l' with a negative RHS.
7: Let $l' : x_i - x_j \le c_{ij}, c_{ij} < 0$, denote this constraint.
8: Replace the variable x_i by the variable $x_i' = x_i - c_{ij}$, in each constraint that x_i occurs.
9: Set $o_i = o_i + c_{ij}$.
10: Let $\mathbf{A}' \cdot \vec{x} \le \vec{c}'$ denote the new constraint system.
11: INCREM-DIFF($\mathbf{A}', \vec{c}', \vec{o}$).

Algorithm 4.1. The Incremental Algorithm for a System of Difference Constraints

We remark that our definition of maximal element is different from the standard definition of maximal element; in the standard definition, an element of a set $\vec{y}$ is declared to be maximal, as long as there is no element $\vec{z}$, such that $\vec{z} \ge \vec{y}$ and $z_i > y_i$, for at least one $i = 1, 2, \ldots, n$. In other words, as per the standard definition, a set could have multiple maximal elements, which are mutually incomparable. We will be using our definition for the rest of the paper.

Lemma 2. *Given a non-empty set S of vectors which is closed and bounded above and a partial order $\ge$ defined on the elements of S, either S has a maximal element $\vec{z}$, or there exists a pair of elements $\vec{u}, \vec{v} \in S$, such that $\vec{u}$ and $\vec{v}$ are incomparable and there is no element $\vec{z} \in S$, such that $\vec{z} \ge \vec{u}$ and $\vec{z} \ge \vec{v}$.*

Proof: Let us say that S does not have a maximal element. Then, as per our definition, there must be at least two elements, say $\vec{u}$ and $\vec{v}$ which are incomparable, since if every pair of elements is comparable, then the elements would form a chain, under the "$\ge$" relationship and every chain which is bounded above, has a maximal element. Now, consider the case in which corresponding to every pair of incomparable elements (say $(\vec{u}, \vec{v})$), there is an element $\vec{z} \in S$, such that $\vec{z} \ge \vec{u}$ and $\vec{z} \ge \vec{v}$. We call $\vec{z}$ the dominator of $\vec{u}$ and $\vec{v}$. Observe that we can create a set of dominators of all pairs of elements that are mutually incomparable; either the elements of this set form a chain or we can create the set of their dominators. As this process repeats, we will be left with a single element, since S is closed and bounded above. This single element is clearly the maximal element of S, violating the assumption that S did not have a maximal element. $\square$

The set $S = \{\mathbf{A} \cdot \vec{x} \le \vec{c}, \vec{x} \le \vec{0}, \vec{c} \text{ integral}\} \ne \phi$, if and only if it contains lattice points, as per the consequences of total unimodularity [6]. The set of lattice points in S is a discrete, closed set which is bounded above by $\vec{0}$. From this point onwards, we shall focus on this set i.e., the set of lattice point solutions only, when the non-emptiness of S is discussed.

Lemma 3. *The set* $S = \{\mathbf{A} \cdot \vec{\mathbf{x}} \leq \vec{\mathbf{c}}, \; \vec{\mathbf{x}} \leq \vec{\mathbf{0}}\}$, *where* $(\mathbf{A}, \vec{\mathbf{c}})$ *is a system of difference constraints, contains a maximal element, if* $S \neq \phi$.

Proof: If S contains a single element, the lemma is trivially true. Assume that S contains more than one element and that it does not have a maximal element. We observe that the elements is S are bounded above by $\vec{\mathbf{0}}$.

Clearly, if every pair of elements in S is comparable, then these elements form a chain under the componentwise "$\geq$" relationship and there must exist a maximal element in S. Since, S does not have a maximal element, as per Lemma 2, it contains at least two elements, say, $\vec{\mathbf{u}}$ and $\vec{\mathbf{v}}$, which are incomparable and further there is no element $\vec{\mathbf{z}} \in S$, such that $\vec{\mathbf{z}} \geq \vec{\mathbf{u}}, \vec{\mathbf{v}}$. We shall now demonstrate that such a vector must exist in S, contradicting the consequences of the hypothesis that S does not have a maximal element.

Construct the vector $\vec{\mathbf{z}}$ formed by taking the componentwise maximum of $\vec{\mathbf{u}}$ and $\vec{\mathbf{v}}$; i.e., $z_i = \max(u_i, v_i)$. We shall show that $\vec{\mathbf{z}} \in S$.

Let $l_1 : x_i - x_j \leq c_{ij}$ denote an arbitrary constraint defining S. Since $\vec{\mathbf{u}}$ and $\vec{\mathbf{v}}$ are in S, we must have:

$$u_i - u_j \leq c_{ij}$$
$$v_i - v_j \leq c_{ij} \tag{1}$$

Without loss of generality, assume that $u_j \geq v_j$; thus $u_j = \max(u_j, v_j)$. Since $u_i - u_j \leq c_{ij}$, it follows that $\max(u_i, v_i) - u_j \leq c_{ij}$, and hence, $\max(u_i, v_i) - \max(u_j, v_j) \leq c_{ij}$.

The constraint l_1 was chosen arbitrarily; we can therefore apply our analysis to every constraint. In other words setting $z_i = \max(u_i, v_i)$, $\forall i = 1, 2, \ldots, n$ gives a solution to the constraint system, i.e., $\vec{\mathbf{z}} \in S$. It follows that the lattice of the elements of S under the componentwise "$\geq$" relationship contains a maximal element, as per our definition of maximal element.   □

Theorem 1. *If* $S = \{\mathbf{A} \cdot \vec{\mathbf{x}} \leq \vec{\mathbf{c}}, \; \vec{\mathbf{x}} \leq \vec{\mathbf{0}}\} \neq \phi$, *then Algorithm 4.1 terminates by returning the maximal element of* S.

Proof: Observe that if $S \neq \phi$, then as per Lemma 3, it contains a unique maximal element, say $\vec{\mathbf{u}} \leq \vec{\mathbf{0}}$. Since $\vec{\mathbf{c}}$ is an integral vector, we are guaranteed that $\vec{\mathbf{u}}$ is integral, by the theory of total unimodularity [6].

Consider the case, in which no recursive calls are made and Line 3 of Algorithm 4.1 is executed on the initial invocation; in this case, $\vec{\mathbf{o}} = \vec{\mathbf{0}}$. Note that if $\vec{\mathbf{c}} \geq \vec{\mathbf{0}}$, then $\vec{\mathbf{0}}$ is clearly a solution to the constraint system and hence belongs to S. Additionally, $\vec{\mathbf{0}}$ is the unique maximal element of S, i.e, $\vec{\mathbf{u}} = \vec{\mathbf{o}} = \vec{\mathbf{0}}$, and hence, the theorem holds.

We now consider the case in which one or more recursive calls are made within Algorithm 4.1.

Let $S_0 = \{\mathbf{A} \cdot \vec{\mathbf{x}} \leq \vec{\mathbf{c}}, \; \vec{\mathbf{x}} \leq \vec{\mathbf{0}}\}$ denote the constraint system when INCREM-DIFF() is called for the first time. Since, $\vec{\mathbf{c}} \not\geq \vec{\mathbf{0}}$, we have a constraint of the form: $x_i - x_j \leq c_{ij}$, $c_{ij} < 0$. Without loss of generality, we assume that x_i is replaced by the variable $x_i'(= x_i - c_{ij})$, in all the constraints of S_0, to get the new constraint system $S_1 = \{\mathbf{A}' \cdot \vec{\mathbf{x}} \leq \vec{c}', \ \vec{\mathbf{x}} \leq \vec{0}\}$. Since $x_j \leq 0$, the constraint $x_i - x_j \leq c_{ij}$ clearly implies that $x_i \leq c_{ij}$, in any solution to the constraint system. Accordingly, replacing the variable x_i by the variable $x_i' = x_i - c_{ij}$, where $x_i' \leq 0$, does not alter the solution space. In other words, S_0 is feasible, if and only if S_1 is. The polyhedron defining S_0 in the initial affine space, has been shifted to a new affine space, in which the ith component of the origin has moved from o_i to $o_i + c_{ij}$. From the mechanics of the translation, it is clear that there is a one-to-one correspondence, between the elements of S_0 and S_1, which preserves the componentwise "$\geq$" relationship. It follows that the maximal element of S_0 is translated to the maximal element of S_1. Hence, Algorithm 4.1 maintains the following invariant: $\vec{o} \geq \vec{y}, \ \forall \vec{y} \in S$.

During each recursive call made by Algorithm 4.1, some component of the origin $\vec{o}$ is decreased by at least unity and hence after at most $n \cdot ||u||_\infty$ recursive calls, we must have $\vec{o} \leq \vec{u}$, where $||u||_\infty$ denotes the largest absolute value over all compoenents of $\vec{u}$. But, by construction, $\vec{o} \geq \vec{y}$, for all $\vec{y} \in S$, and therefore, we must have $\vec{o} = \vec{u}$. $\square$

Observe that if a polyhedron $\mathbf{P}$ has a unique maximal element, then this element is obtained by maximizing the linear function $\vec{p} \cdot \vec{\mathbf{x}}$ over $\mathbf{P}$, where $\vec{p} > \vec{0}$ is an arbitrary positive vector. Therefore, without loss of generality, we can assume that Algorithm 4.1 is in essence, solving the following linear program:

$$\Psi : \max \sum_{i=1}^{n} x_i$$
$$\mathbf{A} \cdot \vec{\mathbf{x}} \ \leq \ \vec{c}$$
$$\vec{\mathbf{x}} \ \leq \ \vec{0} \qquad\qquad (2)$$

Given a network $\mathbf{G} =< \mathbf{V}, \mathbf{E}, \vec{c} >$, it is a straightforward task to construct the constraint system $S' = \{\mathbf{A} \cdot \vec{\mathbf{x}} \leq \vec{c}\}$, such that $S' = \phi$, if and only if $\mathbf{G}$ contains a negative cost cycle [3]. The construction consists of two steps:

(a) Corresponding to each vertex v_i, we create a variable x_i;
(b) Corresponding to the arc $v_j \rightsquigarrow v_i$ with cost c_{ij}, we create the constraint $x_i - x_j \leq c_{ij}$.

It is clear that if $\mathbf{G}$ has m arcs and n nodes, then $\mathbf{A}$ will have m rows and n columns. From Lemma 1, we know that $\mathbf{G}$ contains a negative cost cycle, if and only if, the constraint system $S = S' \wedge \vec{\mathbf{x}} \leq \vec{0}$ is empty. From this point onwards, we shall refer to System (2) as the constraint system corresponding to the network $\mathbf{G}$.

We next observe that Algorithm 2.1 as applied to the network $\mathbf{G}$, is precisely the application of Algorithm 4.1 to the corresponding system of difference constraints, viz., System (2). The STRESS() operation applied to a vertex is equivalent to replacing the variable x_i with the variable $x_i - c_{ij}$, in every constraint in which x_i occurs. Therefore, the phrases "If $\mathbf{G}$ does not contain a negative cost cycle" and "If System (2) is feasible", denote the same state of events. The key differences between the two algorithms are as follows:

(i) Algorithm 2.1 is a actually a terminating procedure, which concludes in exactly $O(m \cdot n)$ steps, whether or not the constraint system corresponding to the input network is feasible, whereas Algorithm 4.1 will not terminate, if the input system is infeasible.
(ii) In Algorithm 2.1, we do not store the value of the origin as is done in Algorithm 4.1. However, in order to simplify the exposition of the proof of correctness of Algorithm 2.1, it is helpful to associate the value o_i of the constraint system (2), with vertex v_i of the network $\mathbf{G}$.
(iii) In Algorithm 4.1, the variable to be replaced is chosen arbitrarily. In Algorithm 2.1, all variables are stressed in each round, if they can be stressed. Further, to stress a vertex, we select the incoming edge of least cost, i..e, there is a greediness to our approach.

Our proof of correctness will establish that stressing every vertex, in each of the $(n-1)$ rounds, is sufficient to detect infeasibility in System (2), in the following sense: If there exists a negative cost edge in the network $\mathbf{G}$, after Line (9 :) of Algorithm 2.1 has been executed for the final time, then System (2) is infeasible. However, this immediately establishes that the network $\mathbf{G}$ contains a negative cost cycle.

We reiterate that a vertex is stressed by Algorithm 2.1, only if it has an incoming edge with negative cost.

Lemma 4. *If System (2) is feasible, then $o_i = 0$ for some vertex v_i.*

Proof: Let $\vec{z}$ denote the solution to System (2), and let $z_i < 0$, $\forall i$. We use Ψ_z to denote the values of the objective function $\sum_{i=1}^{n} x_i$ at this point. Let $k = \max_{i=1}^{n} z_i$. Observe that $\vec{u} = (\vec{z} - \vec{k})$ is also a solution to System (2), and $\vec{u} > \vec{z}$. Hence $\Psi_u > \Psi_z$, thereby contradicting the optimality of $\vec{z}$. The lemma follows.   □

Claim. The o_i values in Algorithm 4.1 decrease monotonically with each recursive call.

Proof: Observe that the only operation, performed on the o_i values, is the addition of a negative number on Line (8 :) of Algorithm 4.1.   □

We use S_i to denote the set of vertices that are stressed during $\mathbf{L_i}$ of Algorithm 2.1. By convention, $S_0 = \mathbf{V}$, i.e., we say that all the vertices of the network are stressed during $\mathbf{L_0}$. Thus, if a vertex is stressed only during $\mathbf{L_0}$, it means that the vertex is not stressed at all.

Definition 2. *A vertex in the network $\mathbf{G}$ is said to be saturated at level i, if it is stressed during $\mathbf{L_i}$, but never afterwards during Algorithm 2.1.*

We use Z_i to denote the set of vertices which are saturated at level i. It is important to note that there could exist a vertex $v_a \in \mathbf{G}$, such that $v_a \in S_i$, but $v_a \notin Z_i$. In other words, a vertex which is stressed during $\mathbf{L_i}$ need not necessarily be saturated at Level i. However, if $v_a \in Z_i$, then v_a is necessarily part of S_i, since v_a is stressed during $\mathbf{L_i}$. We thus have $Z_i \subseteq S_i$, $\forall i = 0, 1, \ldots, (n-1)$.

Lemma 5. *If System (2) is feasible, then there exists at least one vertex v_i which is never stressed by Algorithm 2.1.*

Proof: Observe that when vertex v_i is stressed, o_i drops in value. From Claim 4, we know that o_i can never increase for any vertex v_i. Thus, if System (2) is feasible and all the vertices of **G** are stressed at least once, by Algorithm 2.1, then $o_i < 0, \forall i$ on its termination, thereby contradicting Lemma 4.   □

Lemma 5 establishes that there is at least one vertex which is saturated at level 0.

Lemma 6. *If $S_i = \phi$, then $S_j = \phi$, $j = (i+1), (i+2), \ldots, (n-1)$. Further, System (2) is feasible and **G** does not have a negative cost cycle.*

Proof: If no vertex was stressed during $\mathbf{L_i}$, then no vertex had an incoming edge of negative cost, in the current network and hence the current network does not have a negative edge at all. This situation will not change at the commencement of $\mathbf{L_{i+1}}$ and hence S_j will be empty as well, for $j = (i+1), \ldots, (n-1)$. Let $\mathbf{T} : \mathbf{A'} \cdot \vec{\mathbf{x}} \leq \vec{\mathbf{c'}}$ denote the constraint system corresponding to the network $\mathbf{G_i}$, i.e., the network that results after $\mathbf{L_i}$ completes execution. Since $\mathbf{G_i}$ does not have a negative cost edge, $\vec{\mathbf{c'}} \geq \vec{\mathbf{0}}$. Therefore, $\mathbf{T}$ is feasible. However, as noted in Theorem 1, if $\mathbf{T}$ is feasible, then so is System (2), which is the constraint system corresponding to the network $\mathbf{G_0}$. This immediately implies that $\mathbf{G_0}$ and hence **G**, do not have negative cost cycles.   □

Theorem 2. *Assume that **G** does not have a negative cost cycle. Then,*

$$(S_i \neq \phi) \rightarrow (Z_i \neq \phi), \ i = 0, 1, 2, \ldots n.$$

Proof: By convention, $S_0 = \mathbf{V}$ and by Lemma 5, we know that $Z_0 \neq \phi$; so the theorem is true for $i = 0$.

In order to extend the theorem for higher values of i, we need to understand the structure of Z_0. Let us focus on the constraint system corresponding to **G**, i.e., System (2). We know that the objective function, Ψ, is maximized at a minimal face of the polyhedron $\{\mathbf{A} \cdot \vec{\mathbf{x}} \leq \vec{\mathbf{c}}, \ \vec{\mathbf{x}} \leq \vec{\mathbf{0}}\}$, say F. From the Hoffman-Kruskal theorem, we know that $F = \{\vec{\mathbf{x}} : \mathbf{B} \cdot \vec{\mathbf{x}} = \vec{\mathbf{c'}}\}$, where $\{\mathbf{B} \cdot \vec{\mathbf{x}} \leq \vec{\mathbf{c'}}\}$ is a subsystem of the system $\{\mathbf{A} \cdot \vec{\mathbf{x}} \leq \vec{\mathbf{c}}, \ \vec{\mathbf{x}} \leq \vec{\mathbf{0}}\}$. We have already established that System (2) has a unique maximal point, i.e., the minimal face, F, corresponding to Ψ, is actually a vertex. This implies that the matrix $\mathbf{B}$ in the constraint system describing F is a basis, i.e., $rank(\mathbf{B}) = n$. We shall refer to $\mathbf{B}$ as the optimal basis of System (2).

We reiterate that F is defined by a collection of difference constraint *equalities* along with one or more constraints of the form $x_i = 0$, which we term absolute constraints. Note that all the constraints in the optimal basis cannot be difference constraints, since such a matrix has rank at most $(n-1)$ and $\mathbf{B}$ is a basis. In the network **G**, the arcs corresponding to the difference constraints in $\mathbf{B}$, form a tree, $\mathbf{B_t}$.

We associate a vertex set M_i, $i = 0, 1, \ldots, (n-1)$, with loop $\mathbf{L_i}$; these sets will be populated inductively.

We define the set M_0 as follows:

(i) Vertex $v_j \in M_0$, if the constraint $x_j = 0$ is one of absolute constraints of the basis $\mathbf{B}$.
(ii) If vertex $v_j \in M_0$ and the constraint $x_k - x_j = 0$ is one of the constraints of the basis $\mathbf{B}$, then vertex $v_k \in M_0$.

We shall now establish that $M_0 = Z_0$, i.e., M_0 is precisely the set of vertices that are stressed only during $\mathbf{L_0}$ by Algorithm 2.1.

We focus on $\mathbf{L_0}$. Let v_j denote an arbitrary vertex in M_0, such that $x_j = 0$ is an absolute constraint of $\mathbf{B}$. Note that v_j cannot have an incoming edge of negative cost. To see this, we assume the contrary and arrive at a contradiction. Let us say that there is a negative cost edge into v_j, having cost b, $b < 0$. From the correctness of Algorithm 4.1, we know that $o_j \leq b$ and hence $x_j \leq b$, in any solution of the System (2). But this contradicts the feasibility of the basis $\mathbf{B}$, from which it follows that v_j cannot have an incoming edge of negative cost. We now consider a vertex $v_k \in M_0$, such that $x_k - x_j = 0$ is a constraint of the basis $\mathbf{B}$ and $v_j \in M_0$. Using an identical argument, it is clear that v_k does not have an incoming edge of negative cost either, since x_k must be 0 in the optimal solution. We have thus established that none of the vertices in M_0 have incoming negative edges, at the commencement of Algorithm 2.1. Let v_j be an arbitrary vertex in M_0. Let us say that during some iteration of $\mathbf{L}$, the edge cost of an edge coming into v_j from some vertex v_k becomes negative (say $b < 0$). This would imply that $x_j - x_k \leq b$ is a constraint derived by Algorithm 4.1 and hence $x_j \leq b$ must hold in any feasible solution of System (2), contradicting the feasibility of $\mathbf{B}$. In other words, the vertices in M_0 are *never* stressed by Algorithm 2.1. It is thus clear that $M_0 = Z_0$.

The set M_1 is defined as follows:

(i) Vertex $v_j \in M_1$, if the constraint $x_j - x_a = b$, $b \neq 0$ is a constraint of the basis $\mathbf{B}$, where $x_a \in M_0$.
(ii) If vertex $v_j \in M_1$ and the constraint $x_k - x_j = 0$ is a constraint of the basis $\mathbf{B}$, then $x_k \in M_1$.

We shall show that $M_1 = Z_1$, i.e., M_1 is precisely the set of vertices that are saturated at level 1.

Consider the case in which $M_1 = \phi$. This means that there are no constraints in the basis $\mathbf{B}$, having the form $x_k - x_a = c_{ka}$, where $x_a \in M_0$. Delete all the constraints (rows and columns) corresponding to the vertices in M_0 from $\mathbf{B}$ to get the constraint system $\mathbf{B_1} \cdot \vec{x_1} = \vec{b_1}$. Observe that the deletion of these constraints preserves the basis structure; on the other hand, $\mathbf{B_1}$ is constituted exclusively of difference constraints and hence cannot be a basis. The only conclusion that can be drawn is that there are no constraints in $\mathbf{B_1} \cdot \vec{x_1} = \vec{b_1}$, i.e., $Z_0 = \mathbf{V}$. Thus, $S_1 = \phi$ and the theorem is trivially true.

We now handle the case in which $M_1 \neq \phi$. Focus on $\mathbf{L_1}$ and consider a constraint of the form $x_j - x_a = c_{ja}$ in the basis $\mathbf{B}$, where $v_a \in M_0$ and $c_{ja} \neq 0$.

Since $\mathbf{B}$ is a feasible basis and $x_a = 0$, we must have $c_{ja} < 0$ and $x_j = c_{ja}$. As per the construction of the constraint network, there exists an edge $v_a \rightsquigarrow v_j$ with cost c_{ja}. We now claim that no edge into v_j can have cost lower than c_{ja}. To see this, observe that the cost of edge $v_a \rightsquigarrow v_j$ is altered only through stressing v_j, since v_a is never stressed in the algorithm. Assume that there exists an edge into v_j, with cost strictly less than c_{ja}, say c'_{ja}. During $\mathbf{L_1}$, v_j will be stressed; but this means that $o_j \leq c'_{ja} < c_{ja}$, in any feasible basis, thereby contradicting the feasibility of the basis $\mathbf{B}$. When v_j is stressed during $\mathbf{L_1}$, o_j reaches its correct value, viz., c_{ja}, because we stress v_j using the least cost edge into it. o_j stays at this value over all future iterations, since any additional STRESS() operation on v_j will decrease o_j, contradicting the feasibility of $\mathbf{B}$. In other words, v_j is never stressed again and thus saturated at level 1. Using an identical argument, we can establish that a vertex $v_j \in M_1$, such that $x_j - x_a = 0$ is a constraint of $\mathbf{B}$ and $v_a \in M_1$ will also be saturated at Level 1. It follows that $M_1 = Z_1 \neq \phi$ and the theorem follows.

Now, observe that once the o values of the vertices in M_1 are determined, the constraints in $\mathbf{B}$ having the form $x_r - x_j = c_{rj}$, where $x_j \in M_1$ become absolute constraints, since x_j has been fixed in the current iteration.

This argument can be applied inductively for $i = 2, 3, \ldots, (n-1)$ as follows:

(i) Let M_i denote the set of vertices v_j, such that either there is a constraint $x_j - x_a = c_{ja}$ in $\mathbf{B}$, where $v_a \in M_{i-1}$, or there is a constraint $x_j - x_a = 0$, where v_a is already in M_i.

(ii) If $M_i = \phi$. the elimination of the constraints, in $\cup_{j=1}^{i-1} M_j$ from $\mathbf{B}$, to get $\mathbf{B_i}$, should preserve the basis structure of $\mathbf{B}$; however, a system of pure difference constraints, such as $\mathbf{B_i}$, cannot form a basis and therefore, the resultant constraint system should be empty, i.e., $\cup_{j=1}^{(i-1)} M_j = \mathbf{V}$. This implies that $Z_i = S_i = \phi$ and the theorem is true.

(iii) If $M_i \neq \phi$, then $\mathbf{L_i}$ fixes the o values of the vertices in M_i to their final values, so that these vertices are never stressed again. In other words, $M_i = Z_i \neq \phi$ and hence the theorem holds. □

Theorem 3. *If $\mathbf{G_i}$ has a negative cost cycle, then so does $\mathbf{G_{i+1}}$, $i = 0, 1, \ldots,$ $(n-1)$.*

Proof: Recall that Algorithm 2.1 alters only the cost structure of the input network $\mathbf{G}$ and not its topology. Accordingly, each cycle in $\mathbf{G_i}$ is also a cycle in $\mathbf{G_{i+1}}$ and vice versa. The key observation is that the STRESS() operation, applied to a vertex, preserves the cost of all cycles in $\mathbf{G_i}$. Let R denote an arbitrary cycle in $\mathbf{G_i}$, with cost $c(R)$. Consider the application of a STRESS() operation, to a vertex v_i in $\mathbf{G_i}$. Clearly, the cost of the cycle R is not affected, if v_i is not on R. On the other hand, if v_i is on R, then a negative number is subtracted from the edge into v_i (on R) and the same number is added to the cost of the edge out of v_i (on R). It follows that the sum of the costs of the edges around the cycle R remains the same as before, i.e., $c(R)$ is unaltered. Since the above argument can be applied for each application of the STRESS() operation, the theorem follows. □

Theorem 4. *The Stressing Algorithm executes Line (11:), if and only if the input network* **G** *contains a negative cost cycle.*

Proof: If the input network, **G**, contains a negative cost cycle, then the STRESS() operations executed in lines (1 :) − (9 :) of Algorithm 2.1 are not going to change its cost, as per Theorem 3. Indeed, after Line (9 :) is executed for the last time, **G** will continue to have a negative cost cycle and hence a negative cost edge. This negative cost edge will be detected in Line (10 :) of the algorithm and hence Line (11 :) will be executed.

On the other hand, if **G** does not contain a negative cost cycle, then as argued previously, the constraint system, defined by System (2) is feasible and has a unique optimal solution. We need to consider the following two cases:

(i) $S_i = \phi$, for some $i \in \{1, 2, \ldots, (n-1)\}$
(ii) $S_i \neq \phi$, $\forall i \in \{1, 2, \ldots, (n-1)\}$.

In the former case, there were no vertices to be stressed during L_i, for some $i = 0, 1, \ldots, (n-1)$. This means that all the incoming edges of all the vertices are non-negative. However, this is only possible, if all the edges in the network are non-negative. This implies that Line (11 :) will not be executed.

In the latter case, $|Z_i| \geq 1$, $i = 0, 1, \ldots, (n-1)$ and hence $\cup_{i=0}^{n-1} Z_i = \mathbf{V}$. This means that all n vertices of the network have been saturated at some level, between 0 and $(n-1)$, i.e., the o value corresponding to each vertex has reached its final value. Hence, no vertex can be stressed any longer and all the incoming edges of each vertex are non-negative. This implies that Line (11 :) will not be executed.   □

We observe that in an adhoc networks setting, a node has to merely monitor the cost of its incoming edges to determine that the given set of constraints is infeasibile.

5   Conclusions

In this paper, we introduced a new technique for discovering the existence of negative cost cycles in networks with positive and negative arc costs. The novelty of our approach is that it takes *zero* extra space; all other algorithms that are cited in the literature use $\Omega(n)$ extra space, in the worst case. It must be noted that the Stressing Algorithm is asymptotically optimal, modulo the fastest known algorithm for the NCCD problem. We believe that the arguments used in proving the correctness of the stressing approach will find applications in the design of algorithms for other network optimization problems such as Min-cost flow.

We also note that any label-correcting algorithm to determine negative cost cycles requires that all vertices in the network be reachable from the specified source whereas the Stressing Algorithm does not impose any connectivity requirement on the network.

References

1. R. K. Ahuja, T. L. Magnanti, and J. B. Orlin. *Network Flows: Theory, Algorithms and Applications*. Prentice-Hall, 1993.
2. Boris V. Cherkassky, Andrew V. Goldberg, and T. Radzik. Shortest paths algorithms: Theory and experimental evaluation. *Mathematical Programming*, 73:129–174, 1996.
3. T. H. Cormen, C. E. Leiserson, and R. L. Rivest. *Introduction to Algorithms*. MIT Press and McGraw-Hill Book Company, Boston, Massachusetts, *2nd* edition, 1992.
4. M. R. Garey and D. S. Johnson. *Computers and Intractability: A Guide to the Theory of NP-Completeness*. W. H. Freeman Company, San Francisco, 1979.
5. Andrew V. Goldberg. Scaling algorithms for the shortest paths problem. *SIAM Journal on Computing*, 24(3):494–504, June 1995.
6. Alexander Schrijver. *Theory of Linear and Integer Programming*. John Wiley and Sons, New York, 1987.
7. K. Subramani and L. Kovalchick. A greedy strategy for detecting negative cost cycles in networks. *Future Generation Computer Systems*, 21(4):607–623, 2005.

Cache Placement in Sensor Networks Under Update Cost Constraint

Bin Tang, Samir Das, and Himanshu Gupta

Computer Science Department,
Stony Brook University, Stony Brook NY 11790, USA
{bintang, samir, hgupta}@cs.sunysb.edu.edu

Abstract. In this paper, we address an optimization problem that arises in context of cache placement in sensor networks. In particular, we consider the cache placement problem where the goal is to determine a set of nodes in the network to cache/store the given data item, such that the overall communication cost incurred in accessing the item is minimized, under the constraint that the total communication cost in updating the selected caches is less than a given constant. In our network model, there is a single server (containing the original copy of the data item) and multiple client nodes (that wish to access the data item). For various settings of the problem, we design optimal, near-optimal, heuristic-based, and distributed algorithms, and evaluate their performance through simulations on randomly generated sensor networks.

1 Introduction

Advances in embedded processing and wireless networking have made possible creation of sensor networks [1, 9]. A sensor network consists of sensor nodes with short-range radios and limited on-board processing capability, forming a multi-hop network of irregular topology. Sensor nodes must be powered by small batteries, making energy efficiency a critical design goal. There has been a significant interest in designing algorithms, applications, and network protocols to reduce energy usage of sensors. Examples include energy-aware routing [13], energy-efficient information processing [8, 9], and energy-optimal topology construction [21]. In this article, we focus on designing techniques to conserve energy in the network by caching data items at selected sensor nodes in a sensor network. The techniques developed in this paper are orthogonal to some of the other mentioned approaches, and can be used in combination with them to conserve energy.

Existing sensor networks assume that the sensors are preprogrammed and send data to a sink node where the data is aggregated and stored for offline querying and analysis. Thus, sensor networks provide a simple sample-and-gather service, possibly with some in-network processing to minimize communication cost and energy consumption. However, this view of sensor network architecture is quite limited. With the rise in embedded processing technology, sensor networks are set to become a more general-purpose, heterogeneous, distributed databases

V.R. Sirotiuk and E. Chávez (Eds.): ADHOC-NOW 2005, LNCS 3738, pp. 334–348, 2005.
© Springer-Verlag Berlin Heidelberg 2005

that generate and process time-varying data. As energy and storage limitations will always remain an issue – as much of it comes from pure physical limitations – new techniques for efficient data handling, storage, and dissemination must be developed. In this article, we take a general view of the sensor network where a subset of sensor nodes (called *servers*) generate data and another subset of nodes (called *clients*) consume this data. The data generation and consumption may not be synchronous with each other, and hence, the overall communication cost can be optimized by caching generated data at appropriately selected intermediate nodes. In particular, the data-centric sensor network applications which require efficient data dissemination [4, 6] will benefit from effective data caching strategies.

In our model of the sensor network, there is a single data item at a given server node, and many client nodes. (See Section 6 for a discussion on multiple data items and servers.) The server is essentially the data item producer and maintains the original copy of the item. All the nodes in the network cooperate to reduce the overall communication cost of accessing the data via a caching mechanism, wherein any node in the network can serve as a cache. A natural objective in the above context could be to select cache nodes such that the sum of the overall access and update cost is minimized. However, such an objective does not guarantee anything about the general distribution of enery usage across the sensor network. In particular, the updates always originate from the server node, and hence, the server node and the surrounding nodes bear most of the communication cost incurred in updating. Hence, there is a need to constrain the total update cost incurred in the network, to prolong the lifetime of the server node and the nodes around it – and hence, possibly of the sensor network. Thus, in this article, we address the cache placement problem to minimize the total access cost under an update cost constraint. More formally, we address the problem of selecting nodes in the network to serve as caches in order to minimize the total access cost (communication cost incurred in accessing the data item by all the clients), under the constraint that the total update cost (communication cost incurred in updating the cache nodes using an optimal Steiner tree over the cache nodes and the server) is less than a given constant. Note that since we are considering only a single data item, we do not need to consider memory constraints of a node.

Paper Outline. We start with formulating the problem addressed in this article and a discussion on related work in Section 2. For the cache placement problem under an update cost constraint, we consider a tree topology and a general graph topology of the sensor network. For the tree topology, we design an optimal dynamic programming algorithm in Section 3. The optimal algorithm for the tree topology can be applied to the general graph topology by extracting an appropriate tree from the given network graph. For the general graph topology, we consider a simplified multiple-unicast update cost model, and design a constant-factor approximation algorithm in Section 4.1. In Section 4.2, we present an efficient heuristic for the general cache placement problem under an update cost constraint, i.e., for a general update cost model in general graph topology. In Section 4.3, we present an efficient distributed implementation. Finally, we present simulation results in Section 5, and give concluding remarks in Section 6.

2   Problem Formulation and Related Work

In this section, we formulate the problem addressed in this article. We start with describing the sensor network model.

Sensor Network Model. A sensor network consists of a large number of sensor nodes distributed randomly in a geographical region. Each sensor node has a unique identifier (ID). Each sensor node has a radio interface and can communicate directly with some of the sensor nodes around it. For brevity, we sometimes just use *node* to refer to a sensor node. The sensor network can be modeled as an undirected weighted graph $G = (V, E)$, where V is the set of nodes, and E is the set of edges formed by pairs of nodes that can directly communicate with each other. The communication distance between any two nodes i and j is the number of hops d_{ij} between the two nodes. The network has a data item, which is stored at a unique node called a *server*, and is updated at a certain update frequency. Each sensor node could be a client node. A client node i requests the data item with an *access frequency* a_i. The cost of accessing a data item (*access cost*) by a node i from a node j (the server or a cache) is $a_i d_{ij}$, where d_{ij} is the number of hops from node i to node j.

Problem. Informally, our article addresses the following *cache placement problem* in sensor networks. Select a set of nodes to store copies of the data item such that the total access cost is minimized under a given update cost constraint. The total access cost is the sum of all individual access costs over all clients for accessing the data item from the nearest node (either a cache or the server) having a copy of the data item. The *update cost* incurred in updating a set of caches M is modeled as the cost of the optimal Steiner tree [10] spanning the server and the set of caches. This problem is obviously NP-hard, as even the Steiner tree problem is known to be NP-hard [3]. In this article, we look at the above problem in various stages – a tree topology, a graph topology with a simplified update cost model, a graph topology with the general update cost model – and present optimal, approximation, and heuristic-based algorithms respectively.

More formally, given a sensor network graph $G = (V, E)$, a server r with the data item, and an update cost Δ, select a set of cache nodes $M \subseteq V$ $(r \in M)$ to store the data item such that the total access cost

$$\tau(G, M) = \sum_{i \in V} a_i \times min_{j \in M} d_{ij}$$

is minimum, under the constraint that the total update cost $\mu(M)$ is less than a given constant Δ, where $\mu(M)$ is the cost of the minimum Steiner tree over the set of nodes M. Note that in the above definition *all* network nodes are considered as potential clients. If some node i is not a client, the corresponding a_i would be zero.

Related Work. The general problem of determining optimal cache placements in an arbitrary network topology has similarity to two problems widely studied in graph theory viz., facility location problem and the k-median problem. Both the problems consider only a single facility type (data item) in the network. In the facility-location problem, setting up a cache at a node incurs a certain fixed cost, and the goal is to minimize the sum of total access cost and the setting-up costs for all the caches, without any constraint. On the other hand, the k-median problem minimizes the total access cost under the number constraint, i.e., that at most k nodes can be selected as caches. Both problems are NP-hard, and a number of constant-factor approximation algorithms have been developed for each of the problems [7, 15], under the assumption that the edge costs in the graph satisfy the triangular inequality. Without the triangular inequality assumption, either problem is as hard as approximating the set cover [14], and therefore cannot be approximated better than $O(\log |V|)$ unless $\mathbf{NP} \subseteq \tilde{\mathbf{P}}$. Here, $|V|$ is the size of the network.

Several papers in the literature circumvent the hardness of the facility-location and k-median problems by assuming that the network has a tree topology [19, 22]. In particular, Li et al. [19] address the optimal placement of web proxies in a tree topology, essentially designing an $O(n^3 k^2)$ time dynamic programming algorithm to solve the k-median problem optimally in a tree of n nodes. In other related works on cache placement in trees, Xu et al. [22] discuss placement of "transparent" caches to minimize the sum of reads and writes, Krishnan et al. [18] consider a cost model based on cache misses, and Kalpakis et al. [16] consider a cost model involving reads, writes, and storage. In sensor networks, which consist of a large number of energy-constrained nodes, the constraint on the number of cache nodes is of little relevance.

Relatively less work has been done for caching in sensor networks. Intanagonwiwat et al. [6] propose directed diffusion, a data dissemination paradigm for sensor networks, which adopts a data centric approach and enables diffusion to achieve energy savings by selecting empirically good paths and by caching/processing data in-network. Bhattacharya et al. [4] develop a distributed framework that improve energy consumption by application layer data caching and asynchronous update multicast. In this article, we consider cache placement in sensor network under update cost constraint. As mentioned before, the update cost is typically mostly borne by the server and the surrounding nodes, and hence, is a critical constraint. To the best of our knowledge, we are not aware of any prior work that considers the cache placement problem under an update cost constraint.

3   Tree Topology

In this subsection, we address the cache placement problem under the update cost constraint in a tree network. The motivation of considering a tree topology (as opposed to a general graph model which we consider in the next section) is two fold. Firstly, data dissemination or gathering in sensor networks is typically done over an appropriately constructed network tree. Secondly, for the tree topol-

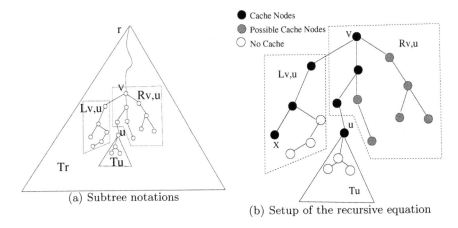

(a) Subtree notations

(b) Setup of the recursive equation

Fig. 1. Dynamic Programming algorithm for the tree topology

ogy, we can actually design polynomial time *optimal* algorithms. Thus, we can apply such optimal algorithms for the tree topology to the general graph topology by extracting an appropriate tree (e.g., shortest-path tree or near-optimal Steiner tree connecting the clients) from the general graph. In Section 5, we show through extensive simulations that such a strategy of applying an optimal tree algorithm to a general graph topology yields heuristics that deliver near-optimal cache placement solutions.

Consider an ad hoc network tree T rooted at the node r. Since the communication edges are bidirectional, any node in the network could be designated as the root; thus, we assume that the root node r is also the server for the data item. The cache placement problem under update cost constraint in a tree topology can be formally defined as follows.

Given the tree network T rooted at r, a data item whose server is r, and an update cost constraint Δ, find a set of cache nodes $M \subseteq T$ ($r \in M$) for storing copies of the data item, such that the total access cost $\tau(T, M) = \sum_{i \in T} a_i \times min_{j \in M} d_{ij}$ is minimized under the constraint that the total update cost $\mu(M)$ is less than Δ, where $\mu(M)$ is the cost of minimum cost Steiner tree over M. Note that the minimum cost Steiner tree spanning over a set of nodes M is simply the smallest subtree connecting the root r to all the nodes in M.

3.1   Dynamic Programming Algorithm

In this subsection, we present an optimal dynamic programming algorithm for the above described cache placement problem under the update cost constraint in a tree topology. We first start with some subtree notations [19] that are needed to describe our dynamic programming algorithm.

Subtree Notations. Consider the network tree T rooted at r. We use T_u to denote the subtree rooted at u in the tree T with respect to the root r (i.e., the subtree rooted at u not containing r); the tree T_r represents the entire tree T. For ease of presentation, we use T_u to also represent the set of nodes in the subtree T_u. We use $p(i)$ to denote the parent node of a node i in the tree T_r. Let $\pi(i, j)$ denote the unique path from node i to node j in T_r, and $d_{k,\pi(i,j)}$ denote the distance of a node k to the closest node on $\pi(i, j)$.

Consider two nodes v and u in the network tree, where v in an ancestor of u in T_r. See Figure 1(a). Let $L_{v,u}$ be the subgraph induced by the set of nodes on the left of and excluding the path $\pi(v, u)$ in the subtree T_v, and $R_{v,u}$ be the subgraph induced by the set of nodes on the right of and including the path $\pi(v, u)$, as shown in Figure 1(a). It is easy to see T_v can be divided into three distinct subgraphs, viz., $L_{v,u}$, T_u, and $R_{v,u}$.

DP Algorithm. Consider a subtree T_v and a node x on the leftmost branch of T_v. Let us assume that all the nodes on the path $\pi(v, x)$ (including v and x) have already been selected as caches. Let $\tau(T_v, x, \delta)$ denote the optimal access cost for all the nodes in the subtree T_v under the *additional* update cost constraint δ, where we do *not* include the cost of updating the already selected caches on the path $\pi(v, x)$. Below, we derive a recursive equation to compute $\tau(T_v, x, \delta)$, which would essentially yield a dynamic programming algorithm to compute $\tau(T_r, r, \Delta)$ – the minimum value of the total access cost for the entire network tree T_r under the update cost constraint Δ.

Let O_v be an optimal set (not including and in addition to $\pi(v, x)$) of cache nodes in T_v that minimizes the total access time under the additional update cost constraint δ. Let u be the leftmost deepest node of O_v in T_v, i.e., the node u is such that $L_{v,u} \cap O_v = \emptyset$ and $T_u \cap O_v = \{u\}$. It is easy to see that adding the nodes along the path $\pi(v, u)$ to the optimal solution O_v does not increase the additional update cost incurred by O_v, but may reduce the total access cost. Thus, without loss of generality, we assume that the optimal solution O_v includes all the nodes along the path $\pi(v, u)$ as cache nodes, if u is the leftmost deepest node of O_v in T_v.

<u>Recursive Equation.</u> As described above, consider an optimal solution O_v that minimizes $\tau(T_v, x, \delta)$, and let u be the leftmost deepest node of O_v in T_v. Note that O_v does not include the nodes on $\pi(v, x)$. Based on the definition of u and possible cache placements, a node in $L_{v,u}$ will access the data item from either the nearest node on $\pi(v, u)$ or the nearest node on $\pi(v, x)$. In addition, any node in T_u will access the data item from the cache node u, while all other nodes (i.e., the nodes in $R_{v,u}$) will choose one of the cache nodes in $R_{v,u}$ to access the data item. See Figure 1(b). Thus, the optimal access cost $\tau(T_v, x, \delta)$ can be recursively defined in terms of $\tau(R_{v,u}, p(u), \delta - d_{u,\pi(v,x)})$ as shown below. Below, the quantity $d_{u,\pi(v,x)}$ denotes the shortest distance in T_v from u to a node on the path $\pi(v, x)$ and hence, is the additional update cost incurred in updating the caches on the path $\pi(v, u)$. We first define $S(T_v, x, \delta)$ as the set of nodes u such that the cost of updating u is less than δ, the additional update cost constraint. That is, $S(T_v, x, \delta) = \{u | u \in T_v \wedge (\delta > d_{u,\pi(v,x)})\}$.

Now, the recursive equation can be defined as follows.

$$
\tau(T_v, x, \delta) = \begin{cases} \sum_{i \in T_v} a_i \times d_{i, \pi(v,x)} & \text{if } S(T_v, x, \delta) = \emptyset \\ \\ min_{u \in S(T_v, x, \delta)} \\ \left(\begin{array}{l} \sum_{i \in L_{v,u}} a_i \times min(d_{i, \pi(v,u)}, d_{i, \pi(v,x)}) \\ + \sum_{i \in T_u} a_i d_{iu} \\ + \tau(R_{v,u}, p(u), \delta - d_{u, \pi(v,x)}) \end{array} \right) \end{cases}
$$

In the above recursive equation, the first case corresponds to the situation when the additional update constraint δ is not sufficient to cache the data item at any more nodes (other than already selected cache nodes on $\pi(v, x)$). For the second case, we compute the total (and minimum possible) access cost for each possible value of u, the leftmost deepest additional cache node, and pick the value of u that yields the minimum total access cost. In particular, for a fixed u, the first term corresponds to the total access cost of the nodes in $L_{v,u}$. Note that for a node in $L_{v,u}$ the closest cache node is either on the path $\pi_{v,x}$ or $\pi_{v,u}$. The second and third terms correspond to the total access time of nodes in T_u and $R_{v,u}$ respectively. Since the tree T_u is devoid of any cache nodes, the cache node closest to any node in T_u is u. The minimum total access cost of all the nodes in $R_{v,u}$ can be represented as $\tau(R_{v,u}, p(u), \delta - d_{u, \pi(v,x)})$, since the remaining available update cost is $\delta - d_{u, \pi(v,x)}$ where $d_{u, \pi(v,x)})$ is the update cost used up by the cache node u. The overall time complexity of the above dynamic programming algorithm can be shown to be $O(n^4 + n^3 \Delta)$ by careful precomutation.

4   General Graph Topology

The tree topology assumption makes it possible to design a polynomial-time optimal algorithm for the cache placement problem under update cost constraint. In this subsection, we address the cache placement problem in a general graph topology. In the general graph topology, the cache placement problem becomes NP-hard. Thus, our focus here is on designing polynomial-time algorithms with some performance guarantee on the quality of the solution.

As defined before, the total update cost incurred by a set of caches nodes is the minimum cost of an optimal Steiner tree over the set of cache nodes and the server; we refer to this update cost model as the Steiner tree update cost model. Since the minimum-cost Steiner tree problem is NP-hard in general graphs, we solve the cache placement problem in two steps. First, we consider a simplified multiple-unicast update cost model and present a greedy algorithm with a provable performance guarantee for the simplified model. Then, we improve our greedy algorithm based upon the more efficient Steiner tree update cost model.

4.1   Multiple-Unicast Update Cost Model

In this section, we consider the cache placement problem for general network graph under a simplified update cost model. In particular, we consider the multiple-unicast update cost model, wherein we model the total update cost incurred in updating a set of caches as the sum of the individual shortest path lengths from the server to each cache node. More formally, the total update cost of a set M of cache nodes is $\mu(M) = \sum_{i \in M} d_{si}$, where s is the server. Using this simplified update cost model, the cache placement problem in general graphs for update cost constraint can be formulated as follows.

Problem Under Multiple-Unicast Model. Given an ad hoc network graph $G = (V, E)$, a server s with the data item, and an update cost Δ, select a set of cache nodes $M \subseteq V$ ($s \in M$) to store the data item such that the total access cost $\tau(G, M) = \sum_{i \in V} a_i \times min_{j \in M} d_{ij}$ is minimum, under the constraint that the total update cost $\mu(M) = \sum_{i \in M} d_{si} < \Delta$.

The cache placement problem with the above simplified update cost model is still NP-hard, as can be easily shown by a reduction from the k-median problem. A number of constant-factor approximation algorithms have been proposed [7, 15] for the k-median problem which can also be used to solve the above cache placement problem. However, all the constant-factor approximation algorithms are based on the assumption that the edge costs in the network graph satisfy the triangular inequality. Moreover, the proposed approximation algorithms for k-median problem cannot be easily extended to the more efficient Steiner tree update cost model. Below, we present a greedy algorithm that returns a solution whose "access benefit" is at least 63% of the optimal benefit, where access benefit is defined as the reduction in total access cost due to cache placements.

Greedy Algorithm. In this section, we present a greedy approximation algorithm for the cache placement problem under the multiple-unicast update cost constraint in general graphs, and show that it returns a solution with near-optimal reduction in access cost. We start with defining the concept of a benefit of a set of nodes which is important for the description of the algorithm.

Definition 1. (Benefit of Nodes) Let A be an arbitrary set of nodes in the sensor network. The *benefit* of A with respect to an already selected set of cache nodes M, denoted as $\beta(A, M)$, is the decrease in total access cost resulting due to the selection of A as cache nodes. More formally, $\beta(A, M) = \tau(G, M) - \tau(G, M \cup A)$, where $\tau(G, M)$, as defined before, is the total access cost of the network graph G when the set of nodes M have been selected as caches. The *absolute benefit* of A denoted by $\beta(A)$ is the benefit of A with respect to an empty set, i.e., $\beta(A) = \beta(A, \emptyset)$.

The *benefit per unit update cost* of A with respect to M is $\beta(A, M)/\mu(A)$, where $\mu(A)$ is the total update cost of the set A under the multiple-unicast update cost model.   □

Our proposed Greedy Algorithm works as follows. Let M be the set of caches selected at any stage. Initially, M is empty. At each stage of the Greedy Algorithm, we add to M the node A that has the highest benefit per unit update cost with respect to M at that stage. This process continues until the update cost of M reaches the allowed update cost constraint. The algorithm is formally presented below.

Algorithm 1. Greedy Algorithm
   Input: A sensor network graph $V = (G, E)$.
               Update cost constraint Δ.
   Output: A set of cache nodes M.
   BEGIN
      $M = \emptyset$;
      while $(\mu(M) < \Delta)$
         Let A be the node with maximum $\beta(A, M)/\mu(A)$.
         $M = M \cup \{A\}$;
      end while;
      RETURN M;
   END. $\Diamond$

The running time of the above greedy algorithm is $O(kn^2)$, where k is the number of iterations and n is the number of nodes in the network. Note that the number of iterations k is bounded by n.

Performance Guarantee of the Greedy Algorithm. The Greedy Algorithm returns a solution that has a benefit at least 63% of that of the optimal solution. The proof techniques used here are similar to the techniques used in [11] for the closed related view-selection problem in a data warehouse. Due to the space limitation, we omit the proof here.

Theorem 1 *Greedy Algorithm (Algorithm 1) returns a solution M whose absolute benefit is of at least $(1-1/e)$ times the absolute benefit of an optimal solution having the update cost (under the multiple-unicast model) of at most that of M.*
                                                                              ∎

4.2   Steiner Tree Update Cost Model

Recall that the constant factor performance guarantee of the Greedy Algorithm described in previous section is based on the multiple-unicast update cost model, wherein whenever the data item in a cache nodes needs to be updated, the updated information is transmitted along the individual shortest path between the server and the cache node. However, the more efficient method of updating a set of caches from the server is by using the optimal (minimum-cost) Steiner tree over the selected cache nodes and the server. In this section, we improve the performance of our Greedy Algorithm by using the more efficient Steiner tree update cost model, wherein the total update cost incurred for a set of cache nodes is the cost of the optimal Steiner tree over the set of nodes M and the server of the data item.

   Since the minimum-cost Steiner tree problem is NP-hard, we adopt the simple 2-approximation algorithm [10] for the Steiner tree construction, which constructs a Steiner tree over a set of nodes L by first computing a minimum spanning tree in the "distance graph" of the set of nodes L. We use the term 2-approximate Steiner tree to refer to the solution returned by the 2-approximation Steiner tree approximation algorithm. Based on the notion of 2-approximate Steiner tree, we define the following update cost terms.

Definition 2. (Steiner Update Cost) The *Steiner update cost* for a set M of cache nodes, denoted by $\mu'(M)$, is defined as the cost of a 2-approximate Steiner tree over the set of nodes M and the server s.

The *incremental Steiner update cost* for a set A of nodes with respect to a set of nodes M is denoted by $\mu'(A, M)$ and is defined as the increase in the cost of the 2-approximate Steiner tree due to addition of A to M, i.e., $\mu'(A, M) = \mu'(A \cup M) - \mu'(M)$.   □

Based on the above definitions, we describe the Greedy-Steiner Algorithm which uses the more efficient Steiner tree update cost model as follows.

Algorithm 2.   Greedy-Steiner Algorithm
Same as Algorithm 1 except μ is changed to μ'.   ◊

Unfortunately, there is no performance guarantee of the solution delivered by the Greedy-Steiner Algorithm. However, as we show in Section 5, the Greedy-Steiner Algorithm performs the best among all our designed algorithms for the cache placement problem under an update cost constraint.

4.3   Distributed Implementation

In this subsection, we design a distributed version of the centralized Greedy-Steiner Algorithm (Algorithm 2). Using similar ideas as presented in this section, we can also design a distributed version of the centralized Greedy Algorithm (Algorithm 1). However, since the centralized Greedy-Steiner Algorithm outperformed the centralized Greedy Algorithm for all ranges of parameter values in our simulations, we present only the distributed version of Greedy-Steiner Algorithm. As in the case of centralized Greedy-Steiner Algorithm, we cannot prove any performance guarantee for the presented distributed version. However, we observe in our simulations that solution delivered by the distributed version is very close to that delivered by the centralized Greedy-Steiner Algorithm. Here, we assume the presence of an underlying routing protocol in the sensor network. Due to limited memory resources at each sensor node, a proactive routing protocol [20] that builds routing tables at each node is unlikely to be feasible. In such a case, a location-aided routing protocol such as GPSR [17] is sufficient for our purposes, if each node is aware of its location (either through GPS [12] or other localization techniques [2, 5]).

Distributed Greedy-Steiner Algorithm. The distributed version of the centralized Greedy-Steiner Algorithm consists of rounds. During a round, each non-cache node A estimates its benefit per unit update cost, i.e., $\beta(A, M)/\mu'(A, M)$, as described in the next paragraph. If the estimate at a node A is maximum among all its communication neighbors, then A decides to cache itself. Thus, during each round, a number of sensor nodes may decide to cache the data item according to the above criteria. At the end of each round, the server node gathers information from all the newly added cache nodes, and computes the Steiner tree involving all the selected cache nodes till the round. Then, the remaining update cost (i.e., the given update cost constraint minus the current update cost of the Steiner tree involving the selected cache nodes) is broadcast by the server to the entire network and a new round is initiated. If there is no remaining update cost, then the server decides to discard some of the recently added caches (to keep the total update cost under the given update cost constraint), and the algorithm terminates. The algorithm is formally presented below.

Algorithm 3. Distributed Greedy-Steiner Algorithm
> **Input:** A network graph $V = (G, E)$.
> > Update cost constraint Δ.
> **Output:** The set of cache nodes M.
> **BEGIN**
> > $M = \emptyset$;
> > **while** $(\mu'(M) < \Delta)$
> > > Let $\mathcal{A}$ be the set of nodes each of which (denoted as A) has the maximum $\beta(A, M)/\mu'(A, M)$ among its non-cache neighbors.
> > > $M = M \cup \mathcal{A}$;
> > **end while;**
> > **RETURN** M;
> **END.** $\Diamond$

Estimation of $\mu'(A, M)$. Let A be a non-cache node, and T_A^S be the shortest path tree from the server to the set of communication neighbors of A. Let $C \in M$ be the cache node in T_A^S that is closest to A, and let d be the distance from A to C. In the above Distributed Greedy-Steiner Algorithm, we estimate the incremental Steiner update cost $\mu'(A, M)$ to be d. The value d can be computed in a distributed manner at the start of each round as follows. As mentioned before, the server initiates a new round by broadcasting a packet containing the remaining update cost to the entire network. If we append to this packet all the cache nodes encountered on the way, then each node should get the set of cache nodes on the shortest path from the server to itself. Now, to compute d, each node only needs to exchange the above information with all its immediate neighbors.

Estimation of $\beta(A, M)$. A non-cache node A considers only its "local" traffic to estimate $\beta(A, M)$, the benefit with respect to an already selected set of cache nodes M. The local traffic of A is defined as the data access requests that use A as an intermediate/origin node. Thus, the local traffic of a node includes its own data requests. We estimate the benefit of caching the data item at A as

$\beta(A, M) = d \times t$, where t is the frequency of the local traffic observed at A and d is the distance to the nearest cache from A (which is computed as shown in the previous paragraph). The local traffic t can be computed if we let the normal network traffic (using only the already selected caches in previous rounds) run for some time between successive rounds. The data access requests of a node A during normal network traffic between rounds can be directed to the nearest cache in the tree T_A^S as defined in the previous paragraph.

5   Performance Evaluation

We empirically evaluate the relative performances of the cache placement algorithms for randomly generated sensor networks of various densities. As the focus of our work is to optimize access cost, this metric is evaluated for a wide range of parameters such as number of nodes and network and network density, etc.

We study various caching schemes (listed below) on a randomly generated sensor network of 2,000 to 5,000 nodes in a square region of 30×30. The distances are in terms of arbitrary units. We assume all the nodes have the same transmission radius (T_r), and all edges in the network graph have unit weight. We have varied the number of clients over a wide range. For clarity, we first present the data for the case where number of clients is 50% of the number of nodes, and then present a specific case with varying number of clients. All the data presented here are representative of a very large number of experiments we have run. Each point in a plot represents an average of five runs, in each of which the server is randomly chosen. The access costs are plotted against number of nodes and transmission radius and several caching schemes are evaluated:

- *No Caching* – serves as a baseline case.
- *Greedy Algorithm* — greedy algorithm using the multiple-unicast update cost model (Algorithm 1).

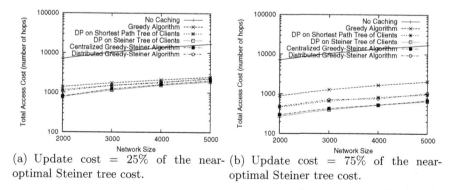

(a) Update cost = 25% of the near-optimal Steiner tree cost.

(b) Update cost = 75% of the near-optimal Steiner tree cost.

Fig. 2. Access cost with varying number of nodes in the network for different update cost constraints. Transmission radius $(T_r) = 2$. Number of clients = 50% of the number of nodes, and hence increases with the network size.

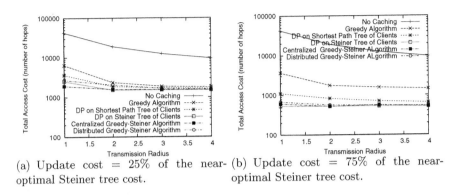

(a) Update cost = 25% of the near-optimal Steiner tree cost.

(b) Update cost = 75% of the near-optimal Steiner tree cost.

Fig. 3. Access cost with varying transmission radius (T_r) for different update cost constraints. Number of nodes = 4000, and number of clients = 2000 (50% of number of nodes).

- *Centralized Greedy-Steiner Algorithm* — greedy algorithm using the Steiner tree-based update cost model (Algorithm 2).
- *Distributed Greedy-Steiner Algorithm* – distributed implementation of the Greedy-Steiner Algorithm (Algorithm 3).
- *DP on Shortest Path Tree of Clients* – Dynamic Programming algorithm (Section 3.1) on the tree formed by the shortest paths between the clients and the server.
- *DP on Steiner Tree of Clients* – Dynamic Programming algorithm (Section 3.1) on the 2-approximate Steiner tree over the clients and the server.

Figure 2 shows the effect of the number of nodes; the transmission radius (T_r) is fixed at 2. Figure 3 shows the effect of T_r; a network of 4,000 nodes is chosen for these experiments and T_r is varied from 1 to 4. The general trend in these two sets of plots is similar. Aside from the fact that our algorithms offer much less total access cost than the no-caching case, the plots show that (i) the two Steiner tree-based algorithms (DP on Steiner Tree of Clients and Centralized Greedy-Steiner Algorithm) perform equally well and the best among all algorithms except for very sparse graphs; (ii) the Greedy-Steiner Algorithm provides the best overall behavior; (iii) the Distributed Greedy-Steiner Algorithm performs very closely to its centralized version.

6   Conclusions

We have developed a suite of data caching techniques to support effective data dissemination in sensor networks. In particular, we have considered update cost constraint and developed efficient algorithms to determine optimal or near-optimal cache placements to minimize overall access cost. Minimization of access cost leads to communication cost savings and hence, energy efficiency. The choice of update constraint also indirectly contributes to resource efficiency. Two models have been considered – one for a tree topology, where an optimal algorithm based on dynamic programming has been developed, and the other for the general graph topology, which presents a NP-hard problem where a polynomial-time approximation algorithm has been developed. We also designed efficient distributed implementations of our centralized algorithms, empirically showed that they performs well for random sensor networks.

References

1. B. Badrinath, M. Srivastava, K. Mills, J. Scholtz, and K. Sollins, editors. *Special Issue on Smart Spaces and Environments,* IEEE Personal Communications, 2000.
2. P. Bahl and V. N. Padmanabhan. Radar: An in-building RF-based user-location and tracking system. In *INFOCOM'00.*
3. P. Berman and V. Ramaiyer. Improved approximation algorithms for the steiner tree problem. *J. Algorithms,* 1994.
4. S. Bhattacharya, H. Kim, S. Prabh, and T. Abdelzaher. Energy-conserving data placement and asynchronous multicast in wireless sensor networks. In *MobiSys'03.*
5. N. Bulusu, J. Heidemann, and D. Estrin. GPS-less low cost outdoor localization for very small devices. *IEEE Personal Communications Magazine,* 7(5), 2000.
6. D. Estrin C. Intanagonwiwat, R. Govindan and J. Heidemann. Directed diffusion for wireless sensor networks. *IEEE/ACM TON,* 11(1):2–16, February 2003.
7. M. Charikar and S. Guha. Improved combinatorial algorithms for the facility location and k-median problems. In *IEEE FOCS'99,* pages 378–388.
8. M. Chu, H. Haussecker, and F. Zhao. Scalable information-driven sensor querying and routing for ad hoc heterogeneous sensor networks. *IEEE Journal of High Performance Computing Applications,* 2002.
9. D. Estrin, R. Govindan, and J. Heidemann, editors. *Special Issue on Embedding the Internet,* Communications of the ACM, volume 43, 2000.
10. E. N. Gilbert and H. O. Pollak. Steiner minimal trees. *SIAM J. Appl. Math,* 16:1–29, 1968.
11. H. Gupta. *Selection and Maintenance of Views in a Data Warehouse.* PhD thesis, Computer Science Department, Stanford University, 1999.
12. B. Hofmann-Wellenhof, H. Lichtenegger, and J. Collins. *Global Positioning System: Theory and Practice.* Springer-Verlag Telos, 1997.
13. Chalermek Intanagonwiwat, Ramesh Govindan, and Deborah Estrin. Directed diffusion: a scalable and robust communication paradigm for sensor networks. In *MOBICOM'00.*
14. K. Jain and V. Vazirani. Approximation algorithms for metric facility location and k-median problems using the primal-dual schema and lagrangian relaxation. *J. ACM,* 48(2), 2001.
15. K. Jain and V. V. Vazirani. Approximation algorithms for metric facility location and k-median problems using the primal-dual schema and lagrangian relaxation. *Journal of the ACM,* 48(2):274–296, 2001.
16. K. Kalpakis, K. Dasgupta, and O. Wolfson. Steiner-optimal data replication in tree networks with storage costs. In *Proceedings of IDEAS,* pages 285–293, 2001.
17. Brad Karp and H.T Kung. GPSR: Greedy perimeter stateless routing for wireless networks. In *MOBICOM'00.*

18. P. Krishnan, D. Raz, and Y. Shavitt. The cache location problem. *IEEE/ACM TON*, 8:568–582, 2000.
19. B. Li, M. J. Golin, G. F. Italiano, and X. Deng. On the optimal placement of web proxies in the internet. In *INFOCOM'99*.
20. C. E. Perkins and P. Bhagwat. Highly dynamic destination-sequenced distance-vector routing (dsdv) for mobile computers. In *SIGCOMM'94*.
21. R. Wattenhofer, L. Li, P. Bahl, and Y.-M. Wang. Distributed topology control for wireless muitihop ad-hoc networks. In *INFOCOM'01*.
22. J. Xu, B. Li, and D. L. Lee. Placement problems for transparent data replication proxy services. *IEEE JSAC*, 20(7):1383–1397, 2002.

A Service Discovery Protocol with Maximal Area Disjoint Paths for Mobile Ad Hoc Networks*

Shihong Zou, Le Tian, Shiduan Cheng, and Yu Lin

State Key Lab of Networking and Switching Technology, P.O.Box 79,
Beijng University of Posts & Telecommunications, Beijing 100876, China
{zoush, chsd, linyu}@bupt.edu.cn

Abstract. Service discovery is a basic requirement for mobile ad hoc networks to provide service efficiently with dynamically changing network topology. In this paper, we propose a service discovery protocol with maximal area disjoint path (SDMAD) for mobile ad hoc networks. The goal of SDMAD is to exploit multiple servers simultaneously and do parallel delivery. One key feature of SDMAD is that there is no contention between the multiple paths discovered by it. Moreover, SDMAD considers service discovery and routing jointly to reduce control overhead. Extensive simulations show that with SDMAD the performance of service delivery is greatly improved over that with other service discovery protocols.

1 Introduction

Mobile ad hoc network (MANET) [1] is a network formed by a group of wireless devices with limited power and transmission range. There is no existing infrastructure in MANET. Communication between two nodes that are not in direct radio range takes place in a multi-hop fashion, with other nodes acting as routers.

Service discovery, which allows devices to automatically discover resources or services with required attribute, is an integral part of the ad hoc networking to achieve stand-alone and self-configurable communication networks. There have been many literatures on service discovery in MANETs [2-5]. In MANETs, there are often duplicated services distributed in multiple nodes. For example, in mobile ad hoc P2P networks [6-7], the duplicated files (such as video files, mobile games, and so on) are distributed over several nodes. When one node wants to download a file, it would like to download from multiple servers simultaneously to shorten the delay. In addition, for some other application, such as Network Time Protocol, Public Key Cryptography and so on, any client must contact several servers at the same time for synchronization or certification. Therefore, one client contacting multiple servers simultaneously is a common scenario in MANETs, which is also the focus of this paper. However, current service discovery protocols adopted the idea similar to anycast [8], ending with one server discovered. Even they can be extended to find several servers, such as manycast

* This work was supported by the National Basic Research Program of China (Grant No. 2003CB314806, No. 2006CB701306) and the National Natural Science Foundation of China (Grant No.90204003, No. 60472067).

V.R. Sirotiuk and E. Chávez (Eds.): ADHOC-NOW 2005, LNCS 3738, pp. 349–357, 2005.

proposed by Carter et al. [11], they did not take underlying routing into account and may result in too more control overhead and route coupling. Service discovery and routing both rely on network wide broadcasting, which is an expensive operation in MANETs. If service discovery and routing are considered jointly, the control overhead can be reduced significantly. In addition, route coupling is another reason for the performance degradation with multipath routing.

When there are multiple servers, there are multiple paths from servers to client. There have been many related works [9-10, 17-20] on multipath routing. These papers all assume that there are only one source and one destination and then proposed protocols to discovery node-disjoint or link-disjoint paths. However, when service is delivered simultaneously on all the paths, node-disjoint and link-disjoint may result in serious performance degradation due to route coupling [18]. Two routes (paths) that have nodes or links in common are considered highly coupled. However, route coupling may occur even if two routes have no nodes or links in common. In the case of multiple channel spread spectrum networks, packet transmission may result in degraded quality of a simultaneous transmission on a neighboring link. In single channel networks, a node's transmission can prevent neighbors from receiving separate transmissions altogether. Therefore we propose a new concept for multipath –area disjoint to avoid the contention between the multiple paths. Then we propose a service discovery protocol with maximal area disjoint paths (SDMAD). It jointly considers service discovery and routing, and discovers multiple paths without route coupling. Extensive simulations show that with SDMAD, intra-contention between the multiple paths is avoided and the completion time for one transaction is reduced significantly.

The remainder of this paper is organized as follows. In section 2, we review related works. In section 3, we introduce the concept of area disjoint and then propose SDMAD. Simulation results and analysis are presented in section 4. Section 5 concludes the paper.

2 Related Work

Kozat and Tassiulas [2] discussed network layer support for service discovery in mobile ad hoc networks. They proposed a distributed service discovery architectures which relies on a virtual backbone for locating and registering available services within a dynamic network topology. It is a directory based system. However, a directoryless system is more suited to most of MANET scenarios.

Helal et al. [3] designed a service discovery and delivery protocol – Konark, which uses a completely distributed, peer-to-peer mechanism that provides each device the ability to advertise and discover services in the network. However, most of this paper is focused on system design and implementation issues.

Mohan et al. [4] studied the problems associated with service discovery by first simulating two well known service discovery techniques – pull and push, and investigated their limitations for large network sizes. Then they proposed a new approach combining the best features of the above two approach. Liu et al. [5] proposed a framework that provides a unified solution to the discovery of resources and QoS-aware selection of resource providers. The key entities of this framework are a set of self-organized discovery agents. These agents manage the directory information of resources using hash indexing. They also dynamically partition the network into domains and collect intra- and inter- domain QoS information to select appropriate providers. However, these two papers both did not consider service discovery with routing jointly and did not address the issue how to improve the performance of parallel delivery.

Klemm et al. [6] presented a special-purpose system for searching and file transfer tailored to both the characteristics of MANET and the requirements of peer-to-peer file sharing, which is named ORION. ORION combines application-layer query processing with the network layer process of route discovery. However, it only allows choosing one server for a request and can not exploit multiple servers.

Duran et al. [7] proposed two efficient search schemes that use query messages filtering/gossiping and adaptive hop-limited search, respectively, for peer-to-peer file sharing over mobile ad hoc networks. However, they focused on reducing the overhead of service discovery mechanism and did not try to improve the performance of service delivery.

Hsieh and Sivakumar [12] proposed a transport layer protocol called R^2CP that efficiently enables real-time multipoint-to-point video streaming in wireline Internet. R^2CP is a receiver-driven transport protocol and exploits the parallel transmission from several peers. Furthermore, it does not address this issue in mobile ad hoc networks.

3 A Service Discovery Protocol with Maximal Area Disjoint Path

In this section, we first discuss the advantage of jointly considering service discovery and routing, and then propose a new disjoint concept – area disjoint. At last, we present SDMAD in detail.

3.1 Jointly Consider Service Discovery and Routing

Service discovery in P2P systems [13] are built on the overlay connections, which are clearly separated from underlying routing mechanism because underlying routing is transparent to the peers. However, in MANETs, every nodes act as not only hosts but also routers. Hence, nodes are aware of the routes. On the other hand, on demand protocols are preferred in MANETs due to their low overhead. When on demand protocols are used, considering service discovery and routing separately will result in significant control overhead. This is because after the client receives the service reply, it still needs to initiate route discovery procedure to find the route to the server. Therefore, we believe that service discovery and route discovery should be done simultaneously and so the client can obtain the name of the server and the route to the server at the cost of one discovery procedure. The discovery procedure can be performed on-demand with many techniques, such as simple flooding, range limited multicast and so on [2].

3.2 Area Disjoint vs. Node Disjoint

Node disjoint has been exploited in many multipath routing protocols of mobile ad hoc networks [9, 10]. It means that two paths have no common nodes. However, when parallel delivery is applied, node disjoint paths may result in significant contention between the multiple paths used by the client. This will induce increasing collisions and longer delay. Hence we propose a new disjoint concept – area disjoint to explicitly avoid such kind of contention. *Area disjoint* is defined as follows: when any node of one path does not locate within the transmission range of any node of another path, we say that these two paths are area disjoint. If two paths are area disjoint, then they are independent of each other and do not contend for channel. In the following, when talking about node disjoint and area disjoint, we do not take the client into account. Fig. 1 shows the difference between node disjoint and area disjoint, where the circle with origin N2 indicates N2's transmission range. Although the paths to server1 and server2 are node disjoint, N1 locates at the coverage of N2 and so will contend for the channel with N2. For one-client-one-server scenario, it is impossible to find area disjoint paths. In one-client-multiple-servers scenario, it is often easy to find area disjoint paths, which is also verified by the simulation. Apparently, the flows coming from multiple servers will contend for the channel around the client. However, according to the paper [21], a flow across 3 hops or more can only achieve 1/4 of single hop capacity in theory or 1/7 of single hop capacity in simulation. Therefore, we can assume that when there is no server 1 hop from the client, the client will not be the bottleneck. When there is any server that is only 1 hop from the client, the client just only request the service from this server. Even if in occasional cases the client becomes the bottleneck, the client can perceive the congestion around it and do congestion control easily.

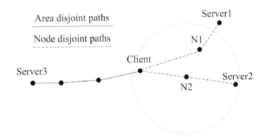

Fig. 1. Area disjoint vs. node disjoint

If two paths are node disjoint but not area disjoint, such as the path [Client, N1, Server1] and the path [Client, N2, Server2] which are shown in Fig.1, the throughput of one session is completely restrained by the other session, which is also called unfairness problem in the paper [15]. When N2 is transmitting to Client, if Server1 send RTS (Request To Send) to N1, N1 can not hear this RTS because it is within the coverage of N2. However, Server1 does not know it and sends RTS to N1 again and again till the retry_limit is reached and then drops the packet. The probability for Server1 to send RTS successfully is very low if N2 always has data to send. Hence the throughput from Server1 is completely forced down, which increases the delay for the client to receive the same amount of data. This is one main motivation to propose SDMAD.

As a byproduct of area disjoint, it is clearly that service reliability will be greatly improved due to increasing path diversity.

3.3 SDMAD

SDMAD is an on-demand service discovery protocol that discovers not only servers but also corresponding area disjoint paths to the servers. When the client needs a service, it floods a Service Request (SREQ) message to the entire network. Because this packet is flooded, several duplicates that traversed through different routes reach multiple servers. These multiple servers then send Service Reply (SREP) messages back to the client via the chosen routes.

In order to prevent route coupling and utilize the available network resource efficiently, the main goal of SDMAD is to discover *maximal area disjoint multiple paths*. To achieve this goal in on-demand routing schemes, the client must know the entire path of all available routes so that it can choose the routes. Therefore, we use the source routing approach [14] where the information of the nodes that consist the route is included in the RREQ (Route REQuest) message. In SDMAD, service discovery and routing discovery are done at the same time, RREQ of DSR (Dynamic Source Routing) [14] is appended one more field *service type* and acts also as SREQ. When a node receives the SREQ, if it can not provide such type of service, it just appends its ID to the SREQ and forwards it, otherwise it replies with a SREP and sends it to the client along the reverse path recorded in the SREQ. RREP (Route REPly) of DSR is appended two more fields to act as SREP. One field is *neighbors ID* to record all the neighbors of forwarding node, the other is *service type* to identify the service the server can provide, which is copied from the SREQ. Each node forwarding the SREP (including servers) should fill in the neighbors ID field.

The pseudo codes of the above procedures are showed as follows:

```
recvSREQ ()//when receiving a SREQ call this procedure
{
If (can-provide-this-service) {sendSREP; }
Else {    append-my-ID-to-SREQ;
          broadcast-SREQ;}
}
sendSREP ()//when I can provide service call this procedure
{
 construct-SREP-with-Reverse-path-in-SREQ;
 append-my-neighbor-list-to-SREP;
 send-SREP-to-next-hop;
}
recvSREP()//when receiving SREP call this procedure
{
 If (destination-is-me)
{record-and-construct-MAD-paths; }
 Else{    append-my-neighbor-list-to-SREP;
          send-SREP-to-next-hop;}
}
```

When there are multiple servers in the network, the client may receive multiple SREPs from different servers. In real deployment, the client can exploit as many as paths to obtain services. However, to simplify the scenario, we limit the number of paths to 2 in this study, which does not loss the essentiality of SDMAD. One of the two paths chosen by SDMAD is the shortest delay path, the path taken by the first SREP the client receives. After the client receives the first SREP, the client can use the route information in the SREP to contact the server immediately. Hereafter, the client waits a certain duration of time to receive more SREPs and learn all possible servers and paths. It then selects the path that is maximal area disjoint to the path which has been already used. If there are more than one paths that are area disjoint with the first one, the one with the shortest hop distance is chosen. If still remain multiple paths that meet the condition, server's Node ID is used to break the tie.

SDMAD judges whether two paths are maximal area disjoint through comparing the neighbors' ID piggybacked in the SREPs. If there is a node in the path of one SREP, which is also in the neighbor set of another SREP, than these two paths are not maximal area disjoint. Otherwise these two paths are maximal area disjoint. The pseudo code is showed as follows:

```
is-MAD-Paths(SREP_a, SREP_b)

{

neighborSet = {all the nodes in the neighbor set of SREP_b};

For (each node i in the path of SREP_a) {

        If (i • neighborSet) Return FALSE;

        Else Continue;

}

Return TRUE;

}
```

With CSMA (carrier sensing multiple access) based MAC (medium access control) protocols, such as IEEE 802.11 [22], which is used extensively in both research and testbed on MANETs, the neighbor ID can be obtained easily by overhearing neighbor's transmission. While with other kinds of MAC protocols, some additional mechanisms are needed to obtain neighbor's ID. In this paper we assume that some kind of CSMA based MAC protocols is used. When SREQ is flooded to the server, the nodes along the path can know all the information of its neighbors assuming there is no collision. Then the nodes can use this information to fill the field in SREP. This means that obtaining the information of neighbors does not introduce additional overhead. If some neighbors are not heard due to collisions, then this may results in choosing two paths which are in fact not maximal area disjoint. In addition, node's movement may change area disjoint paths into node disjoint paths. When this happens, the client can detect it by summing the throughput from each server respectively. If the throughput of one server is significantly lower than that of the other server (which may also be caused by heavy load in the former server), it then simply removes the low throughput server and chooses another maximal area disjoint path. Another approach to reduce such probability is to assign higher priority to two paths not only maximal area disjoint, but also having no common neighbors, which means they are far apart from each other. When a path to one server breaks, the client just chooses an alternate area disjoint path if it can find one from the SREPs it received; otherwise it does nothing. Since the service discovery procedure needs to execute some kind of flooding, which is expensive in MANETs, in SDMAD, only when both paths break does the client initiate the service discovery procedure again.

Each server replies to only a selected set of queries. The queries that are replied to are those that carry a path that is node disjoint from the primary path. This also implicitly controls the total number of the replies, thus preventing a reply storm.

4 Simulation Results and Analysis

We used the well known simulator NS2 to validate the performance improvement of SDMAD and implemented SDMAD based on DSR module in NS2. We simulated multihop ad hoc network scenario with 50 nodes distributed randomly in a network area of 670m*670m. The transmission range is set to 250m and data rate is set to 2Mbps. Packet size is set to 1000 Bytes. We use the random waypoint model with pause time 0s. The maximal speed of node is 5m/s. There are one client (Node 0) and five servers (Nodes 1-5). All the following results are averaged over 20 runs.

When a flow spans multiple hops, the hidden and exposed terminal problems degrade the throughput of the flow seriously [15]. To see clearly the improvement of SDMAD over maximal node disjoint approach (MND), we use DBTMA [16] as MAC protocol to eliminate the performance degradation due to hidden and exposed terminal problems.

Since we focus on service discovery protocol, we assume all the servers generate CBR traffic and do not adapt flow control in transport layer such as R^2CP [12]. One of our future works is to design a transport layer protocol to efficiently exploit parallel delivery in mobile ad hoc networks. We assume we have an ideal transport layer protocol which can ideally schedule and reorder the transmission between multiple paths. We simulated a scenario where client downloads a 1M bytes file which can be downloaded from any of the five servers. At the same time there is no background traffic. We use the transaction completion time (TCT) as the performance metric, which is the time that the client spends on downloading 1M bytes file.

Fig.2 compares the TCT of three mechanisms. SP stands for single path and so SP/2 (the TCT of single path divided by 2) means the ideal performance that two path parallel delivery can achieve. When CBR data rate is 500 kbps, the network is almost saturated due to contention between neighboring hops. Hence, when CBR data rate increases to 1024 kbps, the TCT of SP/2 does not decrease. From Fig.2 we can see that in most of cases SDMAD can obtain the ideal performance. Only when traffic load becomes very heavy, the SDMAD's performance is lower than that of SP/2. This is because that the contention around the client dominates the throughput (some servers are within two hops of the clients). On the other hand, we can see that SDMAD have significant improvement over MND. It is because that with MND the client receives data from only one server at most of time.

We also sum the traffic received from 2 servers respectively and obtained the ratio of big traffic to small traffic, which is given in Fig. 3. From Fig. 3 we can see that with

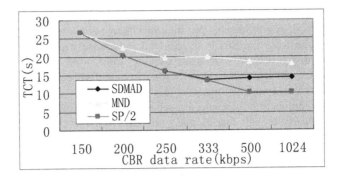

Fig. 2. Transaction completion time

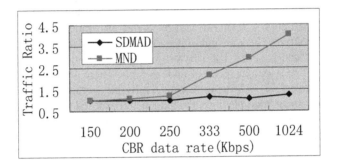

Fig. 3. Traffic ratio of two paths

SDMAD, the bytes receives from two servers are almost same, which indicates the full exploitation of two servers. However, with MND, the ratio becomes much higher than 1 when traffic load becomes heavy. By analyzing the simulation trace carefully, we found that with MND it is often that one server dominates the traffic while the other is refrained to serve the client. Therefore, with MND the two servers are not fully exploited and so the TCT are much higher.

5 Conclusion

In this paper, we first pointed out that one client contacting multiple servers simultaneously is a common scenario in MANETs. Then we proposed SDMAD, a service discovery protocol to find multiple servers with maximal area disjoint paths. SDMAD is an on-demand service discovery protocol which is based on DSR routing protocol. In order to reduce control overhead, it incorporates service discovery into routing. When a server is found, the route to the server is also known to the client. In order to avoid route coupling between the multiple paths, it discovers maximal area disjoint paths. Extensive simulations show that SDMAD fully exploits the servers discovered and shortens the transaction completion time significantly.

The future work includes designing an efficient multipoint-to-point transport layer protocols for parallel delivery after the maximal area disjoint paths were found with SDMAD.

References

1. IETF MANET work group,   http://www.ietf.org/html.charters/manet-charter.html
2. Ulas C. Kozat, Leandros Tassiulas. Network layer support for service discovery in mobile ad hoc networks. IEEE Infocom'03.
3. Sumi Helal, Nitin Desai, Varun Verma and Choonhwa Lee. Konark – a service discovery and delivery protocol for ad hoc networks. IEEE WCNC 03.
4. Uday Mohan, Kevin C. Almeroth and E.M.Belding-Royer. Scalable service discovery in mobile ad hoc networks. Networking, 2004.
5. J. Liu, K. Sohraby, Q. Zhang, B. Li, and W. Zhu, Resource Discovery in Mobile Ad Hoc Networks, in Handbook on Ad Hoc Wireless Networks, edited by M. Ilyas, CRC Press, 2002.
6. A. Klemm, C. Lindemann, and O. Waldhorst, "A special-purpose peer-to-peer file sharing system for mobile ad hoc networks," in Proc. Workshop on Mobile Ad Hoc Networking and Computing (MADNET 2003), Sophia-Antipolis, France, Mar. 2003, pp. 41–49.
7. Ahmet Duran, Chien-Chung Shen. Mobile ad hoc p2p file sharing. IEEE WCNC 2004.
8. C. Partridge, T.Mendez, and W. Milliken. Host anycasting service. RFC1546, Nov. 1993.
9. Sung-Ju Lee, Mario Gerla. Split multipath routing with maximally disjoint paths in ad hoc networks. IEEE ICC 2001.
10. Anand Srinivas, Eytan Modiano. Minimum energy disjoint path routing in wireless ad hoc networks. ACM MOBICOM 2003.
11. Casey Carter, Seung Yi, Prashant Ratanchandani. Manycast: Exploring the Space Between Anycast and Multicast in Ad Hoc Networks. In Proceedings of MobiCom 2003, San Diego, California.
12. Hung-Yun Hsieh, Raghupathy Sivakumar. Accelerating peer-to-peer networks for video streaming using multipoint-to-point communications. IEEE Communication Magzine, 2004. 8.
13. Klingberg T., R. Manfredi. "Gnutella 0.6", June 2002.
14. D. Johnson, D. Maltz, Y.-C. Hu, and J. Jetcheva. The dynamic source routing protocol for mobile ad hoc networks (DSR). Internet Draft: draft-ietf-manet-dsr-09.txt, April. 2003.
15. S. Xu, T. Saadawi. Does the IEEE 802.11 MAC Protocol Work Well in Multihop Wireless Ad hoc Network. IEEE Commun. Mag., June 2001
16. Haas, Z.J. Jing Deng. Dual busy tone multiple access (DBTMA)-a multiple access control scheme for ad hoc networks. IEEE Transactions on Communications, Volume: 50, Issue: 6, June 2002. Pages:975 – 985
17. A. Nasipuri and S.R. Das, On-Demand Multi-path Routing for Mobile Ad Hoc Networks, IEEE ICCCN'99, pp. 64-70.
18. M.R. Pearlman et al, On the Impact of Alternate Path Routing for Load Balancing in Mobile Ad Hoc Network works, MobilHOC 2000.
19. A. Tsirigos, Z. J. Haas, Multi-path Routing in the Present of Frequent Topological Changes, IEEE Communications Magazine, Nov, 2001.
20. Yashar Ganjali, Abtin Keshavarzian. Load Balancing in Ad Hoc Networks: Single-path Routing vs. Multi-path Routing. IEEE INFOCOM'04, 2004.
21. Jinyang Li, Charles Blake, Douglas S. J. De Couto, Hu Imm Lee, Robert Morris. Capacity of Ad Hoc Wireless Networks. ACM MOBICOM 2001.
22. LAN MAN Standards Committee of the IEEE Computer Society. IEEE Std 802.11-1999, wireless LAN medium access control (MAC) and physical layer (PHY) specifications. IEEE, 1999.

Author Index

Lecture Notes in Computer Science

For information about Vols. 1–3629

please contact your bookseller or Springer

Vol. 3677: J. Dittmann, S. Katzenbeisser, A. Uhl (Eds.), Communications and Multimedia Security. XIII, 360 pages. 2005.

Vol. 3676: R. Glück, M. Lowry (Eds.), Generative Programming and Component Engineering. XI, 448 pages. 2005.

Vol. 3675: Y. Luo (Ed.), Cooperative Design, Visualization, and Engineering. XI, 264 pages. 2005.

Vol. 3674: W. Jonker, M. Petković (Eds.), Secure Data Management. X, 241 pages. 2005.

Vol. 3673: S. Bandini, S. Manzoni (Eds.), AI*IA 2005: Advances in Artificial Intelligence. XIV, 614 pages. 2005. (Subseries LNAI).

Vol. 3672: C. Hankin, I. Siveroni (Eds.), Static Analysis. X, 369 pages. 2005.

Vol. 3671: S. Bressan, S. Ceri, E. Hunt, Z.G. Ives, Z. Bellahsène, M. Rys, R. Unland (Eds.), Database and XML Technologies. X, 239 pages. 2005.

Vol. 3670: M. Bravetti, L. Kloul, G. Zavattaro (Eds.), Formal Techniques for Computer Systems and Business Processes. XIII, 349 pages. 2005.

Vol. 3669: G.S. Brodal, S. Leonardi (Eds.), Algorithms – ESA 2005. XVIII, 901 pages. 2005.

Vol. 3668: M. Gabbrielli, G. Gupta (Eds.), Logic Programming. XIV, 454 pages. 2005.

Vol. 3666: B.D. Martino, D. Kranzlmüller, J. Dongarra (Eds.), Recent Advances in Parallel Virtual Machine and Message Passing Interface. XVII, 546 pages. 2005.

Vol. 3665: K. S. Candan, A. Celentano (Eds.), Advances in Multimedia Information Systems. X, 221 pages. 2005.

Vol. 3664: C. Türker, M. Agosti, H.-J. Schek (Eds.), Peer-to-Peer, Grid, and Service-Orientation in Digital Library Architectures. X, 261 pages. 2005.

Vol. 3663: W.G. Kropatsch, R. Sablatnig, A. Hanbury (Eds.), Pattern Recognition. XIV, 512 pages. 2005.

Vol. 3662: C. Baral, G. Greco, N. Leone, G. Terracina (Eds.), Logic Programming and Nonmonotonic Reasoning. XIII, 454 pages. 2005. (Subseries LNAI).

Vol. 3661: T. Panayiotopoulos, J. Gratch, R. Aylett, D. Ballin, P. Olivier, T. Rist (Eds.), Intelligent Virtual Agents. XIII, 506 pages. 2005. (Subseries LNAI).

Vol. 3660: M. Beigl, S. Intille, J. Rekimoto, H. Tokuda (Eds.), UbiComp 2005: Ubiquitous Computing. XVII, 394 pages. 2005.

Vol. 3659: J.R. Rao, B. Sunar (Eds.), Cryptographic Hardware and Embedded Systems – CHES 2005. XIV, 458 pages. 2005.

Vol. 3658: V. Matoušek, P. Mautner, T. Pavelka (Eds.), Text, Speech and Dialogue. XV, 460 pages. 2005. (Subseries LNAI).

Vol. 3657: F.S. de Boer, M.M. Bonsangue, S. Graf, W.-P. de Roever (Eds.), Formal Methods for Components and Objects. VIII, 325 pages. 2005.

Vol. 3656: M. Kamel, A. Campilho (Eds.), Image Analysis and Recognition. XXIV, 1279 pages. 2005.

Vol. 3655: A. Aldini, R. Gorrieri, F. Martinelli (Eds.), Foundations of Security Analysis and Design III. VII, 273 pages. 2005.

Vol. 3654: S. Jajodia, D. Wijesekera (Eds.), Data and Applications Security XIX. X, 353 pages. 2005.

Vol. 3653: M. Abadi, L. de Alfaro (Eds.), CONCUR 2005 – Concurrency Theory. XIV, 578 pages. 2005.

Vol. 3652: A. Rauber, S. Christodoulakis, A M. Tjoa (Eds.), Research and Advanced Technology for Digital Libraries. XVIII, 545 pages. 2005.

Vol. 3651: R. Dale, K.-F. Wong, J. Su, O.Y. Kwong (Eds.), Natural Language Processing – IJCNLP 2005. XXI, 1031 pages. 2005. (Subseries LNAI).

Vol. 3650: J. Zhou, J. Lopez, R.H. Deng, F. Bao (Eds.), Information Security. XII, 516 pages. 2005.

Vol. 3649: W.M. P. van der Aalst, B. Benatallah, F. Casati, F. Curbera (Eds.), Business Process Management. XII, 472 pages. 2005.

Vol. 3648: J.C. Cunha, P.D. Medeiros (Eds.), Euro-Par 2005 Parallel Processing. XXXVI, 1299 pages. 2005.

Vol. 3646: A. F. Famili, J.N. Kok, J.M. Peña, A. Siebes, A. Feelders (Eds.), Advances in Intelligent Data Analysis VI. XIV, 522 pages. 2005.

Vol. 3645: D.-S. Huang, X.-P. Zhang, G.-B. Huang (Eds.), Advances in Intelligent Computing, Part II. XIII, 1010 pages. 2005.

Vol. 3644: D.-S. Huang, X.-P. Zhang, G.-B. Huang (Eds.), Advances in Intelligent Computing, Part I. XXVII, 1101 pages. 2005.

Vol. 3643: R. Moreno Díaz, F. Pichler, A. Quesada Arencibia (Eds.), Computer Aided Systems Theory – EUROCAST 2005. XIV, 629 pages. 2005.

Vol. 3642: D. Ślezak, J. Yao, J.F. Peters, W. Ziarko, X. Hu (Eds.), Rough Sets, Fuzzy Sets, Data Mining, and Granular Computing, Part II. XXIII, 738 pages. 2005. (Subseries LNAI).

Vol. 3641: D. Ślezak, G. Wang, M. Szczuka, I. Düntsch, Y. Yao (Eds.), Rough Sets, Fuzzy Sets, Data Mining, and Granular Computing, Part I. XXIV, 742 pages. 2005. (Subseries LNAI).

Vol. 3639: P. Godefroid (Ed.), Model Checking Software. XI, 289 pages. 2005.

Vol. 3638: A. Butz, B. Fisher, A. Krüger, P. Olivier (Eds.), Smart Graphics. XI, 269 pages. 2005.

Vol. 3637: J. M. Moreno, J. Madrenas, J. Cosp (Eds.), Evolvable Systems: From Biology to Hardware. XI, 227 pages. 2005.

Vol. 3636: M.J. Blesa, C. Blum, A. Roli, M. Sampels (Eds.), Hybrid Metaheuristics. XII, 155 pages. 2005.

Vol. 3634: L. Ong (Ed.), Computer Science Logic. XI, 567 pages. 2005.

Vol. 3633: C. Bauzer Medeiros, M. Egenhofer, E. Bertino (Eds.), Advances in Spatial and Temporal Databases. XIII, 433 pages. 2005.

Vol. 3632: R. Nieuwenhuis (Ed.), Automated Deduction – CADE-20. XIII, 459 pages. 2005. (Subseries LNAI).

Vol. 3631: J. Eder, H.-M. Haav, A. Kalja, J. Penjam (Eds.), Advances in Databases and Information Systems. XIII, 393 pages. 2005.

Vol. 3630: M.S. Capcarrere, A.A. Freitas, P.J. Bentley, C.G. Johnson, J. Timmis (Eds.), Advances in Artificial Life. XIX, 949 pages. 2005. (Subseries LNAI).